Modern As

Modern Astronomy

An Activities Approach

By R. Robert Robbins and Mary Kay Hemenway

With contributions from William H. Jefferys

Illustrations by Patricia Cornelison

University of Texas Press, Austin

Printed in the United States of America
First Edition, 1982

International Standard Book Number 0-292-75064-1
Library of Congress Catalog Card Number 81-70525

For reasons of economy and speed, this volume has been printed from camera-ready copy furnished by the authors, who assume full responsibility for its contents.

Publication of this book was assisted by the Hooks Contingency Fund.

Numerous individuals have made helpful contributions to the development of these materials over the years, but we would like to single out for special appreciation the following astronomers and educators: Paul Vanden Bout, Earle Luck, Paul Makinen, Kathleen O'Sullivan, David Lynch, Lon Clay Hill, Dick Marasso, and Richard White.

Bill Jefferys initiated the use of these materials in a self-paced course of instruction at the University of Texas, and his insight into educational design has resulted in the continuing utility of one of our department's most successful courses. The authors (primarily RRR) have taught the course and developed the materials in their present form over the last decade.

We would also like to thank our editors for unusually extensive support of this project: Archer Mayor, for first believing in the worth of it, and Chuck Arthur and Holly Carver for actually doing all the hard work during the homestretch. Hugs and kisses to our editors—they're super.

One of the authors wishes to dedicate this book to her husband, Paul, for his support and encouragement. The other author (RRR) has no spouse but realizes that he nevertheless owes a similar debt to his calico cats, Lorna Doone and Agatha Bob, who inspected, sat on, and commented on virtually every page in the manuscript.

Contents

Modern Astronomy

Introduction: What This Book Is All About

This book is designed to help you learn about astronomy, its tools and its methods, through actual experimentation and observation. It does not attempt to be an encyclopedic survey of astronomy but concentrates instead on learning a set of important concepts by "doing it yourself." The units of this book were written during a decade of teaching astronomy to student teachers and science education majors at the University of Texas at Austin. They are designed to give the reader an active DISCOVERY approach to astronomy using observational activities. Most of the activities require no prerequisite knowledge or skills, and many are carried out using very simple materials (for example, the self-constructed instruments that use the templates on the insert of the book). Early chapters approach the subject under investigation at an introductory level, and later chapters carry the reader forward to higher levels of skill and knowledge. With increasing sophistication, more advanced activities can be undertaken.

Most students today receive little or no measurement experience and hence have no direct contact with quantitative (as opposed to qualitative) thinking. In dealing with celestial phenomena, it is possible to speculate endlessly and formulate an infinite number of possibilities. It is only the process of measurement -- the capturing of a celestial parameter in the form of a number -- that terminates the speculation and leads toward the one acceptable explanation. This important process is generally overlooked in standard science presentations (and even in many lab manuals, which present data already obtained as "facts" to be explained), and it is for this reason that this volume emphasizes learning-by-doing, through active investigation. There are very few activities in this book where data are supplied; the emphasis is on gathering observations and making proper inferences from them. The instructions for the various units are occasionally somewhat open-ended and go beyond simple "do this, then do that" statements; the reader is challenged to extract all the information possible from the observations and formulate all possible hypotheses for explaining the data. This approach places more weight on the process than on the "bottom line" answer, to underline the fact that science is after all a process for discovering new knowledge. The specifics of the knowledge will change with time, but the process remains the same and understanding it remains the heart of science.

Recent research studies have provided a growing body of evidence that many, perhaps most, college level students do not possess the basic abstract thinking and reasoning skills that most instructors expect. Many educators have interpreted these results in the light of Jean Piaget's theories of the development of cognitive skills and have concluded that the most effective method of fostering the growth of these reasoning skills is to develop actual activities that allow the student to practice them. This is a book of such activities. To state our purpose in the most general way, this volume contains an integrated course of study in developing reasoning skills, using astronomy as a vehicle and a subject to which reasoning may be applied. For a beginning observer, this book is an introduction to astronomical research. While this course of study was originally constructed for prospective secondary school teachers, it is nevertheless ideally suited as an observational introduction to astronomy for anyone with such an interest.

The early units begin with simple measurements and simple equipment, designed to introduce the basic principles of making inferences from observations and of dealing with the all-important fact of measurement errors and their treatment. As understanding progresses, the sophistication of the activities can also progress. Some of the final activities in the book do require specialized equipment.

The readers have the flexibility to design programs suited to their interests by choosing from the various materials in this book. Notice that you do not have to do all the activities nor do you have to do them in sequence. Units 1 and 2 are necessary prerequisite activities to all the subsequent ones, as they give the basic introduction to measurement and the night sky. After completing these two units, your route through the materials depends upon your interests and the time available (see page 6). Certain units do require others as prerequisites, however, and you should note such constraints as a strong recommendation. (If your background experience is strong in certain areas of observational astronomy or physics, you may find that the unit serves as a review.) As a specific example, a reader with an interest in learning celestial photography would probably want to follow unit 2 with the units on lenses, photography, telescopes, and finally celestial photography. Brief descriptions of the contents of the units are given below for the purpose of making a preliminary plan for a course of study. The first page of each unit contains a list of the equipment which is necessary to carry it out. Appendix 3 gives a complete description of all the equipment that the authors have traditionally used for these activities in their course at the University of Texas, as well as the suppliers of that equipment. Appendix 4 gives a detailed description of how these units and materials are structured into a self-paced Keller method course of astronomy. This appendix can be used as a model by any instructor interested in designing a course along such lines.

Comments and feedback on these materials will be welcomed by the authors; they will enable us to further revise the units in the future to function even more effectively.

BRIEF DESCRIPTIONS OF THE ACTIVITIES

REGULAR ACTIVITIES

1. The Principles of Measurement: Using a Cross-Staff and a Quadrant

This is a required activity, prerequisite to all others. It does not require clear weather outside. This unit introduces concepts of angular measurement and errors of measurement and, studies the relationship between angular size and distance. Using the templates given on the insert of the book and a meterstick, a cross-staff and quadrant are assembled and used to measure a variety of angles. In all parts of the activity, the nature of the errors of measurement are emphasized.

2. Mapping the Night Sky and Its Motions

This is a required activity, prerequisite to all others. Parts of it require reasonably clear night skies in which at least the brighter stars are visible. This activity begins with a familiarization with the night sky and the major star groupings above the horizon at various times of the year. It introduces simple star maps. With the devices constructed in unit 1, the relative positions of celestial objects are measured, and by measurements over time, the motions of the sky are revealed. All the observations are correlated and checked with a celestial globe.

3. Lunar Surface Features

Portions of this activity require weather which allows for observation of the moon. In this unit, the moon is drawn at two phases, first by eye and then with a small telescope or binoculars, and the observations are compared to a lunar globe or map. Your closest view of the moon then comes from studying in detail a series of Lunar Orbiter photographs of the surface of the moon, with the objective of reconstructing some portions of the moon's evolutionary history.

4. The Motion and Phases of the Moon

Using the cross-staff or sextant, the motion of the moon is followed with respect to the background stars through a complete cycle. In combination with measurements of the phases, this will lead to an understanding of the earth-moon-sun spatial relationships. The measurements require about 5 minutes on each usable night for approximately one month. (A usable night is defined as one in which the moon and a few identifiable stars near it can be seen.)

5. The Motions of the Planets

This unit has more text than most and covers the historical development of our understanding of the motions of the planets, from the Greek era through modern studies. It concludes with a set of different observing activities that can be carried out on the planets, depending upon the type of equipment available to the user.

6. The Sun: Its Size and Daily Motion
Using a simple gnomon and a pinhole camera, the size of the sun and a study of its daily motion are investigated. The observations in this unit take up most of the hours between noon and sunset on one day and require weather that is clear enough to allow shadows to be seen.

7. The Sun: Its Energy Output and Yearly Motion
Observations of the sun's motion over a longer period of time reveal its seasonal changes. A simple solar collector employing an inexpensive thermometer is used to measure the actual energy output of the sun. These observations require a couple of hours on each of two clear days.

8. Properties of Lenses and Mirrors
Using an optical bench with a selection of lenses and mirrors, studies of how light is focused and directed by optical systems are carried out, and the principle of the telescope is revealed. An indoor activity.

9. Cameras and Photography
In this unit, the operation of a camera and the principles of photography are covered. A roll of film is exposed and developed with proper darkroom procedures. This unit needs just a limited amount of clear weather for one portion where star-trail photographs are taken. Prerequisite: unit 8.

10. Using a Small Telescope
This unit involves actual practice using a small telescope to find, study, and draw celestial objects, using good sky maps and stellar atlases to find suitable objects to observe. It requires clear weather and, if possible, a dark sky location. Image scale and the use of setting circles are covered. Prerequisite: unit 8.

11. Introduction to Spectroscopy
A hand-held spectrometer is constructed using templates on the insert of the book. A variety of light sources will be observed and the three basic types of spectra studied. By measuring the emission features of various chemical elements in the lab, it is then possible to study unknown light sources (such as streetlights) and determine their nature and chemical composition. These studies can be generalized into principles used by astronomers to discover the nature, motion, and composition of celestial sources. This unit does not require good weather.

12. Distances and Fundamental Properties of Stars
A parallax activity is carried out to learn in detail the geometric methods for determining the distances to nearby celestial bodies. This unit includes a considerable amount of reading on the fundamental properties of stars -- their brightness, masses, temperatures, sizes, and sources of energy.

The luminosity-temperature diagram so important to modern theories of stellar evolution is studied in considerable detail. This activity does not require good weather.

13. Components of the Milky Way: Dust, Gas, and Stars
This unit employs research-quality photographs to identify and study the various shapes and forms in which the interstellar material, gas and dust between the stars, can be found. The relationship between the interstellar material and the evolutionary histories of stars is investigated. Indoor activity. Prerequisite: unit 11.

14. Studies of Galaxies
This unit employs research-quality photographs to identify and study the different forms and types of galaxies -- stellar systems outside of our own Milky Way system. Two clusters of galaxies at different distance are studied, and methods for distance determinations to very distant objects are uncovered. Indoor activity.

15. Introduction to Computers and the BASIC Language
In this unit, an introduction to computers is presented, some statements and commands in the popular BASIC language are learned, and two programs useful to this course of study in astronomy are written, tested, and run on the computer. Indoor activity.

ADVANCED ACTIVITIES

16. Using a Solar Telescope to Study the Sun
In this unit, a solar telescope is employed to carry out detailed studies of the surface of the sun. From a large projected image of the sun, the detailed structure of sunspots and their life history can be followed, and the rate of rotation of the sun determined. Variations in brightness over the solar image reveal details concerning the structure of the outer layers of the sun, and the projection of the solar spectrum gives detailed information on its chemical composition. Use of a hydrogen alpha narrow band filter allows the time study of the behavior of prominences on the edge of the sun. This unit requires reasonably good daytime weather. Prerequisites: units 6 or 7, and 11.

17. A Spectral Comparison of the Sun and Beta Draconis
Using detailed tracings of the best research spectra of the sun, the principles of spectral studies of the stars can be understood in depth. The spectrum of the sun is compared to that of Beta Draconis, a star with the same surface temperature as the sun but much larger in size, to determine the effects of differences in density in stellar atmospheres. Indoor activity. Prerequisites: units 11 and 12.

18. Celestial Photography
Here, a 35-mm camera is used in combination with a telescope to photograph celestial objects. The darkroom techniques of enlarging and printing are employed. This unit requires good weather and a dark-sky site if possible. Prerequisites: units 9 and 10.

19. Advanced Astronomical Photography
In this unit, a plate camera is employed at the focus of a professional telescope to take photographs of better quality. The additional darkroom techniques involved in using large sized negatives are explored. Prerequisite: unit 18.

20. Astronomical Spectroscopy at the Telescope
A professional spectrograph-telescope combination is employed to photograph stars for spectral classification and other studies. Prerequisites: units 10 and 11.

21. Photometric Studies with a Telescope
A professional photoelectric photometer is employed on a telescope to obtain very accurate measurements of the brightnesses of stars in different colors. Prerequisite: unit 10.

22. Additional Projects: Where to Go from Here
This unit briefly discusses a variety of other activities that could be carried out, on both the introductory level and advanced levels, including such topics as optical phenomena studied with lasers and advanced computer techniques.

Two Possible "Tracks" Through The Course Materials

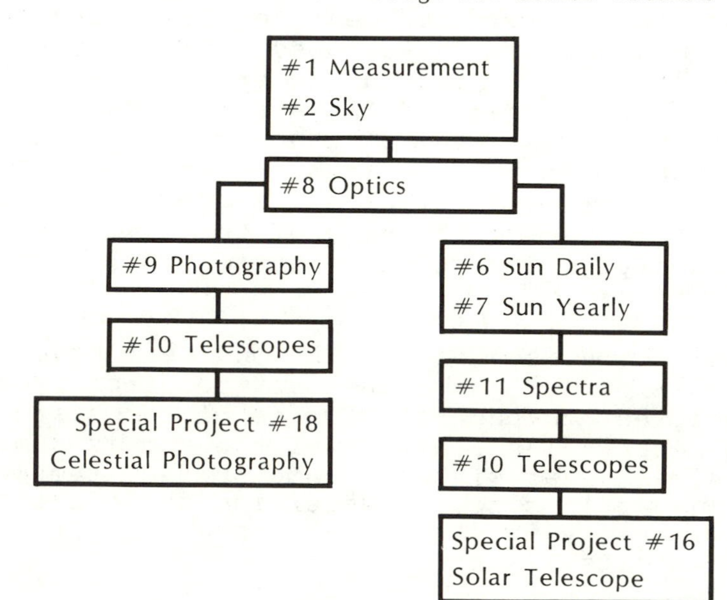

1

The Principles of Measurement: Using a Cross-Staff and a Quadrant

DON BENTZ

From prehistory until 1609, the positions of the stars were measured with naked-eye instruments like the cross-staff.

OBJECTIVES

1. to measure the angular separations of objects with a cross-staff
2. to measure the altitudes of objects with a quadrant
3. to convert angles between different units of measurement, for example, from degrees, minutes, and seconds of arc into fractions of a degree and radians
4. to solve for the actual size of an object whose apparent size and distance are known
5. to improve the accuracy of a measurement by taking the average of a series of observations
6. to distinguish between random and systematic errors and to determine the errors of the cross-staff and quadrant for various sizes of angles
7. to calibrate your hand as a measuring device for angular separations

EQUIPMENT NEEDED

Meterstick, a straight piece of wood about 30 centimeters in length, scissors (or razor blade or Exacto knife) for cutting cardboard pattern, a 10-20 centimeter piece of string, thumbtacks.

In this unit you will construct and employ the cross-staff and quadrant, two simple devices for measuring angles. These instruments probably originated in ancient Greece. Ptolemy, the great encyclopedist of astronomy of the first century A.D., made his measurements of the altitudes of the sun, accurate to fractions of a degree, using a cross-staff type instrument. Following Ptolemy's tradition, the Arabs nurtured the science of astronomy and developed the astrolabe. Except for their work, European astronomy was greatly neglected in the Dark Ages. A renaissance of science began after the invention of the printing press. The fifteenth-century German astronomer Regiomontanus redeveloped several instruments based on his survey of ancient writings. The cross-staff (known under several names, including Jacob's staff or cross-lath) became the most common tool of navigators until the invention of the modern sextant in 1731. By the beginning of the sixteenth century, new series of observations were begun, most having an error of less than one degree. The use of new and accurate instruments greatly speeded the navigational efforts of the new world explorers. The quadrant reached its pinnacle of development with the last great pretelescopic observer, Tycho Brahe (1546-1601). Several fixed quadrants were built by Brahe with radii of about seven feet and an accuracy of one minute of arc. Later astronomers used quadrants with telescopic sights which helped them map the sky and lay the foundations of basic navigation.

You will become familiar with the units of angular measurement, the various types of measuring errors, and how to achieve greater accuracy by averaging repeated measures -- in other words, you will learn the techniques of science as practiced through the centuries. Retain the instruments which you construct for use in other units of this course. For example, in unit 2 you will use them to map parts of the sky and find your latitude.

I. ASSEMBLING AND USING THE CROSS-STAFF

The cross-staff has two pieces -- a meterstick and a sliding crosspiece which fits on it. The pattern for the sliding crosspiece is printed on the insert in this book. Cut the pieces out with a razor blade or scissors, following the assembly instructions above the pattern. The rectangle marked "cutout" is where the crosspiece slides onto the meterstick; after this occurs, staple A to A and B to B.

The following sketches indicate how the cross-staff can be used to measure an angle. See figures 1, 2, and 3. Resting one end of the meterstick lightly against your cheek, you can sight down the stick and line up various objects with the edges of the sliding piece by moving it back and forth. Notice that there are three sets of vertical edges on the sliding crosspiece; they are 4, 2, and 1 inches apart. These edges allow the measurement of a wide range of angles. Suppose that you wanted to measure the angle between two stars and that you used the medium edges. You would slide the crosspiece until the two stars were lined up with the edges as in figure 2 below. Then read off the number of centimeters from your eye to the front of the sliding piece, and, by using the nomograph included in this unit, you can convert the meterstick reading into an angle which is the angle between the two objects you were sighting.

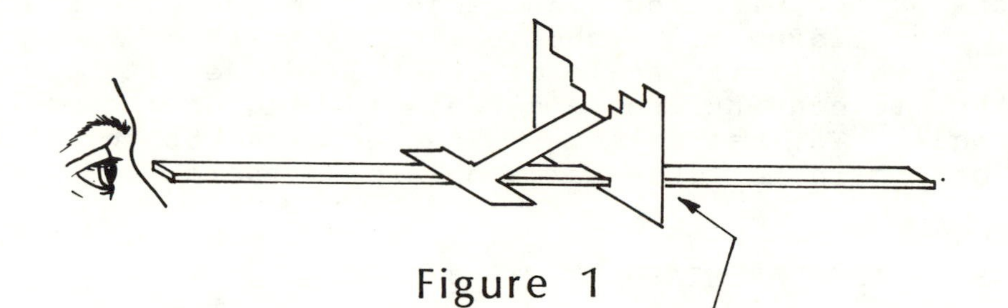

Figure 1

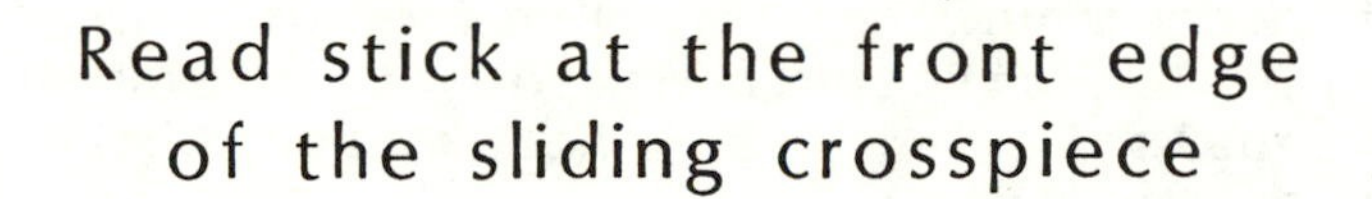

Read stick at the front edge of the sliding crosspiece

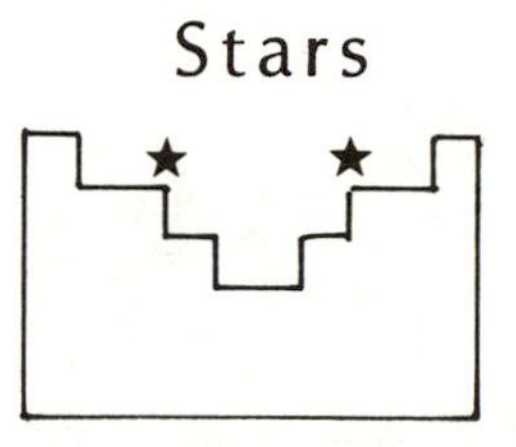

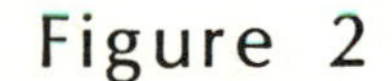

Figure 2

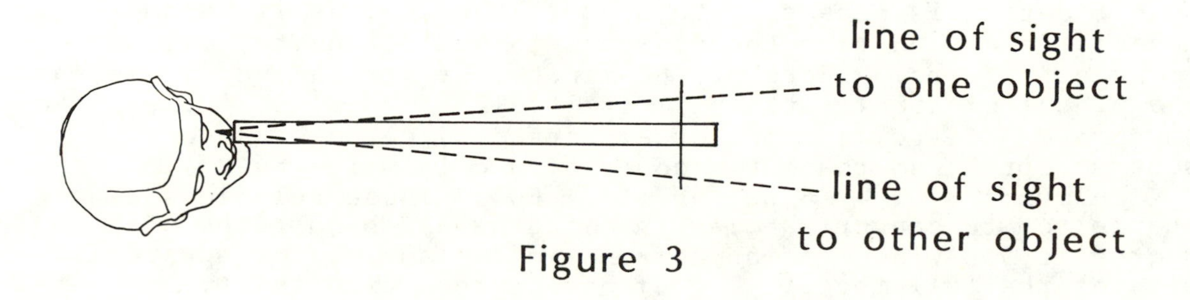

Figure 3

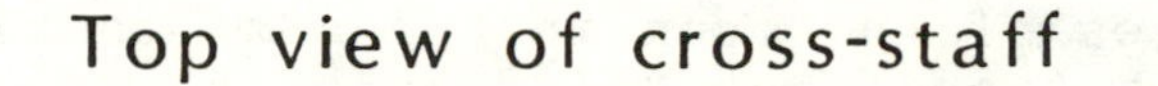

Top view of cross-staff

As a practice exercise with the cross-staff, stand about 4 meters away from an object that is 75 centimeters in size (i.e., three-fourths of a meter) and use the wide separation edges to measure the angle between the two edges of the object. That is, line up the left edge of the object with the appropriate vertical edge on the left side of the sliding piece, and line up the right side of the object with the right edge of the sliding piece. At 4 meters, the sliding piece should line up at about 54 centimeters, which is an angle of about 11 degrees. Be sure that you know how to convert from a certain setting on the meterstick into an angle. You can make your 75-centimeter object from a piece of cardboard, or draw marks on a blackboard.

II. UNITS OF ANGULAR MEASUREMENT

The most common unit of angular measurement is the degree (written °). The circle is divided into 360 degrees, and a right angle equals 90 degrees. A few angles are sketched in figure 4 for examination:

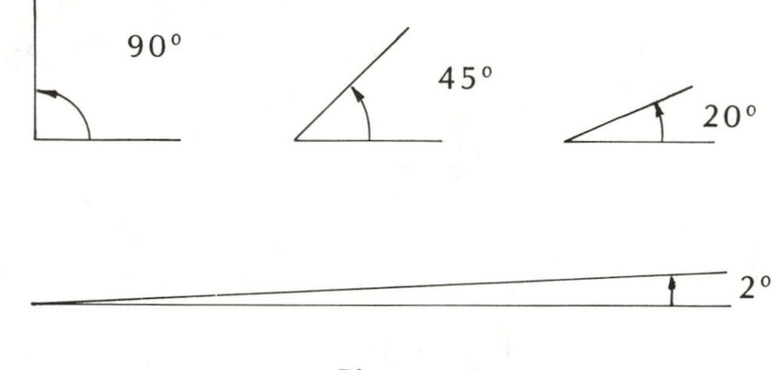

Figure 4

In astronomy, small angles are quite common and telescopic magnification is necessary to measure them. Thus it becomes necessary to subdivide the degree into smaller units, called "minutes" of arc. There are 60 minutes of arc in one degree, and one minute of arc (written 1') is itself further subdivided into 60 seconds of arc (seconds of arc are designated by the superscript ", so that 30 seconds of arc is written 30").

In summary, 1° = 60' and 1' = 60". These relations also enable you to convert angles into tenths and hundredths of a degree if you wish. As an astronomical example, the apparent size of the full moon is 1/2 of one degree, that is, 0.5 degrees. By apparent size, we mean here that, if we measured the angle between the two edges of the moon in the same fashion as we measured the angle between the two edges of the object, the resultant angle would be the angular size, or apparent size, of the moon. Answer each question in your notebook.

1. Compute the angular size of the moon in minutes of arc.

2. Compute the number of seconds of arc in one degree of arc.
3. Convert 31 degrees 25 minutes into decimal form.

While one second of arc may seem to be a tiny angle, astronomers using telescopic magnification frequently deal with angles as small as 0.01". The smallest angle that the naked eye can resolve is about 1'.

A very convenient unit of angular measure is the radian. Figure 5 shows how this angle is defined. If we take a circle and lay off along its perimeter a length of arc AB equal in length to the radius of the circle, by definition the angle indicated at the center of the circle is one radian.

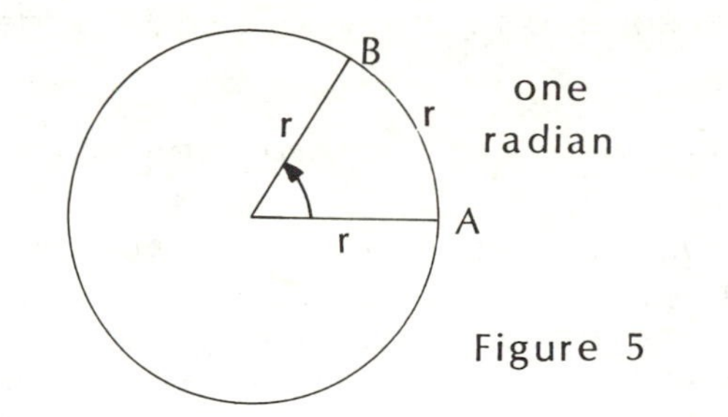

Figure 5

Since the circumference of a circle is 2π times its radius, we can convert between degrees and radians as follows:

$$2\pi \text{ radians} = 360 \text{ degrees}$$
$$\text{so } 1 \text{ radian} = 360/2\pi \text{ degrees} = 57.3 \text{ degrees}$$
$$\text{and } 1 \text{ degree} = 1/57.3 \text{ radians} = 0.01745 \text{ radians}$$

III. VARIATION OF APPARENT SIZE OF AN OBJECT WITH DISTANCE

At this point some practice measurements with the cross-staff are in order. Make two marks on a blackboard 75 centimeters apart (or, if you do this at home, on a piece of paper tacked to the wall). These marks can be thought of as constituting an object 75 centimeters in width. The activity in this section is to measure the apparent size of this "object" at a variety of different distances.

First, stand 1 meter away from the marks (this is as close as you can get with the meterstick of the cross-staff) and measure the angular size of the object. Use the wide sights on the crosspiece of the cross-staff. In your notebook, record your distance from the object and its angular size. Then move farther away and repeat the observation, again recording your distance and the angular size of the object. Make measurements at 2, 4, 8, and 16 meters, continuing to move away from the object until its angular size is quite small (about 3 degrees or so). As you measure smaller and smaller angles, you will want to use the medium and finally the small sights on the crosspiece. Plot all your measurements on a graph with degrees on the vertical axis and distance on the horizontal axis. With a PENCIL, draw a smooth curve connecting your measured points.

The relationship between angular size and distance that you have measured will probably not be a straight line; some type of curve will have resulted from your measurements. Before trying to understand the physical significance of the curve itself, however, it will first be necessary to consider the subject of measurement errors and answer the following question: is the curve you have measured the true relationship between angular size and distance, or could the shape of the curve be explained by observational errors? For example, if the cross-staff measured small angles more accurately than large angles, that would affect the shape of the curve. The cross-staff will have some errors associated with its use, and we would like now to make some estimate of how large these are.

IV. RANDOM ERRORS OF MEASUREMENT

One obvious way to detect the errors of the cross-staff would be to make measurements with it and then make the same measurements with a more accurate instrument. An alternative procedure which can often be used, however, is to employ the device of taking REPEATED OBSERVATIONS and averaging them. If you measure something a number of times, there will inevitably be some variation in your answers. If you average your measurements, the average will be closer to the true answer, since any RANDOM errors of measurement will tend to cancel each other out in the averaging process. For example, those answers that were accidentally too large will be compensated for by those that came out a little too small.

To examine how this works, return to the same 75-centimeter object you measured in activity III and do further measures on it using the wide sight. Stand again 1 meter from the object as you did before. Now measure the angular size of the object ten different times, recording the reading each time. Make sure that each measurement is an independent one; that is, after each measurement move the sliding piece on the cross-staff so that you must resight and reread the setting anew each time.

Now compute your AVERAGE value, which is the sum of all the measurements divided by ten. Plot this value on the graph you drew earlier, and, in addition, also plot the largest and, smallest values you measured for the angular size. Connect all these points with a vertical line, as illustrated in figure 6. Such a plot will give an indication of the RANGE of values through which a single measurement could vary.

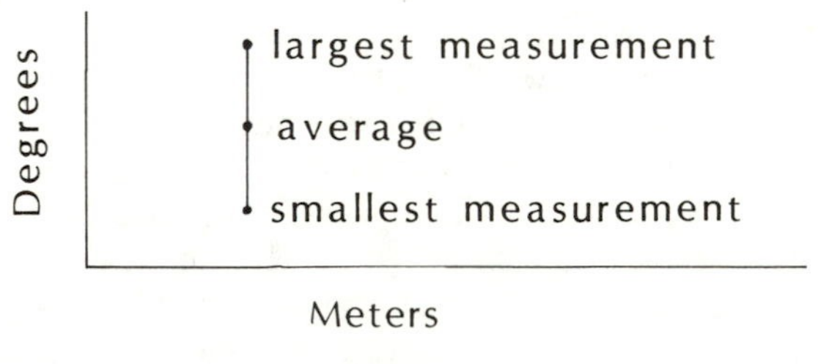

Degrees
largest measurement
average
smallest measurement
Meters

Figure 6

Now do ten independent measurements from distances of 4 and 16 meters and again compute the average, and plot the average, smallest, and largest values.

Scientists generally characterize the variation about an average by a precisely defined quantity called the standard deviation, but for our purposes here it will be sufficient to adopt an approximate rule called Snedecor's Rough Check. This rule states that, if you have ten independent measurements of a quantity, the standard deviation of one measure is approximately equal to the RANGE of the measurements (the largest minus the smallest value) divided by three. If you have five independent measurements, divide the range by two. If you have three independent measurements, the standard deviation is approximately equal to the range itself.

For the measurements you have just taken (ten times each from 1, 4, and 16 meters) evaluate the standard deviation and also a quantity we will call the percentage error, namely:

percentage error = (standard deviation/average) X 100

Are the errors bigger for larger or smaller angles, or are they approximately the same for all angles? What might cause the differences in accuracy for different-size angles? Do you think that the overall shape of the curve you plotted in section III might actually be due to your measurement errors, or are the errors so small in comparison with the changes in the curve that the curve probably does reflect the true relation between distance and angular size?

V. ABSOLUTE ERRORS

Examining the range of your measurements gives you some indication of your CONSISTENCY. It tells you how much confidence you could place in a single measurement, if that was all you were able to make. It does not necessarily tell you about your ABSOLUTE error, however. For example, suppose that you consistently measured angles using the medium sights on the crosspiece but, on the nomograph, accidentally used the "narrow" mark to convert to angles. You might be very consistent in this process and achieve a small range in your measurements yet be systematically off a factor of two from the correct values. This type of error is called a SYSTEMATIC error. Try to think of some other possible sources of systematic errors in using the cross-staff and enter them into your notebook.

While the device of repeating measurements can give some indication concerning random errors, there is no such simple remedy to check for the possibility of systematic errors in measurements. If you can redo the measurements with another, more accurate instrument, this will serve as a check of the less accurate one, but scientists typically try to use the most accurate devices available, so this stratagem is usually not available. If you have a formula to predict what you are measuring, you can check against the formula, but again, on the frontiers of research, such a known relationship is usually not available. The subject of discovering systematic errors in measurements is a difficult one, one with which all scientists must continually grapple by examining every aspect of their measurement sequence and checking it carefully for flaws, no matter how sophisticated and expensive their equipment is.

Similarly, every measurement you make in this course will always have an error associated with it, and you should always attempt to attach an error estimate to every quantity you measure. You should get into the habit of thinking of a measurement WITHOUT an attached error as practically useless, since there is no clue as to how much confidence one can attach to such a measurement.

Therefore, every measurement you make should be expressed either in the form

$x = 120.2 \pm 8.5$ if 8.5 is the estimated error or
$x = 120.2 \pm 7\%$ since $8.5/120.2 = 7\%$

It may not always be obvious how to produce the error estimate, but you must always try. Clearly, you should do repeated measurements when you can and also examine each facet of your measurement process for possible systematic and random errors. Sometimes your error estimate may be just that -- an estimate.

VI. RELATION BETWEEN ACTUAL AND APPARENT SIZE OF AN OBJECT

Using your repeated measurements of angular size at various distances (part IV) and their associated errors, you should now be able to determine whether the shape of the curve you measured in part III might be due to measurement errors or whether the curve describes a real relation. If the changes in the curve are larger than your measurement errors, then they are presumedly real.

Examine the curve you have plotted of angular size versus distance. Is it true that, as the object gets more distant, its angular size gets smaller? How does the shape of the curve change as the angular size gets smaller? How small does the angular size need to be, in your estimation, before the curve is negligibly different from a straight line -- say to within 2 or 3 percent? Answer these questions in your notebook.

Most celestial objects are relatively distant and hence their angular size is small. The long, skinny triangle in figure 7 illustrates the geometry usually encountered by astronomers. When the angular size is small (a few degrees or less), the relation between angle and distance is straightforwardly given by the equation

Calculated Size = (Distance) x (Angular Size in radians)

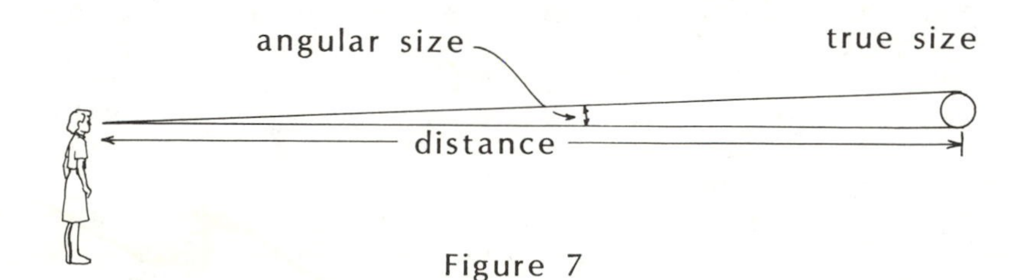

Figure 7

This equation says that, if we take an object whose size is fixed but increase its distance from us, then its angular size must get proportionately smaller. Is this what your observations show?

We may use this formula to determine the absolute measuring errors involved in using the cross-staff to determine angles of various sizes. To do this, take the observations that you made in part IV (ten measurements of the angular size of a 75-centimeter object from 1, 4, and 16 meters) and substitute your AVERAGE values into the formula. Calculate the size of the object you were measuring. If your observations were perfectly accurate, you would get 75-centimeters, but there will generally be some measuring error. Compute the percentage error by calculating

$$\frac{\text{calculated size minus actual size}}{\text{actual size}} \times 100$$

for each of your angular size measurements. How well do your calculated sizes agree with 75 centimeters? Is the cross-staff more accurate for small, medium, or large angles?

Note that in the equation above, the units of length must agree on the two sides of the equation. If you measure a distance in kilometers, then the actual size that you would compute from the equation is also in kilometers. As another example of how this equation works, use it to compute the true size of the moon: assume that the apparent size of the moon is 1/2 degree, and that its average distance is 384,400 kilometers.

Now let us examine the effect of measuring an angle with different sights. In part IV, you measured the angular size of a 75-centimeter object from a distance of 4 meters using the wide sights on the cross-staff, and you have now checked the absolute error in the average of your ten measurements using the formula of this section. Return now to that object and from a distance of 4 meters again measure its angular size with the medium sights on the cross-staff. Repeat this procedure ten times. Use the formula again to evaluate the accuracy of your average value. Was the medium sight more or less accurate than the wide sight? Can you think of any reason for this?

VII. ASSEMBLING THE QUADRANT

On the insert in this book you will find the necessary pattern for the quadrant, which you will recognize to be a simple protractor. Attach it to a straight piece of wood with a staple or thumbtack, being careful that the straight side of the protractor is aligned perpendicular to the length of the stick. Attach a weighted thread or string to another thumbtack stuck through the "x" on the protractor, and you are ready to begin. See figure 8.

The quadrant is a device for measuring the altitude of an object; that is, its angular distance above the horizon. To use the quadrant, sight along the stick at the object whose altitude you wish to measure. Let the weight hang down freely under the influence of gravity until it stops swinging. Then rotate the

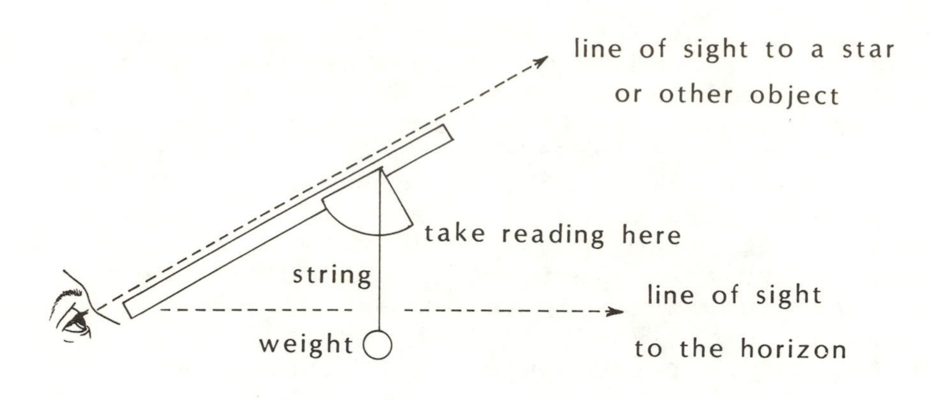

Figure 8

stick (keeping it pointed at the object) until the thread is laying against the protractor. Then with your finger you can hold the thread against the protractor while you move the quadrant away from your eye and read the altitude off the scale. To check that the protractor is properly mounted on the stick, you should find someplace with a clear horizon and sight the stick toward the DISTANT horizon; it should then read zero, since zero degrees is indeed the altitude of the horizon. If it does not, loosen the thumbtacks on your quadrant and reset the protractor.

Find an object which is just a little taller than yourself, for example, the top of a door, and, standing well back from it, measure its altitude ten times, making each measurement independently of the others. The quadrant is being used to measure a fairly small angle here: compute your average value, find the range, and compute the percentage error (standard deviation divided by the average). Now measure a large altitude with the quadrant; that is, choose some object to measure such that its altitude is 60 degrees or more. Measure its altitude ten times independently, and find the average, the standard deviation, and the percentage error. Using the percentage error as a measure of accuracy, is the quadrant more accurate in measuring large or small angles? Why do you think this is so? Do you think the quadrant measures angles more or less accurately than the cross-staff?

VIII. USING THE QUADRANT TO MEASURE A TALL BUILDING'S HEIGHT

Go outside and walk away from a tall building, pausing occasionally to sight the top of the building with the quadrant (i.e., to measure the altitude of the top of the building). Continue moving away from the building until you get an angle of 45 degrees on the quadrant.

The geometry of the situation will then be very simple, as figure 9 shows. AB will then equal BC; that is, the height of the building will equal your distance from it. Now measure your distance from the building in some way (e.g., by pacing it off and measuring the length of ten paces) and compute the height of the building. How does your own height affect this measurement and what should you do about it? What other errors probably enter into the accuracy of your final answer?

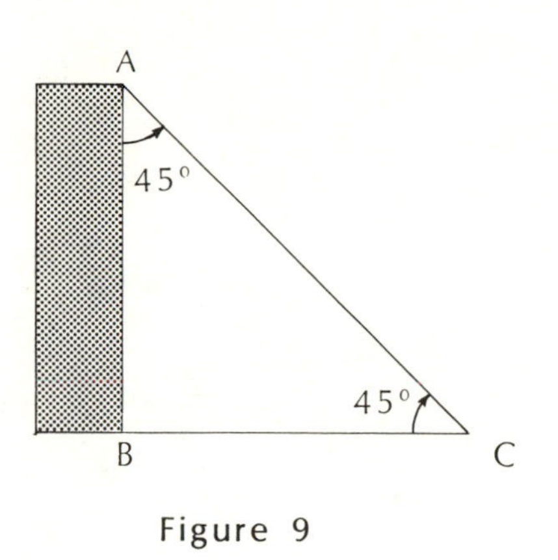

Figure 9

IX. USING YOUR HAND AS A MEASURING DEVICE

In this unit you have learned to measure angles accurately with a cross-staff and quadrant. A less accurate, but more convenient, method is to calibrate something which is always available for use -- your hand. If you extend your arm and make a fist, its angular size is about 10 degrees. With your hand spread, the angle subtended is about 18 degrees, while the finger subtends about 1 degree. See figure 10.

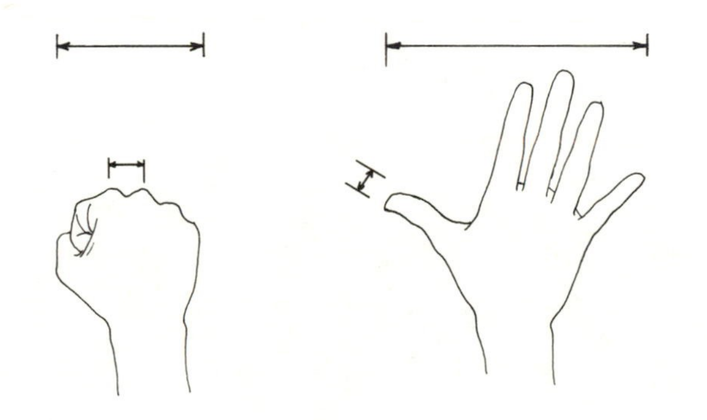

Figure 10

These measurements are approximate only. Now calibrate your own hand. Mark off a paper in either centimeters or inches. Extend your arm. Stand at various distances to compare your fist, hand, and finger against the marker. Then use the formula in section VI to compute the angles in radians. Convert this angle to degrees.

Estimate the amount of error present in this method and its causes. Do you expect this method to work for people of all sizes? Explain your answer in your notebook.

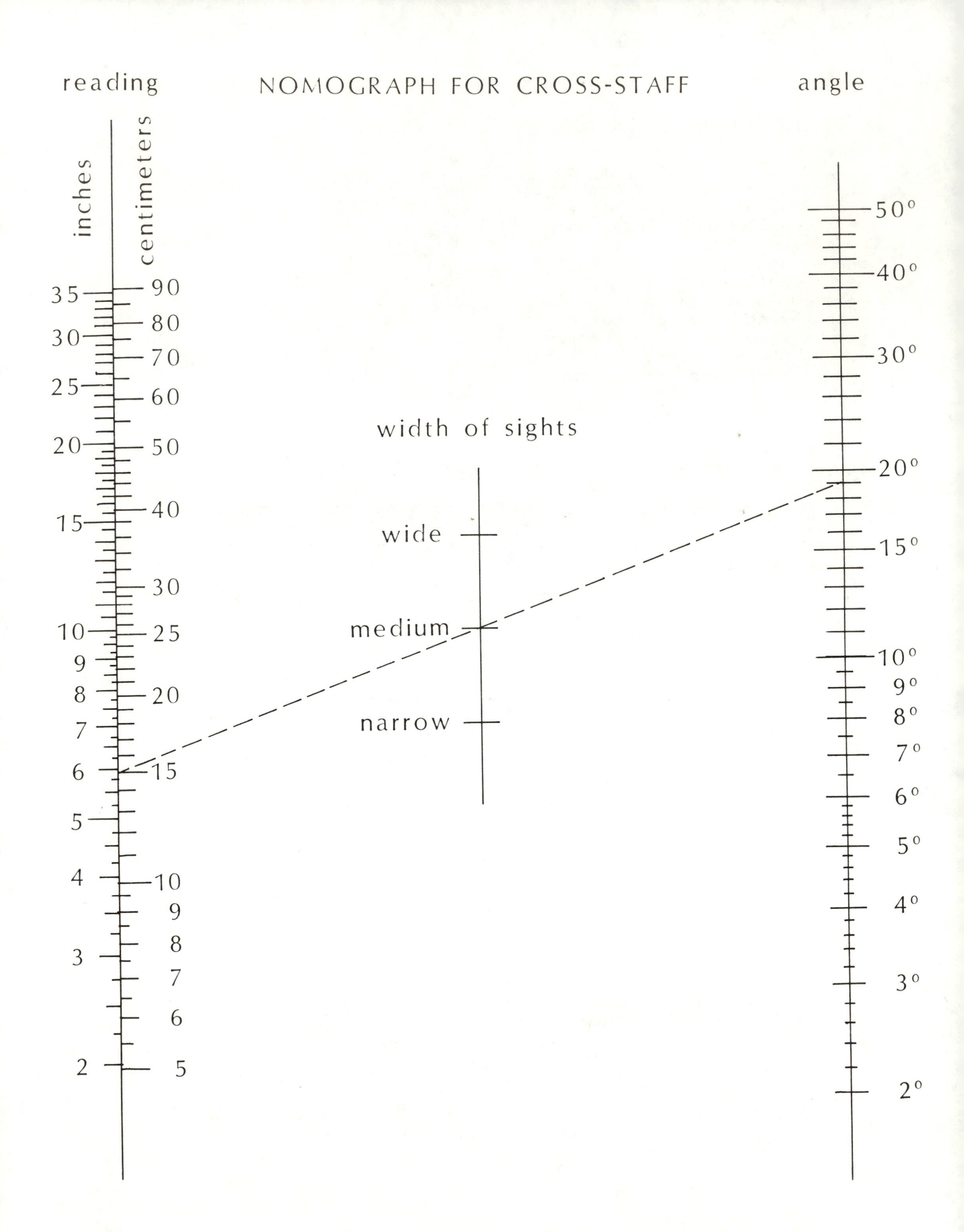
reading
NOMOGRAPH FOR CROSS-STAFF
angle
inches
centimeters
35
30
25
20
15
10
9
8
7
6
5
4
3
2
90
80
70
60
50
40
30
25
20
15
10
9
8
7
6
5
width of sights
wide
medium
narrow
50°
40°
30°
20°
15°
10°
9°
8°
7°
6°
5°
4°
3°
2°

Mapping the Night Sky and Its Motions 2

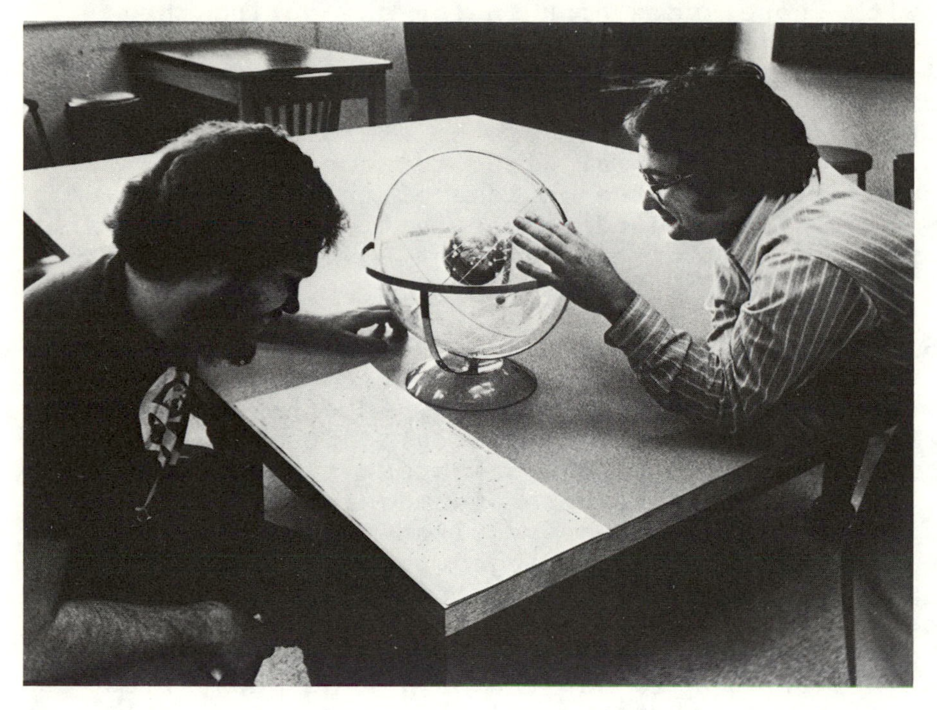

The sky, a star map, and a celestial globe are all you need to know to become an experienced observer.

OBJECTIVES

1. to use the cross-staff and quadrant to measure the angular separations between stars and their altitudes with respect to the horizon
2. to determine the direction and rate at which the sky appears to rotate
3. to draw a diagram illustrating why different stars are seen at night at different times of the year
4. to determine your latitude from measurements of the star Polaris
5. to set up a celestial globe for any latitude on earth and determine the location of any star (with respect to the meridian and the horizon) for any time of the year
6. to find a star on the celestial globe and on a standard star map, given its coordinates from a standard star catalog

EQUIPMENT NEEDED

A cross-staff, quadrant, nomograph, SC1 star chart, protractor, compass, ruler, celestial globe, star map for the month, terrestrial map.

This unit is intended both to familiarize you with the north polar stars and with those stars high in the sky during the evening of the semester in which you take the course and to enable you to establish a correspondence between the sky and conventional star maps. By observing the stars, you will determine the apparent daily motions of the stars across the sky. You will also use a small celestial globe to understand the relation between the earth and the motions of the sky. This unit is prerequisite to all future outdoor observations, since it will serve as your basic orientation to the night sky.

I. OBSERVATIONS OF THE NIGHT SKY

Early observers in Greece viewed the sky as a transparent sphere which surrounded the earth. They divided the stars into six categories of brightness with the brightest stars called first magnitude, the next brightest second magnitude, and the faintest stars visible to the naked eye sixth magnitude. The Greeks, as well as other peoples, arranged the stars into groupings which we call constellations. Most of the names of the eighty-eight constellations we use today are based on Greek mythology and their Latin translations. Modern astronomy uses the constellation designation to map the sky. The boundaries between constellations, set by international convention, are imaginary lines on the celestial sphere. With the invention of the telescope, many fainter stars and objects were discovered. These are also measured on the magnitude system, with a LARGER number denoting a FAINTER object. On maps, the size of the dot is usually related to the brightness of the object.

The maps used in this unit (the star map, the SC1 chart, and the celestial globe) show only the brightest stars. They are named in each constellation from brightest to faintest in order of the Greek alphabet (alpha, beta, gamma,...) used with the constellation name in the genitive form to denote the star; e.g., Alpha Cygni is the brightest star in Cygnus. Many stars also have Arabic names such as Betelgeuse, Algol, and Arcturus; Alpha Cygni is named Deneb. Either designation may be used.

This unit directs you to do five sets of measurements of the night sky. It is important that you be able to identify the stars you are observing in order to interpret the observations correctly.

You will:

1. draw several constellations in parts of the sky not near the north horizon, and measure their angular separations from each other
2. map the stars above the north horizon
3. repeat number 2 at least two hours later
4. measure the stars for determining the horizon on the SC1 chart
5. repeat number 4 at least one hour later

It is suggested that you do numbers 2 and 3 on the same night, followed by their reduction procedure. If you are able, all five sets of observations can be done on the same night, in the order 2,4,1,5,3. This takes a long unbroken observation period. If it is necessary to break a sequence (either 2 3 or 4-5) into two nights, do not let more than three days elapse between the first measure and the repeat measure. If you do, other long-term motions of the night sky will begin to come into play and affect your measurements. But, if only a few days elapse, you may safely assume that the positions of the stars at a certain time of night (to within the accuracy of the measurements) will be the same on the two nights.

A. First Exercise: Drawing constellations

Locate in the sky some of the brighter constellations in the sky at this time of year. Use the star map as a guide. The star maps are given in the Appendix. The schedule below indicates the prominent features to be seen during each semester; after each constellation name in parentheses comes the name of the brightest star in that constellation.

1. January 15 to February 30 evenings: Orion (Betelgeuse, Rigel), Taurus (Aldebaran), Canis Major (Sirius), Auriga (Capella), Gemini (Castor, Pollux), and the Pleiades star cluster are all fairly high in the sky.
2. Late in the spring semester (or past midnight in the beginning of the semester) look for the stars of number 1 in the west, Leo (Regulus) overhead, Bootes (Arcturus) and Virgo (Spica) in the east.
3. Early June: Leo (Regulus) in the west, Bootes (Arcturus) and Virgo (Spica) overhead, Lyra (Vega) Cygnus (Deneb), and Aquila (Altair), in the east, and Scorpius (Antares) in the south.
4. For an early evening late in summer, shift the description on number 3 westward, and look additionally for Pegasus in the east and Sagittarius in the southeast.
5. September evenings: Cygnus, Lyra, Aquila, and Pegasus are all fairly high in the sky.
6. After midnight, or late in the fall semester: look for stars of number 1 shifted eastward a bit.

As a test for the brightness of the night sky, look for the Milky Way. In the winter, it can be found along the line from Capella to Betelgeuse; in the summer, it runs from Cygnus toward Scorpius. If you cannot see it, it means the sky is hazy, or bright, or both.

If you see a bright star and it does not appear to be on the star map, it may be a planet. Mark the positions of such objects on your star map. Draw a sketch of at least FOUR star groupings (brightest stars of a constellation), using your cross-staff to measure the angular separation of some of the stars in the constellation. Use your calibrated hand, fist, and finger to measure the angular separations between the various constellations themselves or between the brightest stars in them when the constellation is large.

B. Second Exercise: Mapping stars in the northern sky

On a night which is clear enough that a fair number of stars are visible, go to a place where you have a clear northern horizon; take your cross-staff and quadrant. Use your instruments to make a drawing, or map, of the stars above the northeast, north, and northwest horizons. Plot your observations on a piece of graph paper. You may prefer to write down your observations and then draw the map, or you could choose to draw the map directly as you make your observations.

Mark a piece of graph paper off in degrees east and west of north. The vertical axis of your graph should be degrees of altitude. You will want to measure the altitudes of the stars and their angular separations, a discussion of some aspects of the measurements is given below, but it is still basically up to you to work out specifically how you want to do the observations.

Note the following USEFUL HINTS. Do not choose stars which are very distant from north. Results will be better if all stars lie within 30 degrees of the north pole. Measurements of angular separations of 40 degrees or more are subject to more error. Be sure to measure stars both east and west of north. There is no set number of stars to be measured, but you should realize that when you return to repeat the observations you will want to measure the SAME set of stars. Thus you should measure enough stars so that you will recognize their patterns and be able to measure the same stars the next time. Further, be sure to keep track of your SEQUENCE of observations, and do them in the same sequence. Note also that haze and dust in the atmosphere make observations near the horizon more difficult, especially in the city where there is a bright sky to contend with. If possible, avoid stars which are quite close to the horizon. In general, all the night observing projects will be more easily done outside the city where the sky is darker, although it may be true that various practical considerations make it difficult to get outside of town very often. To the extent that you can, however, you will find your results improved. Note finally that when the moon is up, the sky is significantly brighter and faint stars are consequently harder to see. Marginal observations should be attempted when the moon is below the horizon, if possible.

MEASURING TECHNIQUES: There are a variety of ways you could make the measurements for the north polar map. In general you will measure the angles between stars with your cross-staff and their altitudes (angles above the horizon) with your quadrant. In figure 1, you would measure the indicated angles with the indicated instrument.

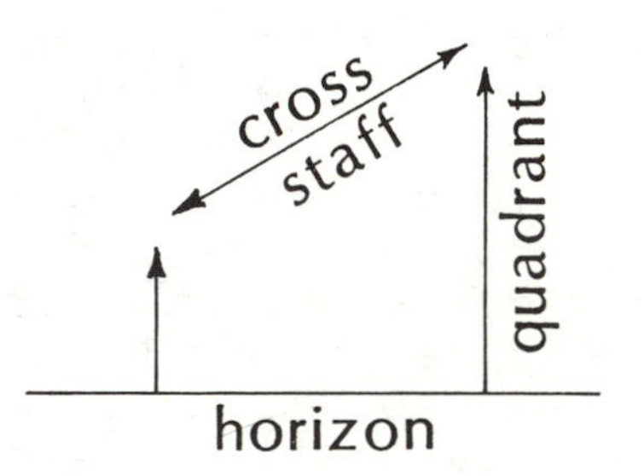

Figure 1

You may also use techniques similar to those used by surveyors to construct maps, as follows: Suppose that you have the positions of two stars, say A and B as shown in the figure, and you want the position of a third star to be 18 degrees from A and the angle between B and the third star to be 12 degrees. Setting a compass at 18 degrees by using the axes on your graph paper as a scale, strike an arc around star A, and then another intersecting arc 12 degrees around star B. The intersection determines the position of the third star, as shown in figure 2.

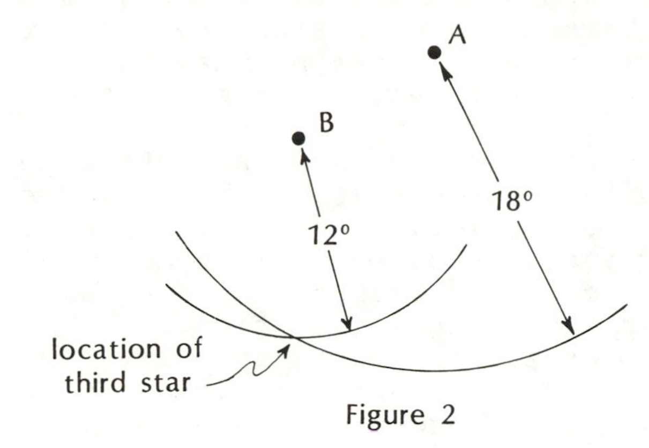

Figure 2

The third star may be between the other two, as is illustrated in figure 3. Generally, the two arcs intersect in two places but it should be obvious which is the correct position. If there is any ambiguity, measure the position of the object with respect to a third star.

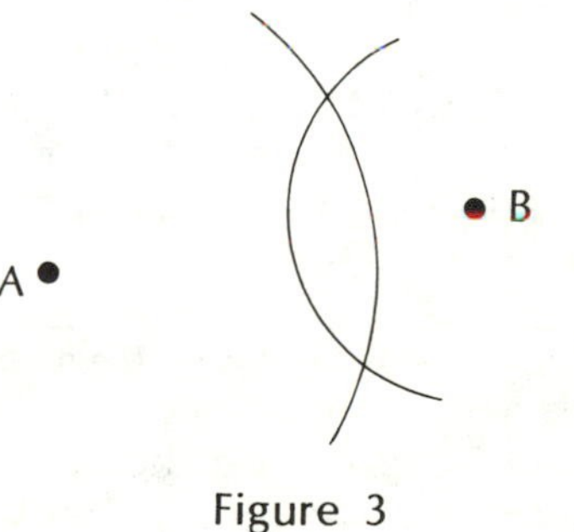

Figure 3

C. Third Exercise: Remeasure the north polar stars

If at least two hours have passed, return outside and again map the stars above the northern horizon in the same fashion as you did before, noting carefully the time you begin and end. Be sure that you measure the SAME STARS as you did previously, and in the same order. You may do this on another night, if within three nights, but make sure your clock time is at least two hours later than the first set of measurements.

Now, turn to the section on reductions and interpretation for the north horizon.

D. Fourth Exercise: Determining the horizon

Select a set of easily identifiable stars (ones which you can find on your star map) which are not too far from the horizon (stars ranging from 10 degrees to 40 degrees in altitude). Find a star in the south, southeast, east, northeast, north, northwest, west, and southwest if you can; that is, find objects all around the horizon in all the compass directions. Choose stars which appear on the SC1 chart.

For example, in the early evening at the beginning of the fall semester, some appropriate objects could be Arcturus in the west, Antares in the southwest, a star from Sagittarius in the south, a star in Cassiopeia in the northeast, and a star in Pegasus in the east. (Stars which are quite high in the sky, such as Vega, Altair, and Deneb, are not very useful for this part of the project at this time of year.) Late in the fall evening (and later in the semester), other brighter stars will be appearing in the east to view. There are no extremely bright stars in the early evenings of the fall semester in the east, so you will have to be careful to choose a star you can definitely locate on your star map in this direction.

Having chosen your set of stars carefully, measure their altitudes. Note the time of your observations and get your measurement procedure well organized in order to measure all the altitudes as quickly as possible, to minimize changes in the sky while you're measuring. Then, with a compass, mark on your SC1 chart the altitude angles you have measured for each star by striking an arc around the star of the proper size. Use the vertical axis of the SC1 chart as a scale for your compass. Your horizon can then be drawn in on the star chart as a smooth curve which is TANGENT TO ALL THE ARCS you have drawn. Note that the shape of the curve is similar for observations from any latitude, but for some latitudes the entire curve will not fit on the SC1 chart.

E. Fifth Exercise: Remeasuring the horizon

If at least one hour has passed since your first horizon measurements, you may begin this exercise. Since the stars move with respect to your horizon, your horizon will appear to change with time on the SC1 chart. The time difference is very important. Remeasure the same set of stars (or as many of them as are still above the horizon) in the same order as your first set of measurements.

If your western stars have set, don't worry about it; just do the eastern stars, or choose a few NEW stars in the west. By comparing the two sets of horizon observations, determine the rate at which the sky is rotating and the direction of rotation.

II. INTERPRETATION OF THE OBSERVATIONS

A. The Star Map

Compare your drawings of the various constellations with the patterns on the star map. Note that on the map one degree is one millimeter, unless you cross the outline. Using a millimeter rule, measure on the map the angular separations between the various constellations and the stars within them, and compare your map readings to the measurements that you made. Make a table of comparisons in your notebook. The star map is an unusual shape in order to eliminate the type of distortion found in the SC1 chart.

B. The North Horizon

Do your observations show the apparent rotation of the celestial sphere? If so, in what direction is the rotation? Were there any stars that didn't move at all? If so, what was the position in the sky of a stationary star? (Answer all questions in your notebook.)

It is clear that the apparent rotation of the sky is simply a reflection of the earth's own daily rotation on its axis. The pole star Polaris, at the present time, is located very near the north pole of the sky or the point where the earth's north rotational pole would intersect the celestial sphere if it were extended out indefinitely.

The diagrams in figure 4 are intended to convince you that the altitude of the pole star depends upon the observer's location on the earth. Each diagram shows the earth, its equator, and the line of sight to Polaris for an observer standing at a different latitude.

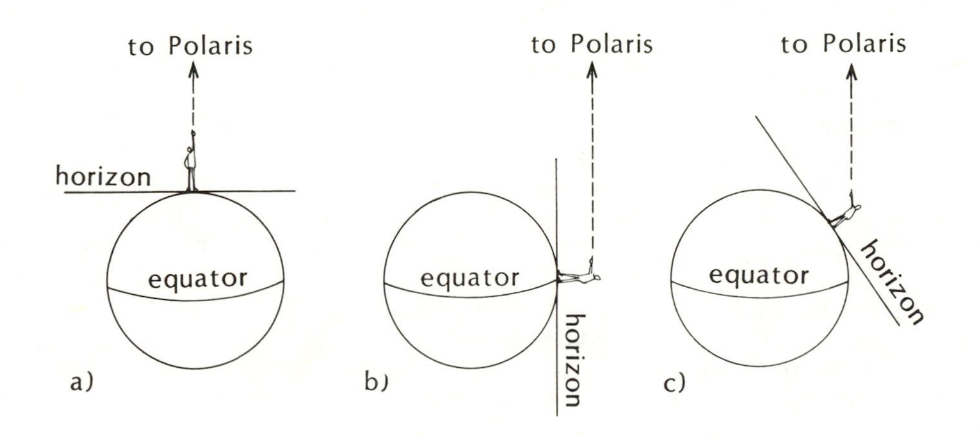

Figure 4

The first diagram is for an observer standing at the north pole. In this case, Polaris would be directly overhead (at the zenith), and its altitude would be 90 degrees. The second diagram shows an equatorial observer, for whom Polaris would be right on the north horizon and thus would have an altitude of zero degrees. Thus, in both cases, the altitude of the pole star has equaled the latitude of the observer. The third diagram shows the situation for an observer at an intermediate latitude. From studying these diagrams it should not be difficult to convince yourself that the altitude of the pole star is equal to the latitude of the observer. Notice that in each diagram the observer's horizon is shown; it is represented by a tangent plane to the earth's surface at the point where he or she is standing. Objects below the observer's horizon are not visible. Notice also that the direction to the pole star is indicated as being the same on each diagram, even though the observers are at different locations on the earth. This is because the stars are so enormously distant compared to the size of the earth that all observers anywhere on the earth see a particular star by looking in the same direction in space. Alternately phrased, one could say that the light from any given star falls in very closely parallel rays over the entire surface of the earth.

Calculate your latitude from your measurements, and write it in your notebook.

Note that if you happen to be going on a trip which involves travel either north or south, the altitude of Polaris will change. If you take your quadrant and measure the change, and if you also know (from an atlas or the car odometer) how many kilometers north or south have been traveled, then you could easily compute the circumference of the earth. The change in the altitude of Polaris would be the same fraction of 360 degrees as the number of kilometers you traveled was of the earth's circumference.

Look at a map to see which cities lie north or south of you.

C. The Rate of Rotation of the Celestial Sphere

By examining your north polar observations, determine how long it takes for the celestial sphere to rotate 360 degrees. Use a protractor to determine how much rotation took place during the time interval between your two sets of observations and then determine a rate in degrees per hour at which the sky moves. Then determine how long it will take to move 360 degrees. Note: do not worry if you don't get exactly twenty-four hours -- you won't. The difference is the observational error of this particular method. Write your answer for the time to rotate 360 degrees (in hours) in your notebook.

Evaluate your observational error as a percentage and then see if you can make some suggestions as to how your error could be reduced if you redid this project.

D. The Horizon

Measure the separation of your horizon lines at several points, and take an average. What is the motion of the horizon?

Compare this rate with that determined from the polar measurements. How could the errors in this determination be reduced?

Notice that the north-south position of a star (its distance from the celestial equator) is given in degrees, while the east-west position is given in units of time. This is convenient because of the earth's eastward rotation, which makes the sky appear to rotate westward. The sky rotates 360 degrees in twenty-four hours, so the rate is 15 degrees per hour. The zero point reference for the east-west coordinate is by convention taken to be at the intersection of the CELESTIAL EQUATOR (which is the intersection between the plane of the sky and the earth's equatorial plane) and the ECLIPTIC (the earth's orbital plane around the sun, projected onto the plane of the sky). Find this intersection and note that it is marked 0. This reference point is now between Pisces and Aquarius.

You may notice as you compare your observations with the SC1 chart that the two agree well along the celestial equator but that there are some divergences when you get far from the equator. This is because the chart actually begins to distort (with respect to the apparent sky) as you go farther from the equator. Such distortion is inevitable when you take a spherical surface such as the sky and turn it into a flat map. The mapmaker has chosen to represent the equatorial regions most like the apparent sky and let the distortion show up progressively as you go away from the equator. Because of this distortion, the chart goes only to 60 degrees above and below the celestial equator. You have no doubt noticed similar distortion on maps of the earth with certain types of projections, where Greenland looks bigger than all of North America put together. The equator is plotted as a straight line on this map, which makes all other great circles on the sky (e.g., the ecliptic) appear as waves.

E. The Celestial Globe (Indoor Exercise)

The celestial globe is a handy device for visualizing the apparent motions of the sky and sun and for determining what stars will be up at a certain time of the year. The directions below are intended to lead you into a familiarity with it.

Note the earth in the center, the celestial sphere around it, and the movable sun. Although the earth in fact circles the sun, from the earth it appears as if the sun moves around us once a year, and this is what the globe represents, that is, its viewpoint is entirely GEOcentric. The plastic globe does actually misrepresent one aspect of the earth-sky relationship, however -- the scale is quite different from the impression given by the device. The scale of the globe suggests that the distance from the earth to the stars is only slightly larger than the distance from the earth to the sun and not too much larger than the size of the earth itself. In reality, the diameter of the earth is 12,750 kilometers (8,000 miles), whereas the distance to the sun is 150,000,000 kilometers (93,000,000 miles). Further, the stars themselves are enormously more distant (as you will see in unit 12); the NEAREST star is approximately 700,000 times more distant than the sun. Thus, if any question of scale arises with the celestial globe, it is well to remember that a proper perspective would require the size of the earth to be shrunk down to an infinitesimal point. If you do imagine the earth shrunk down in this way, then it is not difficult to realize why all observers on the earth will see a given star in the same direction in space. (See figure 5.)

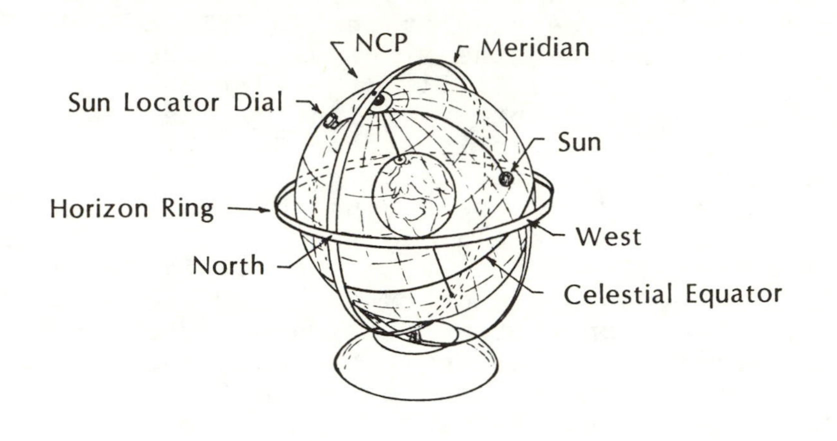

Figure 5

Since the earth circles the sun once a year, the sun will appear to move with respect to the background stars and, from a geocentric viewpoint, it will circle the sky once a year in addition to its daily motion. Figure 6 illustrates why the sun appears to move with respect to the background stars.

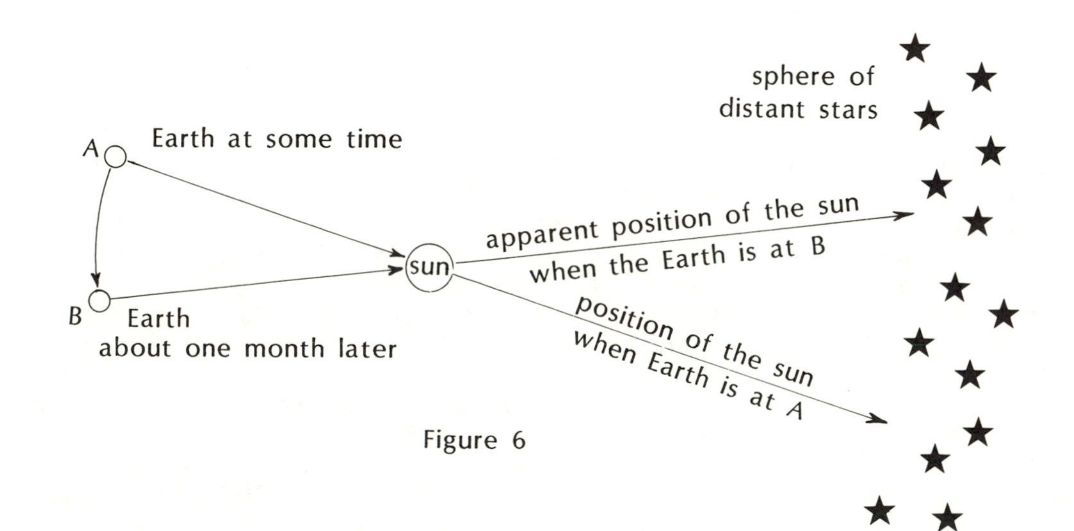

Figure 6

The sun's apparent path across the sky is called the ECLIPTIC; this great circle on the sky is marked on the celestial sphere. Move the sun around and note the dates on the ecliptic, indicating that the sun is found in that part of the sky at that time of year. (The stars in the opposite part of the sky will then be the nighttime stars.)

Notice also on the globe the celestial equator, the extension of the earth's equatorial plane upon the plane of the sky. The celestial equator is marked off by the east-west angular coordinate which is measured in hours of rotation. This coordinate is called the RIGHT ASCENSION. Notice again the zero point where the ecliptic and the celestial equator intersect. The north-south coordinate (called the DECLINATION) is marked off in degrees. This coordinate system is similar to that of the earth, where the east-west measurement is in degrees of longitude and the north-south in degrees of latitude.

The ecliptic and the celestial equator do not coincide, because they are great circles that relate to different motions of the earth. The ecliptic is essentially a reflection of the earth's orbital motion around the sun, whereas the celestial equator is determined by the rotation of the earth upon its axis. Since the earth's rotational axis is actually tilted at an angle of 23 1/2 degrees with respect to the perpendicular to the earth's orbit, the ecliptic and the celestial equator also cross one another at an angle of 23 1/2 degrees. One crossing point, the vernal equinox, serves a function similar to Greenwich, England, the zero point in the earth's longitude.

Notice that because of this, as the sun moves around the ecliptic at a rate of approximately one degree per day, it is sometimes north of the celestial equator and sometimes south. From the celestial globe, answer this question: What is the farthest north of the celestial equator the sun can be, and on what date does this occur? (See figure 7.)

Units 6 and 7 will cover the motion of the sun in more detail.

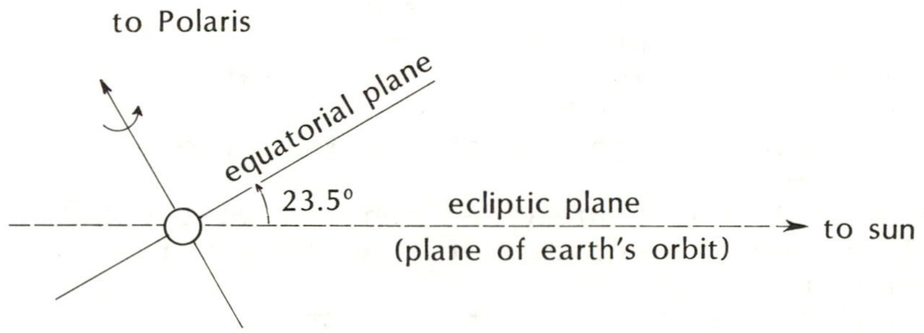

Figure 7

SETTING UP THE CELESTIAL GLOBE:

Rotate the earth until your position is "on top," set the horizon ring so that it is horizontal, and place the sun at today's date on the ecliptic. Finally, adjust the globe so that Polaris is at the angular distance above the north horizon equal to your latitude. The globe is now set up to reproduce the sky as seen today.

Rotate the sky until the sun is as high in the sky as it will go; this is local NOON. At this time, the sun is located on the MERIDIAN, an imaginary line on the sky connecting the overhead point (or ZENITH) of an observer to the north and south points on the horizon. Notice that observers at latitudes north and south of you will have the same meridian as you, although their zeniths will differ. At any given moment, however, observers east and west will have different meridians.

Now rotate the sky until the sun is passing the horizon ring in the west; the globe now represents the position of the sky at sunset. When the sun is as far below your position as it can be (180 degrees from its noon position), it is local midnight. Finally, rotating the sky and sun around 270 degrees from its noontime position will make the sun rise on the eastern horizon.

Locate on the celestial globe the constellations you drew in the first exercise. Check some of the angles between the constellations you measured using the globe and measuring the separation of the same constellations on it with a flexible ruler.

Set up the celestial globe for the time of night and date when you did your first observations of the north polar stars. Look through the globe past your location on the earth and see if the north sky stars resemble the drawing you made of them. Rotate the sky until it looks similar to your observations the second time you measured the north polar stars. Does the time implied by the position of the sky agree with the time you did your second set of observations?

Now set up the globe for the times when you did your observations of the horizon, and see if it agrees with what you saw at those times.

Using the celestial globe, answer the following questions:

1. What are the coordinates of Vega?
   Right ascension =
   Declination =
2. At midnight tonight, what constellations will be found along the meridian?
3. At midnight tonight, what constellations will be closest to the zenith?
4. On June 23 one hour after sunset, what constellations will be on the meridian for an observer at your location?
5. On August 15 at midnight, can an observer at your location see Sirius (the bright star in Canis Major)? Can an observer see Vega (in Lyra)?

Now set up the celestial globe to reproduce the sky as seen by an observer at the north pole (keep the horizon ring horizontal and rotate the sky until the pole star is straight up). Rotate the sky to see how the stars move during the course of a day and answer the following question: Is it true that an observer at the north pole could see half the stars in the sky at some time during the year but could NEVER see the other half?

Move the sun to different dates of the year and notice its behavior. During the day on June 23, how would the sun appear to move for an observer at the north pole? What happens on September 23? On December 23?

Set up the globe for an observer on the equator. Where is the north star now? Spin the globe and examine the motions of the sky that result. Is it true that an observer at the equator, at some time or another during the year, should be able to see all the stars in the sky? Why don't astronomers build observatories on the equator? Note that, at the equator, the stars rise perpendicular to the horizon. How do stars behave with respect to the horizon as they rise and set at your latitude?

Note: You do not necessarily have to proceed on to unit 3 after completing unit 2. You are now free to plan your progress through the rest of the book. Your choice of units should be importantly influenced by the special projects you wish to do and by their prerequisites.

Lunar Surface Features

3

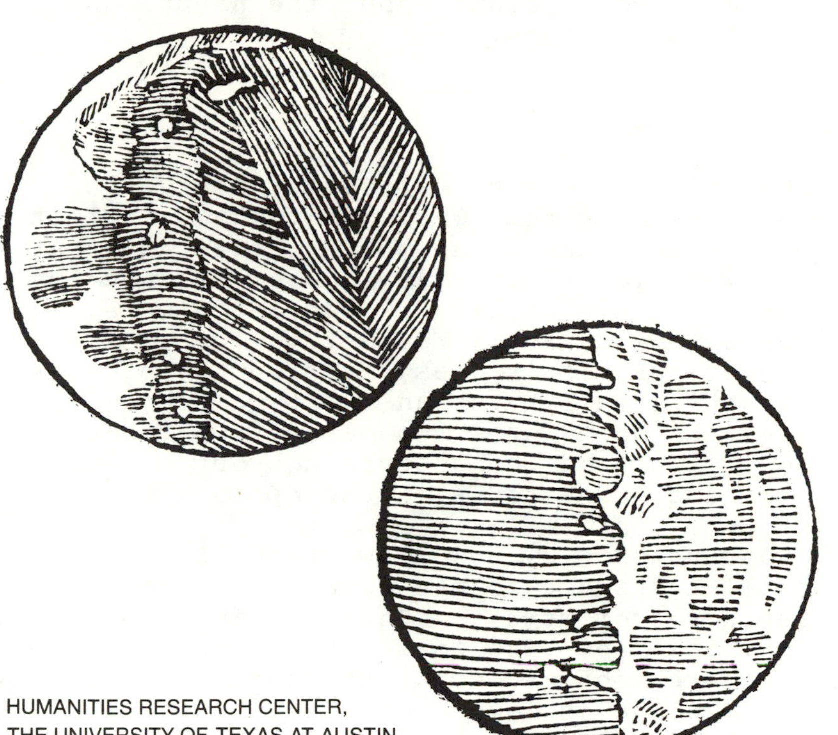

HUMANITIES RESEARCH CENTER,
THE UNIVERSITY OF TEXAS AT AUSTIN

It is not difficult to draw better moon maps than Galileo did in 1610. Surprisingly, even a very small hand-held telescope can detect features as small as 10 miles in scale.

OBJECTIVES

1. to draw lunar features and identify them using a lunar globe
2. to classify lunar features from photographs and estimate their order of formation
3. to calculate the physical size of features as viewed with the naked eye, telescope or binoculars, and photographs

EQUIPMENT NEEDED

A small telescope or binoculars, a lunar globe, "The Lunar Orbiter Photographic Atlas of the Moon" (optional).

In this unit you will investigate the surface features of the moon in several ways and draw conclusions about the nature and evolution of the visible lunar surface.

I. OBSERVING THE MOON VISUALLY

For this part of the unit you are requested to observe the visual surface of the moon at more than one phase. The simplest and best situation would be to observe the moon at first quarter and then at full moon, a week later, or at full moon followed by third quarter a week later. (This is less convenient as the third quarter moon rises somewhat late.) Weather may prevent such tidy scheduling of observations; if it does, observe the moon at two different phases, starting no later than first quarter.

Each time you observe the moon, draw two maps of what you see. (Use your notebook. Don't be shy concerning your artwork; if you enjoy drawing, try the suggestions in the optional appendix.)

Draw one map with your naked eye, and draw the other by looking through your small telescope or binoculars. Do not attempt to make the drawings from memory. They should be as detailed as practical or possible. The following hints may help: do not try to draw the entire moon at once, but construct your sketch in pieces. First examine the moon for a while without drawing anything, noting the light and dark areas on the surface, and also noting that specific features, even if unidentifiable, can be seen. Draw a circle in your notebook and sketch onto it the outlines of the pattern of light and dark on the lunar surface. Do not draw the entire surface at once, but examine one part and sketch it before going on to the next. Then go back over the surface a piece at a time looking for specific features -- lines, circles, rays, etc. Use this procedure for both your naked eye and telescopic drawings.

When your drawings are completed, take time to compare them with the moon one last time. Then compare both of them with the lunar globe. Can you identify what part of the moon you have been seeing by using the globe? Note in your observing book the specific features on the globe you were able to see with your eye and with the small telescope. List their names. Find the smallest object or detail of an object that you were able to see in each case. Using the fact that the moon is 1/2 degree in apparent size and 2,000 mile in true diameter, estimate the apparent and the true physical size of the smallest feature you observed in each drawing.

Repeat this procedure at another phase. At which phase were you able to see more detail? Why is this so? Compare your drawings to those of Galileo. Galileo's drawings, made with the first known telescope, appear on the first page of this unit.

II. EXAMINATION OF LUNAR ORBITER PHOTOGRAPHS

Five Lunar Orbiter missions were flown by NASA in 1966 and 1967 to provide information for the Apollo lunar landings. The pictures were taken and developed in the spacecraft and scanned for transmission back to earth. The motion of the film in the scanner device produced the lines visible in the photos. These lines delineate the areas known as framelettes. Since the altitude varied in the orbits, the physical size of the framelettes varies. Two cameras operated simultaneously, one at high resolution and one at medium resolution. A complete set of Lunar Orbiter photographs consists of approximately 3,100 prints, so obviously a selection has been made for this unit. The detail seen is better than that obtainable by any telescope located under the earth's atmosphere.

Use plates 7, 68, 94, 149, 176, 192, 204, and 622. (These numbers appear on the plates in this book, and on the corresponding page in "The Lunar Orbiter Photographic Atlas of the Moon" where the scale is slightly larger.) Examine the photographs carefully. Note that there are a number of different features visible. After your examination make up a classification scheme for the features. You will find that approximately ten distinct types are needed. List your classification scheme in such a manner that someone else could use it to classify features. Give examples of each feature by plate number and location in the plate. Notice that some features are raised and others are depressed. In some cases it is difficult to tell. (Spend some time considering plate 149.) Can you suggest a criterion for telling whether a feature is raised or depressed?

Due to its airless nature, we see the moon unmolested by the effects of atmospheric erosion. We see the combined effects of over four and one half billion years of lunar history. Some features can clearly be shown to be older than others. What criteria can you use to decide whether one feature was formed before or after another? Find several specific examples of age difference in the photographs and list their positions in your notebook.

Obtain a copy of the "astronomically accepted" classification scheme. Do not worry if your scheme differs from it ... there are many possible ways to classify lunar features. Compare the two schemes. Which seems more complete? Does either scheme include types of features that the other one misses? On the "astronomically accepted" sheet, there is a column indicating the plate number upon which an example of that feature can be found. Locate an example of each of these features and indicate on that sheet the position of the feature.

Calculate the size of the smallest feature visible on plate 149, using the framelette size as listed on the plate.

The Lunar Orbiter photography has been provided by the National Space Science Data Center.

APPENDIX

The Shading-Erasure Technique for Drawing Lunar Features

As stated previously, line drawings are sufficient for this unit, but one rather simple way to add an artistic touch involves using the shading-erasure technique.

1. Uniformly gray in an area about 3 inches in diameter on smooth white paper in the following way: obtain some powdered graphite by filing a pencil lead with a fine file, or purchase powdered graphite where artists' supplies are sold. Place a small amount of graphite on a ball of cotton (or soft tissue paper) and rub the cotton over the paper. Try to make the background uniformly gray.

2. Take this prepared paper to the telescope and trace in the outlines of the features which you wish to draw. It is at this point that you must make sure that the shapes and relative sizes look right.

3. Wherever you see a bright area reflecting sunlight, erase the gray background.

4. Wherever you see dark shadows, shade that area. Add finishing touches and details. Record the date, time, and phase.

"Astronomically Accepted" Classification System

Name/Description	Plate	Location
1. CRATERS, chain - a line of craters adjacent to each other	622	______
2. CRATER, peaked - crater with central peak(s) near the center	7	______
3. RAYS - bright streaks radiating from craters	192	______
4. MARIA - relatively smooth flat areas of darker color caused by lava flows	176	______
5. MOUNTAINS - steeply sloped features surrounding maria basins	94	______
6. HIGHLANDS - rugged higher regions composed of basalt with high aluminum content, thus lighter color than maria basalts	204	______
7. ISOLATED PEAKS - mountains partially covered by maria	176	______
8. DOMES - low rounded raised structures on maria surface	169	______
9. WRINKLE RIDGES - ropelike raised features on maria surface, often sinuous in shape	68	______
10. RILLES - narrow valleys or long narrow canals or cracks	94	______

REFERENCES

Bowker, David E. and Hughes, Kenrick. "The Lunar Orbiter Photographic Atlas of the Moon," NASA SP-206. 1971.

Kosofsky, L. and El-Baz, Farouk. "The Moon as Viewed from Lunar Orbiter," NASA SP-200. 1970.

Schultz, Peter H. "Moon Morphology," University of Texas Press. 1976.

PLATE 7

PLATE 68

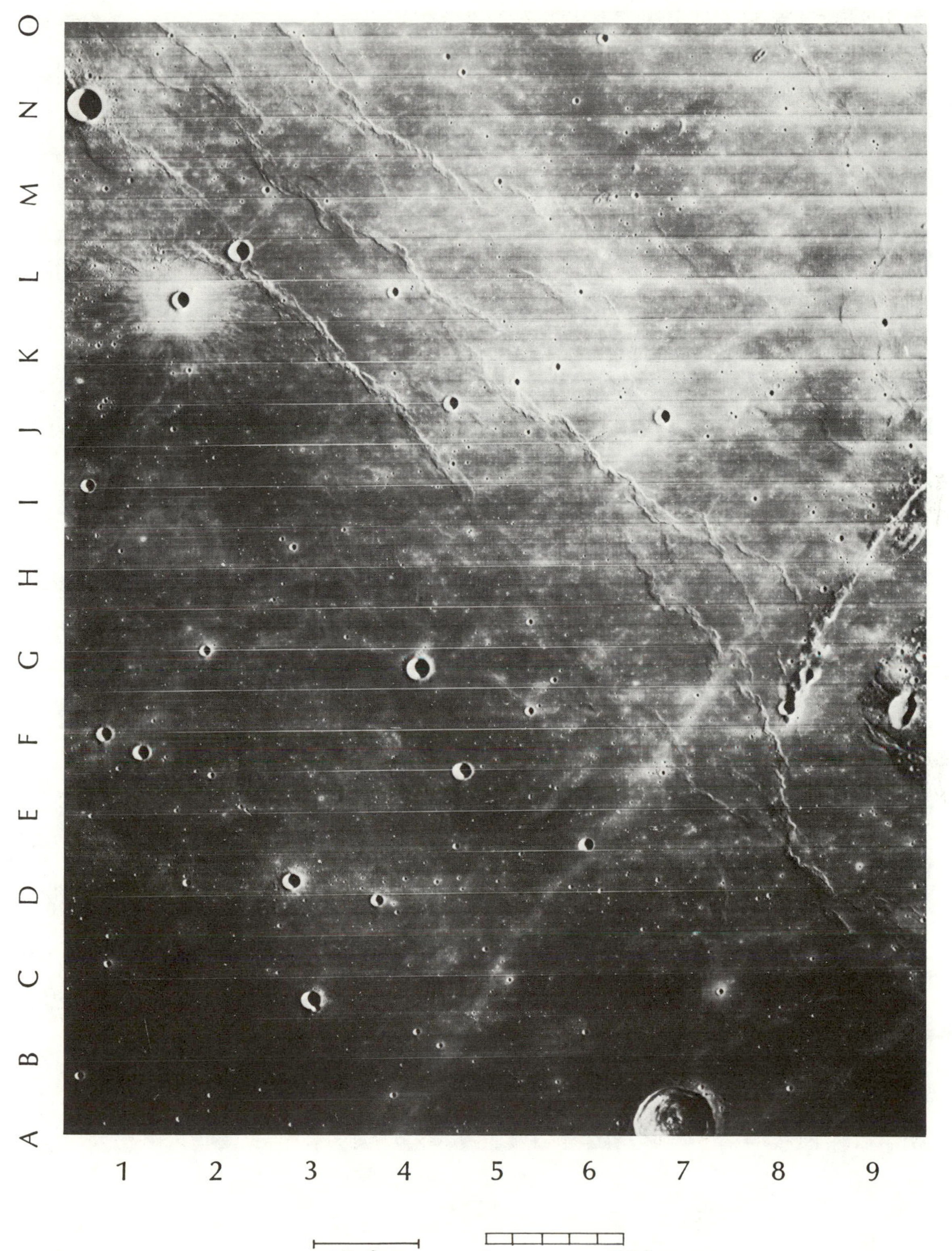
O
N
M
L
K
J
I
H
G
F
E
D
C
B
A
1
2
3
4
5
6
7
8
9

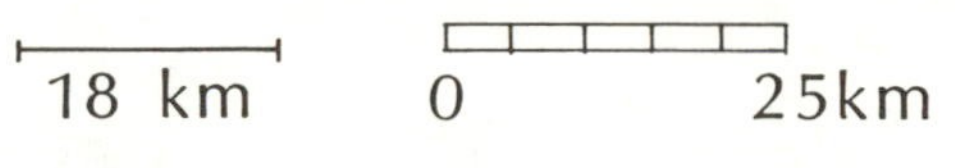
18 km
0
25km

PLATE 94

PLATE 149

PLATE 169

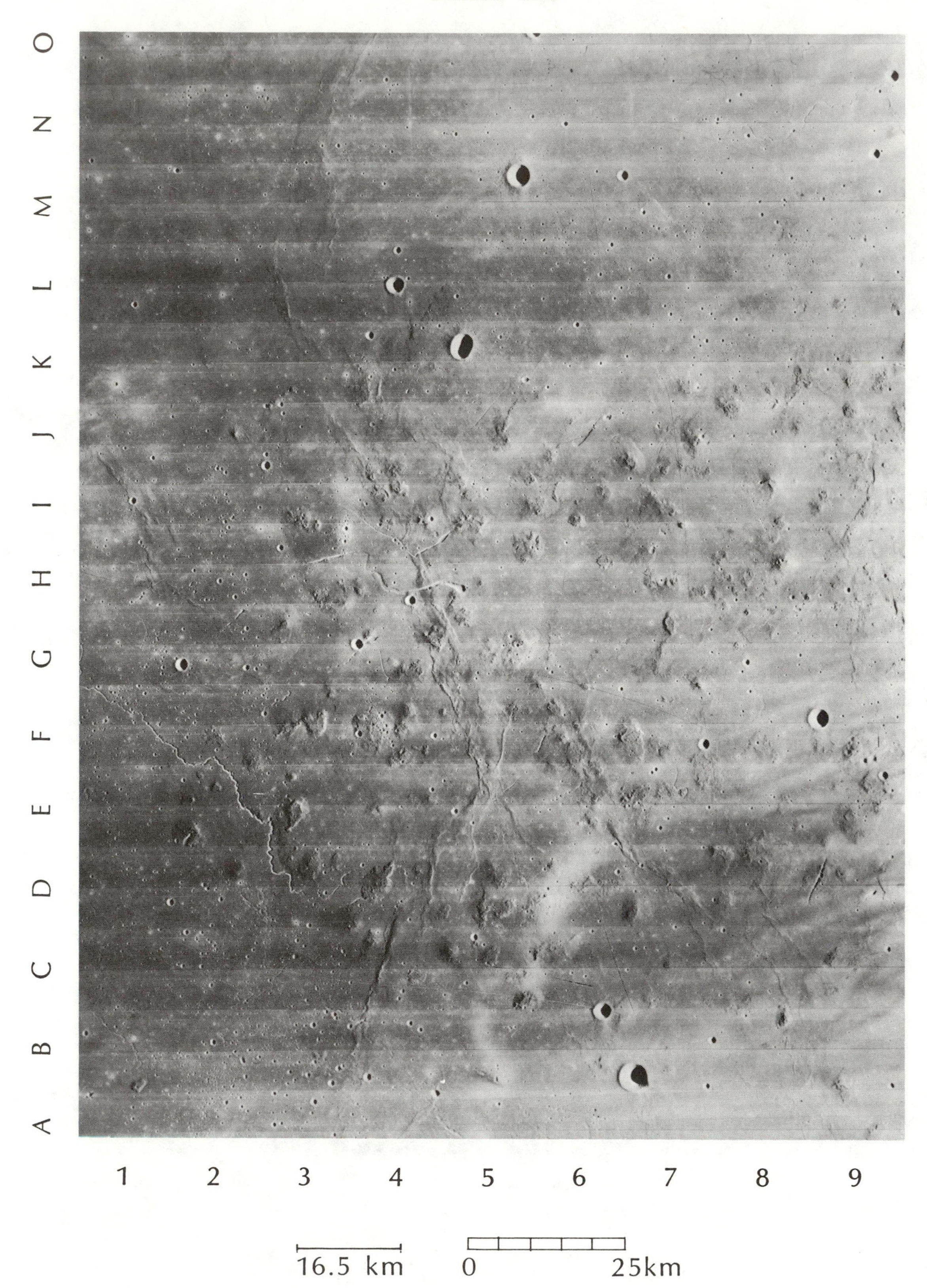

PLATE 176

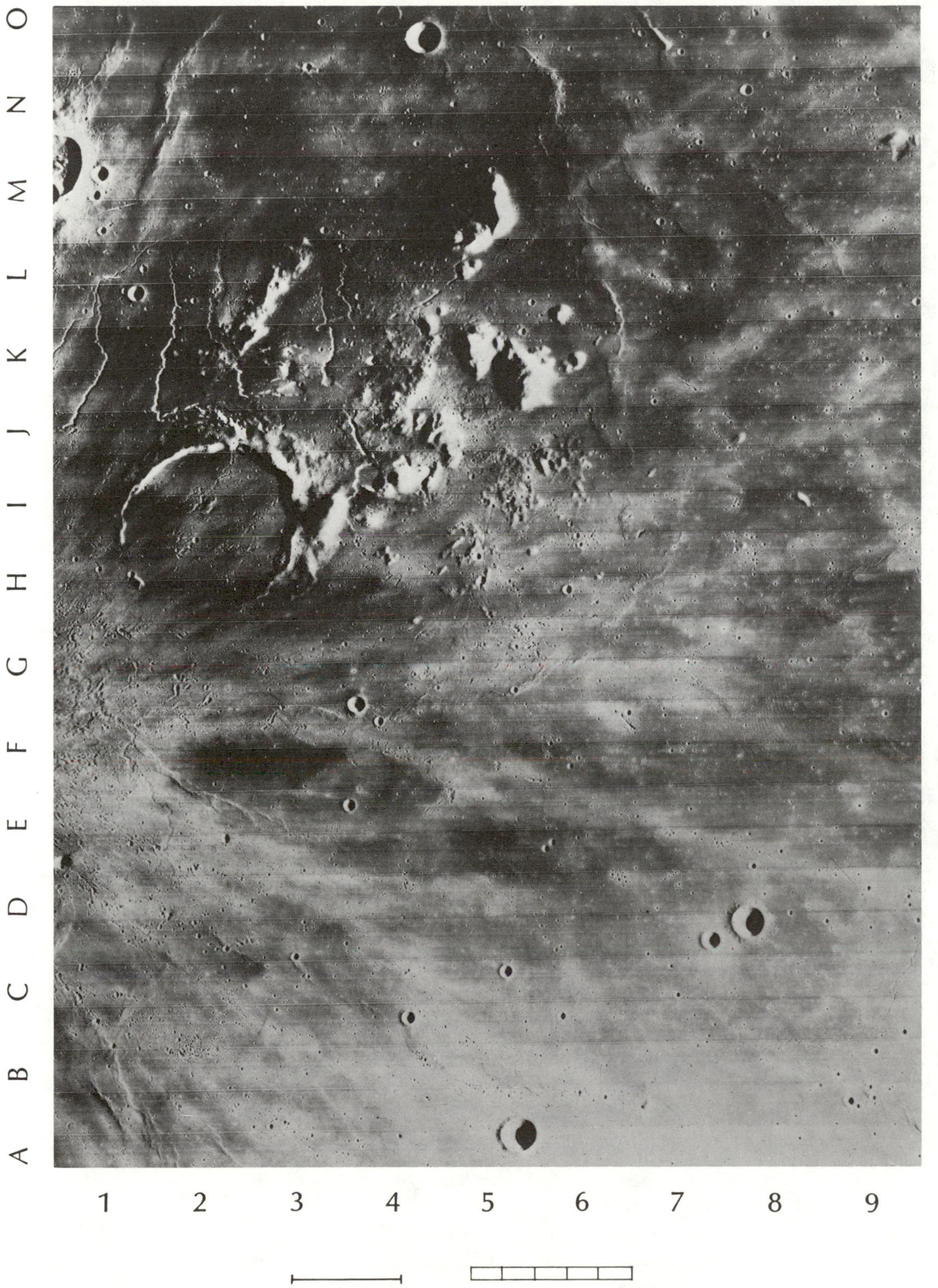

PLATE 192

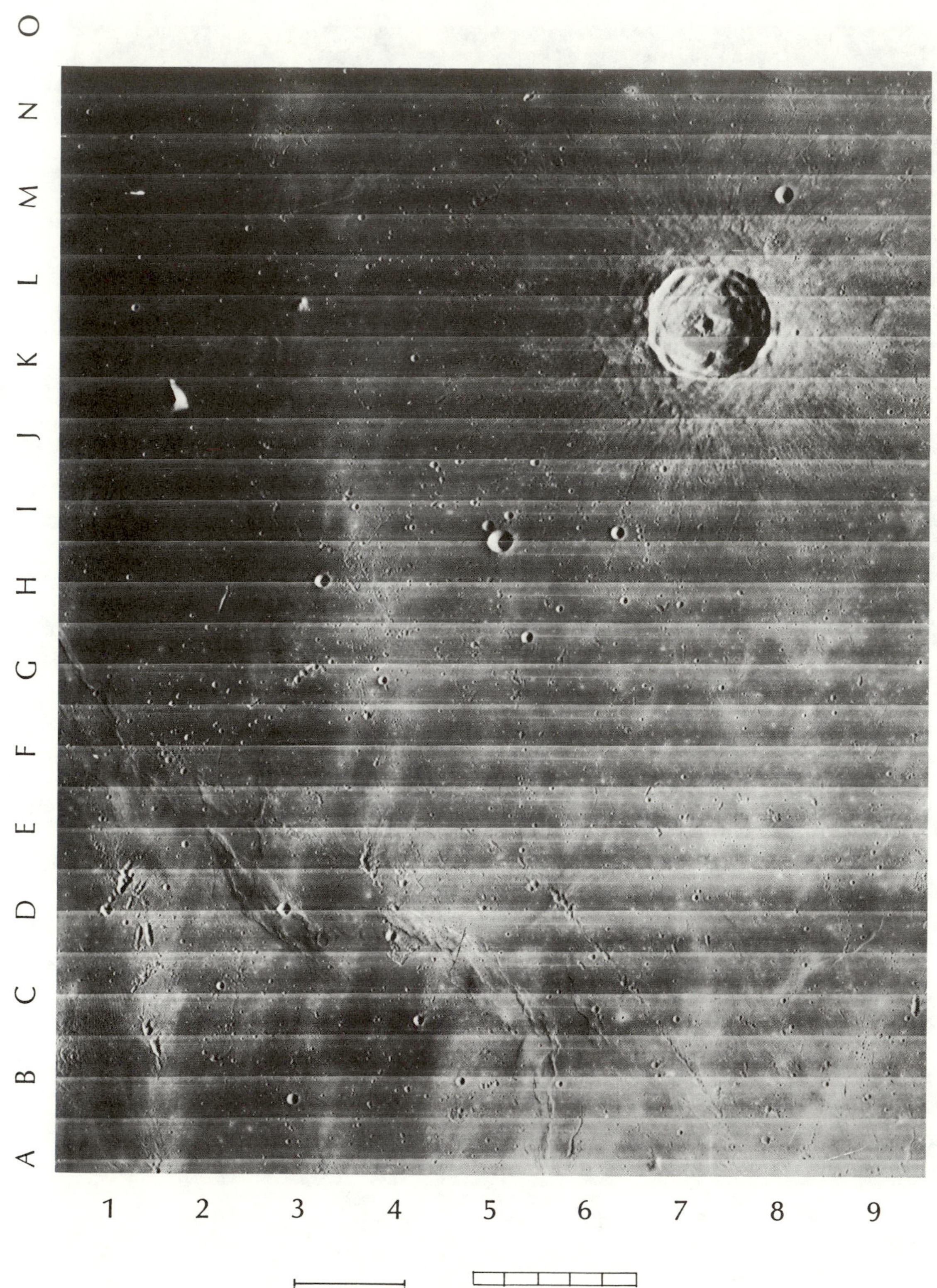

PLATE 204

PLATE 622

The Motion and Phases of the Moon

4

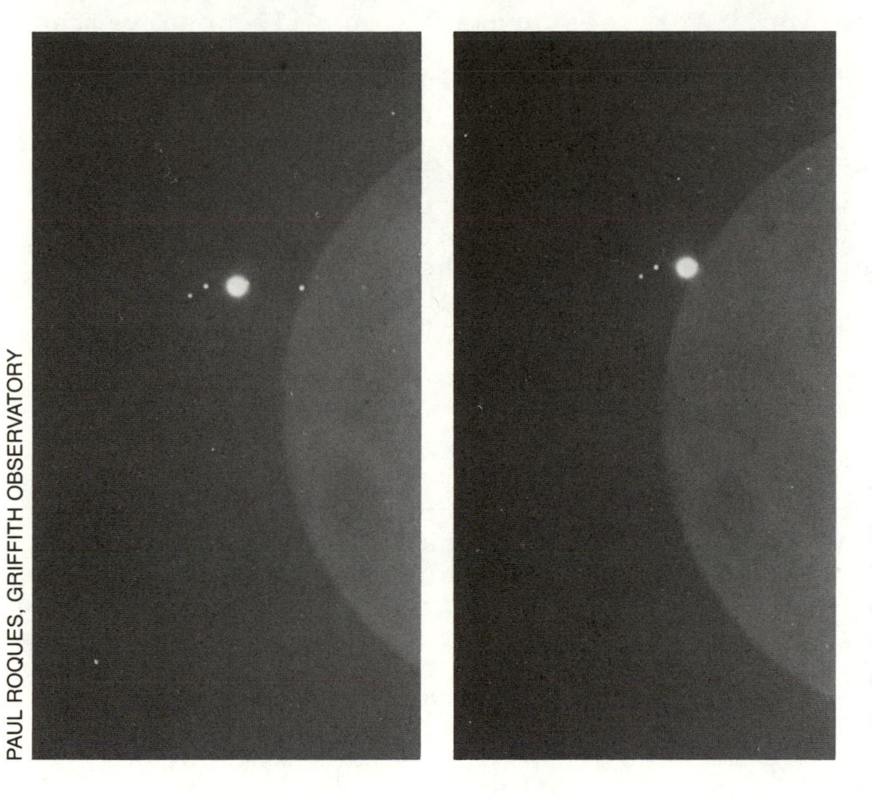

PAUL ROQUES, GRIFFITH OBSERVATORY

The motion of the moon, here shown occulting Jupiter and its satellites, is now determined so precisely by laser ranging that it can be used as a test of the subtle predictions of Einstein's theory of general relativity.

OBJECTIVES

1. to predict the phase of the moon, given the angle between the sun and the moon as seen from the earth, and to predict the phases of the earth, given the angle between the sun and the earth as seen from the moon
2. to determine the angular rate of motion of the moon across the sky with respect to the background stars
3. to determine how long it takes the moon to travel 360 degrees around the sky with respect to the background stars
4. to determine how long the moon takes to go from full moon to full moon, or from any given phase back to that same phase again
5. to be able to explain by using diagrams why the two time intervals determined in 3 and 4 are not the same

EQUIPMENT NEEDED

The cross-staff from unit 1, an SC1 sky chart.

Perhaps as you observed the stars of unit 2, you noticed the moon and its rapid motion. For this project you will observe the phases of the moon and its position in the sky with respect to the background stars, every two or three nights for about three weeks. Any given night's observing should require at most five minutes of time, but, since this project does continue for a longer period than the others, it is an exception to the suggestion that you work only on two projects at a time. Note also that if you are interested in doing unit 3, it is more efficient to work on it and this one together. It is best to begin soon after the new moon, when the night sky is least bright. However, it can be started at any time of the month.

I. OBSERVATIONS

Each night that you observe you should (1) find the moon and determine its position carefully with respect to the stars in its vicinity; indicate in pencil on your SC1 chart where the moon was found, (2) draw the phase of the moon in your observing notebook, and (3) MEASURE, or estimate if the angle is too large, the angular separation of the moon and sun in the sky. Be sure to note the date and time of night next to each observation you record. Note that, as the moon moves through its cycle, you may have to adjust the time of night at which you observe it; this is the reason you should not let too many nights pass without at least finding the moon -- you might otherwise lose it.

To locate the moon with respect to the background stars and record it on your star chart, you may either visually estimate the position of the moon or measure the angle between it and some bright background stars with your cross-staff. Unless you are an experienced observer, the latter method is usually preferable. This procedure would then be the same as that used in unit 2 to determine the position of a third star from the known positions of two other stars. You may also want to use any bright planets you can find as "landmarks" to locate the moon, provided you have carefully located the planet itself on your star chart before beginning. You should not use Mercury, Venus, or Mars for this purpose, as they themselves have a fairly rapid motion with respect to the background stars and thus they will not hold still during the lunar cycle. Jupiter and Saturn are so distant that their motion across the sky is slow, and for the purposes of this exercise they can be considered as fixed objects.

Although this project can be started at any time of the month except in the few days around new moon, the moon rises inconveniently late after the third quarter (past midnight), an awkward time to begin. Full moon is the time when the moon is at its brightest. It is thus more difficult to see the background stars. At this time you will be able to measure only the brightest stars in the sky, even if some of them are at a considerable distance from the moon.

Probably the most convenient time to begin the project is two or three days after new moon, at which time the moon is easily found in the western sky just after sunset. An additional benefit of this position is that it is then easiest to measure the angle between the moon and the sun, since they are close together in the sky. Later on, as they appear farther apart, you may have to estimate the angle between them or measure it in sections and piece it together. But for your first measurement, you will get off to the best start if you go out at sunset and measure the angle between the sun and moon when the sun is just setting below the western horizon. Can you measure the angle between the sun and the moon if the sun has set? It will be convenient at times to do so -- devise a procedure that will give you this angle.

You should continue to observe the moon until it returns to the SAME POINT IN THE SKY that it occupied when you made your first observation. The moon will move with respect to the stars, and your objective is to determine the rate at which it moves. Observe it every two or three nights, until it becomes inconvenient to follow it any longer (this will depend upon how late you habitually stay up -- night owls can follow it longer).

II. CALCULATIONS

After completing your observations calculate the RATE at which the moon has been moving with respect to the stars. Read the total number of degrees from the star chart that the moon has traveled and divide by the number of days between your first and last observations. You will obtain a rate in degrees per day, and with this rate you can calculate how long it will take the moon to make a full 360 degree circuit with respect to the stars. This is called the SIDEREAL period of the moon. You should then go back outside near the time of your prediction (perhaps starting a little before the predicted date, to allow for errors) and reobserve the moon as it returns to the same point in the sky it occupied when you started observing it.

Determine how long it takes the moon to return to the same point in the sky from your observations. When it does return, examine the phase of the moon at that time. Does it have the same phase as it did when it was last at that same point in the sky? If not, write a paragraph in your observing book explaining why it does not. Include drawings if necessary to your explanation. Is the phase cycle of the moon (the time to go from a certain phase back to that same phase, called the SYNODIC period) shorter than or longer than the sidereal period of the moon (the time to go from one point in the sky back to that same point with respect to the stars)?

What is the angle between the moon and the sun (as seen from the earth) when the moon is at first quarter? When the moon is full? When it is at third quarter? Draw diagrams illustrating the arrangement of the earth, moon, and sun at each of these times and indicate in the diagram the angle you measured.

III. AN ALTERNATE PROCEDURE

Finally, note that instead of observing the moon as it moves from a certain point among the stars back to that same point, one can carry out this project in a slightly different but entirely equivalent way that may prove more convenient. You could observe the moon at a certain phase (e.g., start at first quarter or at full moon) and then follow the moon until it returned again to that SAME PHASE. The rest of the directions will be exactly the same -- you can still determine the rate at which the moon moves across the sky by following it as long as you can, predict when it will be back to around a similar phase, reobserve it when it returns to the same phase, and then determine the period of the phases. Then answer the following question: when the moon returns to the same phase as it had, has it returned to the same point in the sky? If not, write a paragraph in your observing book explaining why. Which is longer, the phase cycle or the time to return to the point among the stars? This approach to the project is entirely equivalent to the first approach.

The Motions of the Planets

5

NASA

Until the magnificent photos from the space probes began returning in the 1970s, our information on the planets came entirely from ground-based telescopic observations, distorted and blurred by the earth's turbulent blanket of atmosphere.

OBJECTIVES

1. to describe the Ptolemaic model of planetary motions, defining its various terms, and to explain why the geocentric model persisted for so many years
2. to describe the Copernican model of planetary motions and indicate what its shortcomings were
3. to define an ellipse and its characteristics
4. to describe and explain Kepler's three laws of planetary motion and do simple calculations
5. to describe Galileo's contributions to the acceptance of the heliocentric theory and astronomy
6. to write down Newton's gravitational law, do simple calculations with it, and describe its impact on the development of astronomy
7. to show how Newton's form of Kepler's third law allows us to determine the masses of many celestial bodies
8. to explain several methods by which the actual motion of the earth through space may be demonstrated
9. to carry out a follow-up observational activity

EQUIPMENT NEEDED

A small telescope, cross-staff, sextant, SC1 chart, a camera (optional), string, two thumbtacks.

This unit will cover about 2,000 years of astronomical history, describing the development of our theories of the solar system from ancient Greece through the time of Isaac Newton. During this era, astronomy was basically concerned with the sun, moon, and planets; no one was aware that there was really anything else, except of course the "fixed stars" which served as an unchanging tableau against which the motions of the planets were seen. The word "planet", in fact, comes from the Greek word for wanderer.

After studying the written material, you will perform one of the observational activities described in the last section of this unit.

I. THE PTOLEMAIC MODEL OF THE SOLAR SYSTEM

With their strong feeling for mathematics and geometry added to their interest in the universe, the Greek philosophers in the centuries before Christ produced the first really solid reasoning on celestial problems seen in the western world. It would require a book to describe all the contributions made by such great thinkers as Thales, Pythagoras, Aristarchus, and Eratosthenes; we will concentrate primarily on those lines of thought that led to descriptions of the planetary motions.

The question of whether the sun moved around the earth or the earth orbited the sun was approached by the Greeks in a very pragmatic fashion. They reasoned that if the earth changed its position in space, then we ought to be able to see changes in the apparent positions of the stars with respect to each other due to the "parallax" effect. Just as examining your thumb at arms length with one eye closed and then the other (try it) results in an apparent displacement of your thumb with respect to the background against which you are viewing it, a movement of the earth in space might well be expected to produce such effects on the relative positions of the stars (see figure 1). The Greeks sought evidence of such effects through careful observations of the stars over the course of the year and found none. They concluded, therefore, that the earth was stationary in space and that it was the sun which moved.

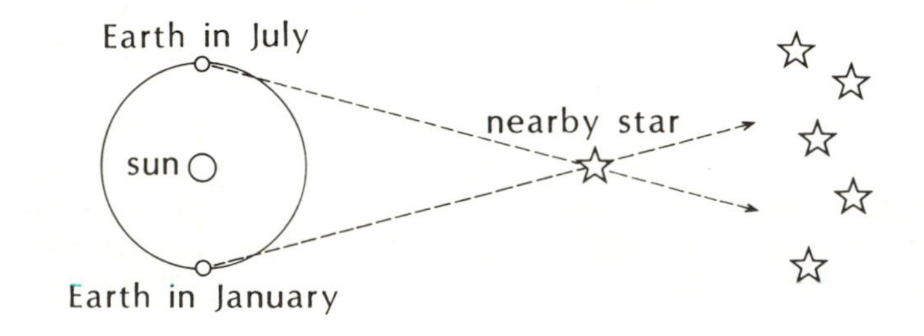

Figure 1 (not to scale)

Before proceeding, can you comment on this line of reasoning and figure out where they went astray?

The difficulty was that the stars were much further away than the ancient mind had ever imagined. Even the nearest star, Alpha Centauri, is 4.5 light-years away. At these vast distances, the parallax effects are simply too small to be seen with the naked eye. In fact, such effects were not measured until 1838. Even the nearest star shows a shift with respect to background stars of only 0.75 seconds of arc, about 200 times smaller than the finest detail the unassisted eye can resolve.

Starting with a stationary earth (that is, a geocentric model), what are the observed motions of the planets that must be explained? Most of the time, the planets are observed to move eastward with respect to the background stars, as illustrated in figure 2. The Greeks correctly assumed that the planets that showed the largest apparent motion on the sky were the closest, just as a nearby car moves more rapidly across our line of sight than a distant car does. Occasionally, however, a planet would be observed to slow its normal eastward motion, stop, and take up a westward (or RETROGRADE) movement. After a period of this backward motion, it would again slow, stop, and then resume its normal eastward motion.

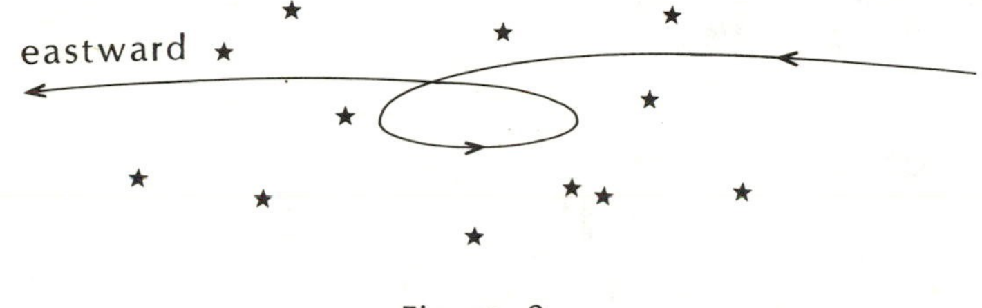

Figure 2

Two planets, Mercury and Venus, exhibited special features in their motion that required an extra explanation. These two planets appeared to be rather closely tied to the sun, in the sense that they would be observed to move away from the sun for a while but would never stray too far from it. After moving outward for a while, they would reverse and move back toward it, as illustrated in figure 3. Mercury never gets more than 28 degrees from the sun, while Venus may wander out to a limit of 47 degrees.

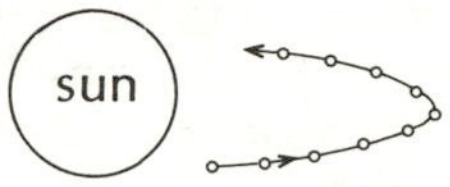

sun

Figure 3

And, of course, the sun and moon also moved with respect to the background stars and thus must be orbiting the earth. The sun traveled full circle (360 degrees) with respect to the stars in one year, for a rate of very close to 1 degree per day. Its apparent path across the sky always traced out the same great circle on the celestial sphere, which came to be called the ECLIPTIC. The moon and the planets in their motions were observed to also stay quite close to this "ecliptic plane," clearly the fundamental reference plane for the solar system. The region near the ecliptic in the sky where all the moving objects seemed to contain themselves came to be called the ZODIAC and figured prominently in all the various systems of astrology that have developed in many cultures throughout the ages.

The Greeks imposed one further requirement on their model of the solar system. Since celestial bodies were thought to be "perfect," they could as a consequence only engage in "perfect" motion, which to the Greek mind meant either a circle or a combination of circles. To explain the retrograde motion, the Greeks imagined an ingenious combination of two circular motions, shown in figure 4. The planet is considered to move around a small circle called the EPICYCLE, at a constant rate. But the center of the epicycle moves around the circumference of a larger circle, called the DEFERENT, centered on the earth. By adjusting the sizes of the circles and the rates of the two motions, the Greeks could explain the apparent movement of any planet and its retrograde intervals. This same combination of deferent and epicycle even worked for Mercury and Venus, simply by adding the condition that the centers of their epicycles had to remain tied to the line between the earth and sun and move as it moved. The moon and sun, of course, showed no westward motions in the sky and hence needed no epicycles. The complete picture that evolved is summed up in figure 5. Although only Mars is shown, the outer planets have similar modelling, and the "sphere of stars" is farther out yet.

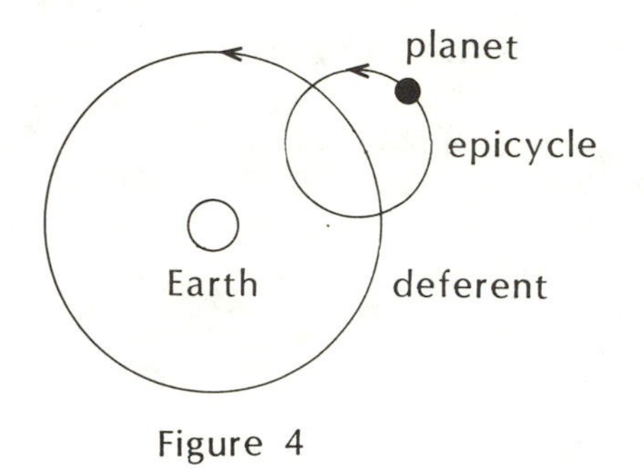

Figure 4

The Greek astronomer Ptolemy added a number of small refinements to this basic picture. As observations accumulated, it became apparent that small discrepancies sometimes emerged. Ptolemy found that he could improve the agreement between theory and observation by slight movements of the epicycle and deferent, such that the deferent was not quite centered on the earth and the epicycle was offset a bit from the circumference of the deferent. As this line of Greek thinking was finally summarized by Aristotle and passed on through the dark ages to the medieval church through his writings, the name of Ptolemy came to be identified with this Greek geocentric model of the solar system.

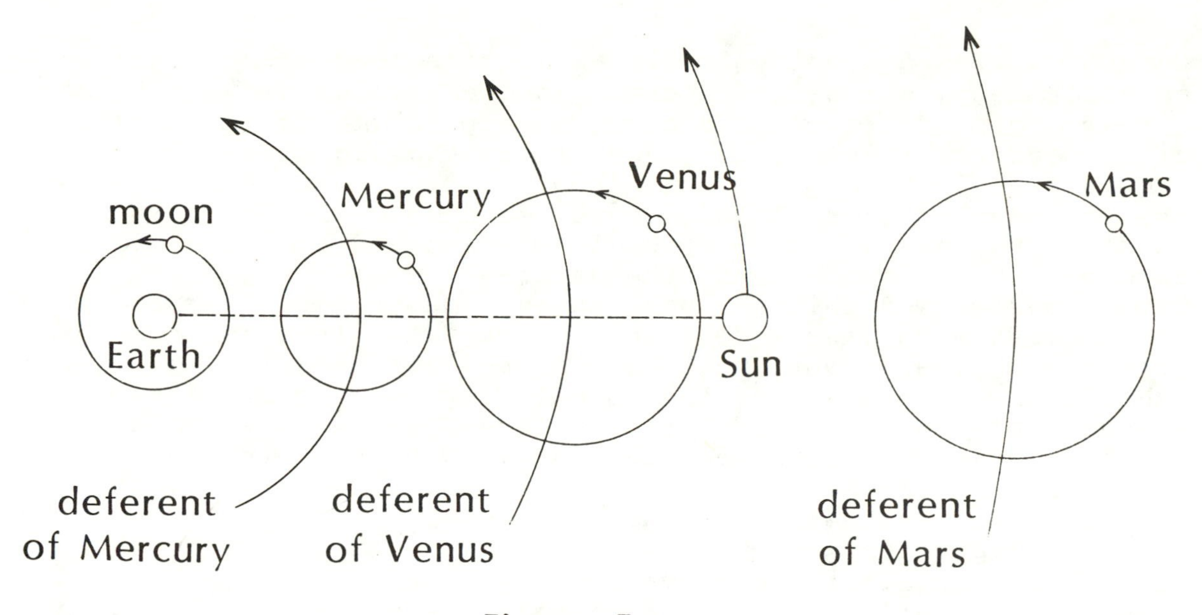

Figure 5

In an age of space travel, when all students are told from early childhood the real motions of the planets, it is frequently difficult to appreciate that the now obsolete Greek model was in fact based on some good thinking. That they went astray on a couple of points is no surprise, because even today correct lines of thinking occasionally go astray. With free and open inquiry, essential to science, such errors are finally detected and corrected.

During the middle ages, however, free and open inquiry did not exist. In the fifteen hundred years following Christ, the church became the most dominant institution, and church theology adopted the Ptolemaic model of the heavens as dogmatic truth. The Greeks regarded these ideas as mathematical models which described the motions that they saw and allowed them to predict future motions; this is exactly the philosophy of modern science, which gladly replaces any scientific model if another comes along that better describes and predicts the observations. The medieval church maintained its authority through the absolute acceptance of rigid dogma. It was only in the Moslem cultures of the Middle East that astronomical research and thinking were kept alive, with many of the ideas of ancient Greece preserved there and later reintroduced into Europe.

II. THE COPERNICAN HELIOCENTRIC MODEL

The first seeds of a new world picture were planted by a timid Polish canon, Nicolaus Copernicus. Copernicus received the printed version of his historic work, "On the Revolutions of the Heavenly Spheres," on his deathbed in 1543. Although he had begun his research in 1506, he was reluctant to publish. A preliminary outline of his theory, published in 1530, attracted great attention and earned the approval of Pope Clement VI.

Copernicus was not motivated by any desire to shake up the establishment, but he had become genuinely concerned about the widely accepted Ptolemaic theory. One problem was that small discrepancies between theory and observation continued to accumulate; it turned out that it was possible to further adjust the Ptolemaic picture by adding yet more epicycles. Thus, it was possible to have a little epicycle which moved around a big epicycle which moved around the deferent, etc; etc; etc. Copernicus viewed this patching as extremely objectionable, since it produced a complex model that was very far from the clean and pure geometrical beauty that he imagined the heavens should exhibit. He wanted to restore simplicity and beauty to the heavens and thus proposed a return to a picture in which the earth moved and the sun was stationary, with the planets orbiting in circular motions. He thus proposed a heliocentric hypothesis for basically aesthetic reasons.

His straightforward theory suffered from the difficulty that it did not explain the observations very well, and as a consequence even serious medieval scholars could rightly object to it. To make it agree with the data using epicycles added on would require even more epicycles than the Ptolemaic picture. Copernicus himself was certainly aware of these problems and of the fact that there were still no hard data demonstrating that the earth moved through space, and these considerations doubtless figured in his reluctance to publish his theoretical speculations. (Copernicus and the other medieval and Renaissance astronomers that followed him have been thoroughly and most enjoyably described in Arthur Koestler's book THE SLEEPWALKERS, a controversial, psychological history of Renaissance science. THE COPERNICAN REVOLUTION, by Thomas Kuhn, places this discovery in a historical frame, relating it to concurrent intellectual revolutions in other areas.)

III. TYCHO BRAHE: ORIGIN OF MODERN OBSERVATIONAL ASTRONOMY

Three years after the death of Copernicus in 1543, Tycho Brahe was born in Denmark. He was from a noble, wealthy family, and early in life he developed a passion for astronomy. He gained enough of a reputation from observations in his youth to acquire the patronage of Frederick II, and, he convinced the king to build him the finest observatory the world had yet seen. The telescope had not yet been invented, so Brahe's observations were naked-eye sightings with devices not unlike the cross-staff and quadrant constructed in unit 1. However, his instruments were quite large and accurately built, so that, with persistent and careful data gathering, Brahe was able to obtain observations that approached the limit of the resolution of the eye in accuracy. Even today, without telescopic assistance, it would be difficult to improve on his data.

Brahe was obsessive about his astronomical observations; he not only compiled data with great care, averaged his numbers so that random errors canceled out, but he also observed persistently and systematically over a long period of time. This combination of factors resulted in the best body of observational

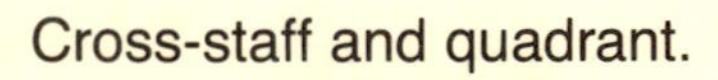

Cross-staff and quadrant.

Cut on solid lines. Fold *away* from you on dashed lines. Rule dashed lines heavily with a ball-point pen before folding.

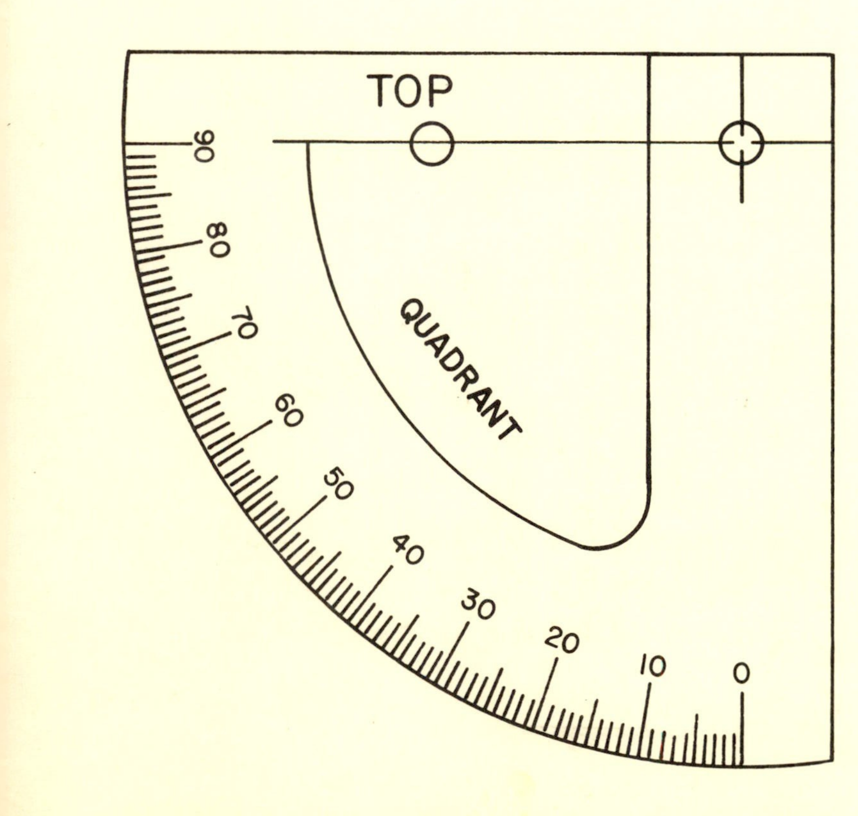

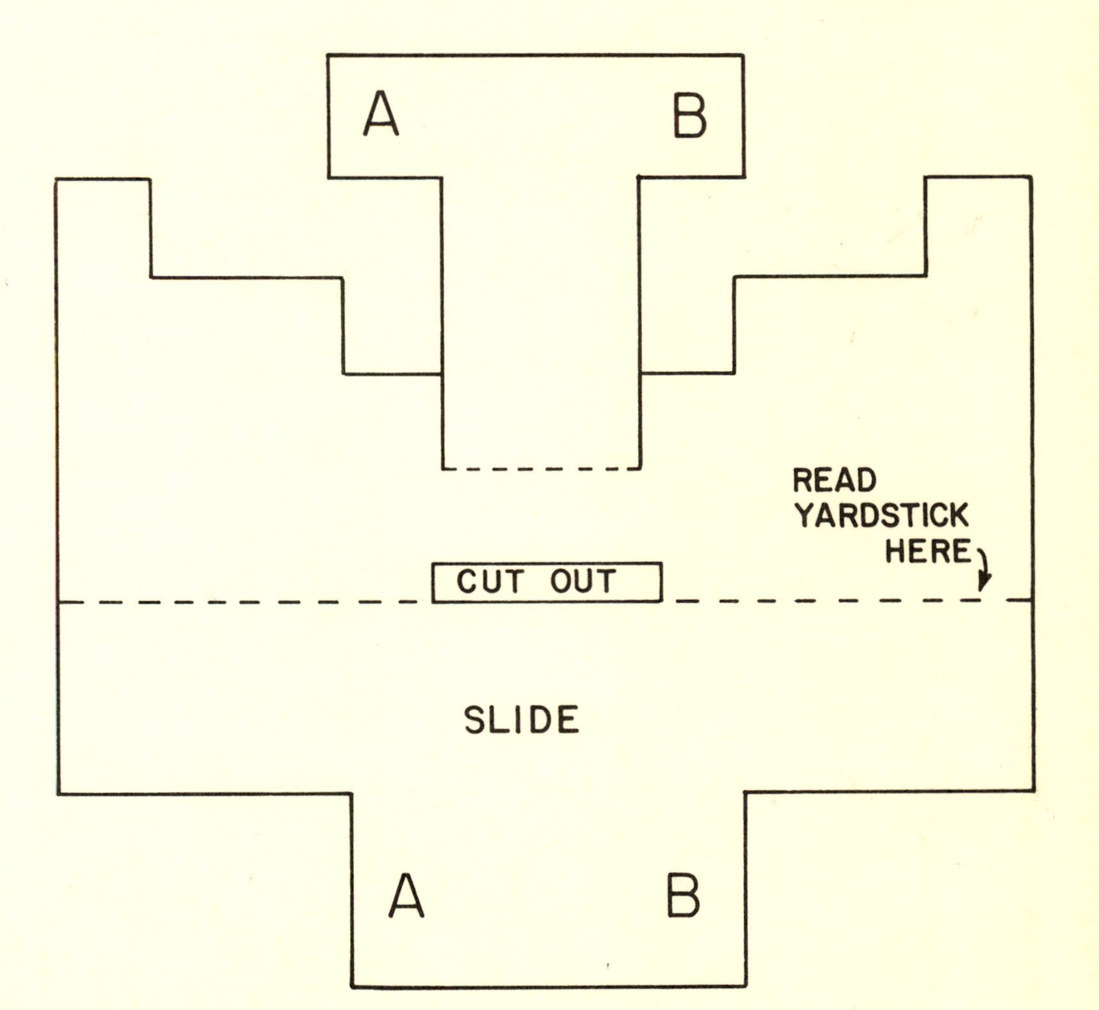

data on the motions of the planets ever accumulated, and these data led the way to an improved model of the solar system. It is probably accurate to call Brahe the first modern observational astronomer.

Brahe himself was not a mathematician, and, although he collected data of great accuracy, he was not able to fit a theoretical model to it. It was apparent to him from his observations that neither the Ptolemaic nor the Copernican models were satisfactory. The greatest deviations between theory and observation existed for the planet Mars, and of course Brahe also attempted to measure stellar parallaxes and failed, again reinforcing the notion that the earth was stationary. (Brahe was basically anti-Copernican.) Near the end of his life, Brahe acquired a collaborator in the form of the young German mathematician Johannes Kepler, who proved to be a thinker of sufficient genius and boldness to overthrow thousands of years of sterile thinking and come up with the correct model for the motions of the planets.

IV. KEPLER AND HIS THREE LAWS OF PLANETARY MOTION

In "The Sleepwalkers," Arthur Koestler attempts to explain why it was that Kepler proved to be the one with the persistence and boldness to come up with the correct answer. Kelper himself commented in his journals that his audacity in challenging the hallowed models passed down from the Greeks was personally terrifying to him. He kept extensive daily journals of his labors, so historians are able to follow his progress in great detail. He also filled these journals with his daily horoscope and hourly comments on his health, since he was a compulsive hypochondriac.

It is hard to appreciate today how difficult and tedious Kepler's job was. He had to fit Brahe's observations to a curve of some sort (of shape unknown), using only paper, pen, and longhand calculations. Not only were there no calculating machines, logarithms hadn't been invented yet. The fitting process took him eight years, and many authors have wondered retrospectively what made him carry on through many discouragements.

In the end, Kepler was successful in describing the motions of the planets with three so-called laws of motion. They are not laws of motion in the modern sense, because as yet there was not even the concept of "force" to be applied in a coherent body of physical law. Kepler's laws are in essence three descriptions.

A. Kepler's First Law

Kepler bravely discarded the hallowed notion of a circular orbit and, after trying many different types of curves, finally discovered that the orbit of Mars was best fit by a curve called an ELLIPSE. The definition of an ellipse is illustrated in figure 6; it is a curve such that the sum of the distances AB and BC is always a constant.

In fact, to draw an ellipse one need only tie a piece of string to two thumbtacks and run a pencil around the inside of the string. The two tacks define the "focal points" of the ellipse, and the shape of the ellipse depends upon how close together these two focal points are relative to the long dimension of the ellipse, called its major axis. If the two focal points are made to coincide, the string and pencil technique will clearly draw a circle, but, if the two points are almost AB plus BC apart, a very elongated curve will result. These various shapes for an ellipse can be defined by a parameter called the ECCENTRICITY, which is defined as the distance between the focal points divided by the major axis. E = 0 gives a circle, while E = 1 would give an ellipse so elongated that it would be a straight line.

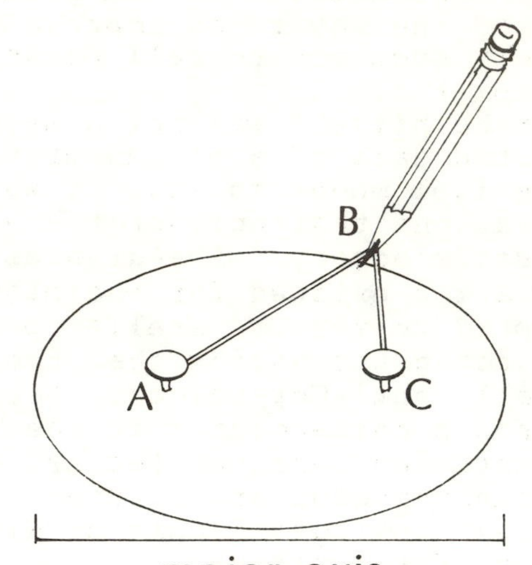

Figure 6

If the sun is imagined to be placed at one of the focal points of an ellipse (and the other one empty), the motions of the various planets can be seen to be following an elliptical path about the sun. Each planet has an orbital path of a different eccentricity, but all are fairly close to zero. If you looked down on the solar system from above its north pole in space, the orbits would resemble circles (with the possible exception of Pluto).

Take a pencil, string, tacks, and paper. Draw several ellipses of various eccentricity. Calculate the eccentricity for each ellipse.

In an elliptical orbit, therefore, the distance of a planet from the sun is constantly changing. The earth in its orbit varies its distance from the sun from about 91 to 95 million miles, coming closest in January. (Is this distance variation the cause of the seasons?) The average distance of the earth from the sun is 93 million miles; this distance has come to be called the Astronomical Unit (and abbreviated AU) and is used as the standard yardstick for describing distances in the solar system. Mathematically, the average distance of a point on an ellipse from either focus also turns out to be one-half of the major axis of the ellipse.

B. Kepler's Second Law

Kepler also discovered that the planets do not move at a constant speed in their orbital paths. The closer they come to the sun, the faster they move. He described this variation in his "law of areas," which says that a planet moves at a rate such that its radius vector (the line from the sun to the planet) sweeps out equal areas in equal times.

This areal description is illustrated in figure 7 for an elliptical orbit more exaggerated than that of any planet. If our hypothetical planet travels from 1 to 2 in one day, the shaded area indicated in figure 7 will be swept out by the radius vector. When it is closer to the sun, it must move faster to sweep out the equivalent area. Thus, it travels from 3 to 4 in one day, and from 5 to 6, such that all the shaded areas in figure 7 are equal in area.

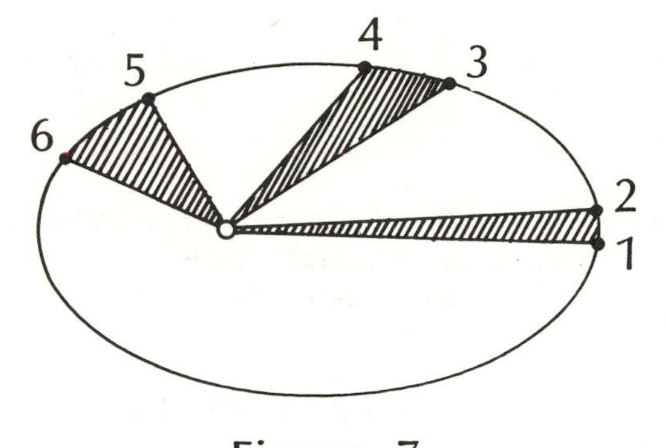

Figure 7

Kepler's law of areas can be related to a physical quantity of considerable importance in modern physics and astronomy, the angular momentum of the orbiting body. The orbital angular momentum of a planet can be defined as the product of its mass, its orbital speed, and its distance from the sun.

$$\text{Angular Momentum} = M \cdot V \cdot D$$

Angular momentum is of interest in physics because it is constant during the motion of a body (the terminologies "invariant" and "conserved" are also used to describe it). Thus, if an orbiting planet has constant angular momentum yet its distance from the sun decreases, it must increase its speed to keep the product MVR constant.

An everyday illustration of the constancy of angular momentum is seen in the familiar figure of an ice skater spinning slowly and then speeding up by pulling the arms in toward the body. To maintain a constant value of angular momentum, speed must increase.

Angular momentum is also important in astronomy during the star formation process when a cloud collapses under the influence of its own gravitational force. As it becomes smaller, it spins faster and the forces generated in the equatorial region of the cloud tend to flatten it out. This process helps us to understand the flattened form of the solar system and the galaxy, and why the planets in the solar system and the stars in the galaxy tend to follow a common traffic pattern and direction in their motion.

C. Kepler's Third Law

Kepler's last law describes a relationship between the average distance of a planet from the sun and its orbital period, the time it takes to complete a full orbit. Kepler discovered that the squares of the periods were proportional to the cubes of the average distances. This law takes its simplest form mathematically if we measure the periods in years and the distances in AU:

$$P^2 = D^3$$

Taking the square root of both sides of this equation yields:

$$P = D\sqrt{D}$$

A planet with an average distance of 4 AU would have a period of eight years. Saturn has a thirty year period about the sun -- what is its average distance from the sun?

V. GALILEO'S CONTRIBUTIONS

Galileo, a contemporary of Kepler working in Italy, angered his professors by refusing to take statements given in lecture for granted. His questioning mind was to lead him to many great discoveries, especially in physics; he began by performing many basic and fundamental experiments on falling bodies, sliding bodies, and pendulums, which soon led to a proper understanding of the laws of their motion. Galileo became internationally famous, however, for his discoveries in astronomy. In 1609, he became the first man to turn a telescope upward toward the sky. His most important observation, judged by its impact on the Ptolemy-Copernicus controversy, was his study of the phases of Venus. When viewed through a telescope the planets show visible disks and the variation of their illuminated surfaces can be followed. In the Ptolemaic model for Venus, the planet could not physically locate itself on the other side of the sun from the earth and, as a consequence, could show only a crescent phase when viewed from earth. However, Galileo observed that Venus could be seen in a "gibbous" phase, that is, with more than half of its disk illuminated. You should draw both the Ptolemaic and Copernican pictures of the motion of Venus, visualizing how Venus would appear from the earth in both models. Convince yourself that the observation of a gibbous phase for Venus means that at least one planet, Venus, must orbit the sun and not the earth.

Galileo also observed four moons revolving about the planet Jupiter. Anti-Copernicans had argued that the earth could not move, since if it did the moon would be left behind. Jupiter acted like a mini-solar system.

VI. THE NEWTONIAN REVOLUTION IN PHYSICS AND ASTRONOMY

The brilliant work of Galileo and Kepler laid the foundations for a real science of physics and astronomy. What was now required was someone of consummate theoretical genius to fit it all together in a set of real "laws." Such a genius did appear in the form of Isaac Newton, and his brilliant conclusions concerning the motions of objects, force laws, gravity, and optics still remain with us today as the underpinnings of modern science. It was not until the twentieth century that significant refinements in his magnificent new theories were finally imagined by scientists.

Students in physics classes most frequently encounter Newton through his three laws of motion for bodies acting under forces. Many scientists prior to Newton had struggled to formulate the concept of a force (e.g., Kepler and Galileo), but Newton was the one who came up with precise adequate definitions amenable to mathematically stated laws of motion. Newton first postulated that if no forces are acting on a body, it retains its current state of motion. If it is at rest, it stays at rest. If it is moving, it continues to move uniformly (that is, in a straight line with a constant speed). This simple statement introduced enormous simplification into physical thought, since from the time of Aristotle it had been customary to assume that, if something was moving, it must be because something was forcing it to move. The end result of such thinking was an almost animistic natural world, in which every object doing anything at all had to be explained by its own special force acting on it. Newton argued that if a body is set in motion (for example, a planet), it will continue its motion even if no further forces are present.

Newton also proposed the idea of forces in equilibrium -- for every force that could be identified, an equal and opposite counterforce could also be understood as acting. Thus, when you stand on the surface of the earth and are pulled downward by the force of gravity, you also are exerting an equal but oppositely directed force on the earth. For a stationary object (for example, a building), the force downward is balanced by the upward push of the rigid ground; such straightforward thinking remains the foundations of much of engineering science today. Newton's most famous law was his proposal that force could be equated to the product of mass times acceleration; that is, $F = ma$. This simple law remains to this day the foundation of everyday (nonrelativistic) motions in the macroscopic world we experience and in the microscopic world of gases and the motions of atomic particles. This law leads to many immediate insights about the motions of bodies that can be applied to astronomical bodies.

As a simple application of this last law, consider the game of "kick the can" and imagine that you kick two cans, each with the same amount of force. Suppose, however, that one can is empty and the other is filled with concrete. It is intuitively clear that the empty can (the one with less mass) is the one that will receive the greater acceleration. Similarly, in a binary (double) star system, the less massive star is the one that will be given the greatest acceleration, move the fastest and have the largest orbit, even though the gravitational force between the two stars is equal and opposite.

In seeking to explain why objects moved as they did on earth and why objects such as the moon orbited about the earth as they did, Newton proposed a simple law of gravitational attraction that worked. It was this law that had an immediate and enormous impact on astronomical thinking; withing a few years it was applied to resolve many situations that had long seemed difficult to analyze. Newton proposed very simply that, between two bodies of masses M and m separated by a distance r, the gravitational force was given by

$$F = GMm/(\text{square of } r).$$

That is, the force was given by the product of the masses, divided by the square of the distance, and multiplied by some numerical constant G (a number that depends upon the units chosen for the quantities in the equation). The force between the two bodies is of course equal and opposite, but because of $F=ma$, their response to the force is different--smaller bodies are accelerated more.

In explaining the motion of the moon, Newton reasoned that, since all objects fell toward the earth's gravity, the moon must also do so, and he computed the rate at which it fell toward earth. But it did not fall into the earth because it also had its own motion (stemming presumably from the days when it was created) that tended to direct it in a straight line at a constant speed. Without the force of gravity and its acceleration, the moon would simply speed away from the earth. Figure 8 illustrates how the combination of the moon's own motion and its falling motion toward earth creates an orbit.

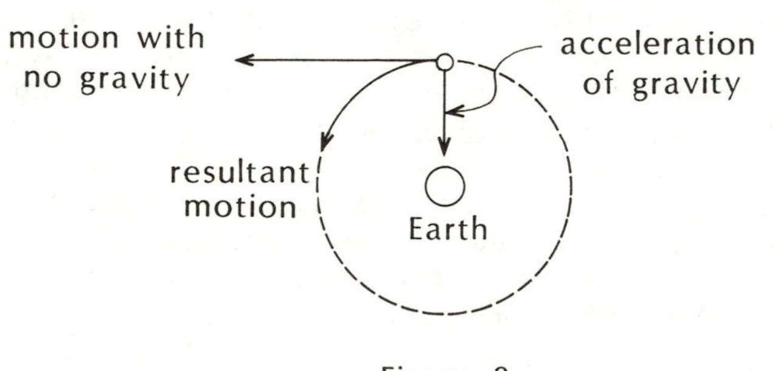

Figure 8

Carry out the following two calculations with the law of gravity to increase your familiarity with it:

1. If the distance between two bodies is decreased by a factor of 10, by how much will the force between them increase?
2. If the distance between the earth and sun remained the same but the sun were suddenly replaced by a black hole of 1 solar mass, compute the change in the gravitational force felt by the earth.

The law of gravity suddenly made most of the motions seen in the solar system understandable. It explained the tides on the earth, the motion of the moon, and the planetary orbits, and astronomers soon noted that it explained the motions of stars bound up together in double systems. Indeed, to this day, the law of gravity has served admirably to explain the fundamental motions of astronomical bodies out to the limits that we have been able to penetrate, that is, several billion light-years. This law is so solidly established that, when the motion of a body does not correspond exactly to the law's prediction, it is assumed that nongravitational forces are presumed to be acting. An example of this was seen in the Voyager flyby of Saturn, in which the particles in the so-called F-ring were found to have a braided appearance. Since gravity must be acting, the search began for other types of forces (e.g., electromagnetic) that might also be present, to explain the deviations from the type of orbit gravity alone would give.

Historically the most famous example of this type of argument was observed in the discovery of the planet Neptune. After William Herschel discovered the planet Uranus by naked-eye observations, its motion through space was followed for a number of years to see if it obeyed the predictions of Newton's gravitational laws. It did not, but this apparent failure turned into one of gravity's greatest triumphs, since it was assumed that the deviations from an elliptical orbit exhibited by Uranus were probably due to an additional gravitational attraction from a more distant and yet undiscovered planet. In a famous problem in the history of mathematics, the German mathematician Karl Gauss computed where the new planet should be and it was discovered on the first night's search.

Newton's physical laws were published in his famous volume "Principia" in 1687, even though it was many years before an actual physical proof of the earth's motion was produced, the theories he proposed explained the solar system motions so naturally that acceptance of the heliocentric hypothesis and Kepler's laws became widespread. In fact, given F=ma and the law of gravity, it is a very short and simple calculation to derive Kepler's laws mathematically. Elliptical orbits, the law of areas, and the law relating the periods to the distances are immediate mathematical consequences of Newton's laws.

Newton was able to observe, through such derivations, that there was in fact a small error in Kepler's third law, and he produced a modified version of the law that has enormous usefulness for astronomy. Including the constants in the equation, Newton showed that the law relating the periods and distances should actually read (for the solar system):

$$P^2 = \frac{4\pi^2 D^3}{G\,(M_{sun} + m_{planet})}$$

Notice that the masses of both the sun and the planet appear in this equation. Kepler's version did not include the mass of the planet, and when he took the ratio between the orbits of different planets (as one is doing when years and AU are used as the units) the mass of the sun cancels out.

But, with the mass of the orbiting body in the equation, it now became possible to determine the masses of many astronomical bodies. For example, consider the planet Jupiter and write this equation for one of its Galilean satellites, say, Ganymede. If the subscript G refers to Ganymede and J to Jupiter, we have:

$$P^2 = \frac{4\pi^2 D^3}{G\,(M_J + m_G)}$$

The mass of Ganymede is very small compared to that of Jupiter and we will make only an infinitesimal error if we neglect it. When we do, we have an equation that relates the mass of Jupiter, the period of Ganymede, and the distance of Ganymede, from Jupiter. But P and D can be observationally determined, and we can then solve for the mass of Jupiter. Similarly, we can solve for the mass of any planet which has an observable satellite, solve for the mass of the sun from the planetary motions, and in general solve for the mass of any massive body that has a small observable body in orbit around it. The equation above remains the main source of information in astronomy concerning the masses of astronomical bodies, information which is usually critical in their understanding. In fact, for studying the structure and equilibrium, energy output, evolution, and life history of a star, the mass is basically all that is needed. If two bodies are in mutual orbit and their masses are comparable, this equation allows us to solve for the sum of the masses, and sometimes additional information can be obtained from other data that enables us to separate out the two components.

VII. THE HISTORICAL DEMONSTRATION OF THE EARTH'S MOTION

It was not until 1725 that observational evidence for the motion of the earth through space was actually obtained. James Bradley noted that, in order to properly observe the positions of stars located in certain parts of the sky, it was necessary to tilt the telescope slightly away from an apparent straight line toward the star. Before this effect was understood, it was called the aberration of light. This turns out to be a misnomer, since there was nothing peculiar about the light itself and the effect was due to the motion of the earth through space. To understand this, first consider the following analogy to the problem: imagine that it is raining on a windless day, so that the raindrops are falling straight down, as in figure 9. If a section of stovepipe were held perpendicular to the surface of the earth, the raindrops would fall straight through the pipe.

But, if the pipe were in motion, it would be necessary to tilt the pipe in the direction of the motion in order to get a raindrop to pass all the way through the pipe. The analogy to the earth in motion through space is exact if we imagine the incoming light from a celestial object as analogous to the falling raindrops, because light does in fact travel at a finite speed (very high, 300,000 kilometers per second, but finite). In order for the light from a star to pass properly through the telescope, the telescope must be slightly tilted in the direction of the earth's motion through space.

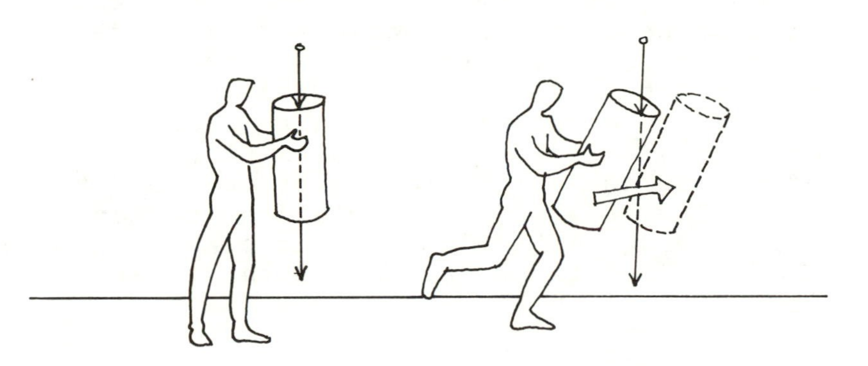

Figure 9

The amount of the tilt depends upon the speed of the earth relative to the speed of light, and when these facts were understood it was possible to observe the tilt and determine that the speed of the earth equaled 0.0001 the speed of light.

Can you explain why there are some stars for which no tilt is observed?

Almost a century later, in 1838, the first stellar parallax was observed by the astronomer Friedrich Bessel and the deliberations of the ancient Greeks were finally resolved -- after 2,000 years. Happily, scientific progress is somewhat more rapid today!

VIII. FOLLOW-UP OBSERVATIONAL ACTIVITIES

To complete this unit it is necessary to carry out an observational activity relating to the solar system. To give flexibility and allow for a variety of interests, we present here a number of different options -- some can be tackled and carried out in a period of continuous activity, while others require observations over a period of time (e.g., to be able to observe the motion of a planet). The total effort involved in each of the activities should be about the same in terms of hours required.

A. Observing the Motions of a Planet

Consult a monthly reference (such as "Sky and Telescope," "Astronomy," "McDonald Observatory News") to find out which planets are conveniently located for observation. If a long enough interval of time is available and the planet Mars is favorably located, its motion can be followed by means of the cross-staff. A month interval will show motion easily visible to the naked eye, and several months of following Mars will reveal large movements. The procedure to follow would be identical to that described in unit 4 for locating the moon with respect to background stars and plotting its changing position on the SC1 constellation chart. Read and follow those instructions, and, at the end of your sequence of observations, calculate the rate of movement of the planet on the sky.

If Jupiter or Saturn is chosen to study, the smaller motions require more accurate observations. In this case, observe the planet with an astronomical sextant, binoculars, or any small telescope and follow the same plotting procedure outlined for the moon in unit 4. The greater accuracy of the sextant will allow you to determine a rate of motion for the planet in a fairly short period of time. If you have not already used the sextant, follow the instructions in the appendix.

B. The Phases of Venus

If a small telescope is available of sufficient power to enable you to resolve the disk of Venus, you can reproduce the observations of Galileo that began the downfall of the Ptolemaic system. You should draw the apparent phases of Venus and the variation in the size of its apparent disk over a period of several months. Then, present your observations with suitable diagrams explaining why you saw what you saw.

C. The Motions of the Moons of Jupiter

The Galilean satellites of Jupiter can be observed with the smallest of telescopes, and their motion with respect to the giant planet is quite rapid. Begin a series of observations of their relative positions with respect to the planet, and use the size of the apparent disk of Jupiter to maintain the proper scale in your drawings of what you see. This activity would best be carried out over two or three successive nights, with a long enough period of observation on each night to detect motions (several hours). In all probability, you will observe events such as one of the satellites going into eclipse behind Jupiter, an event that occurs regularly! In your observations try and keep track of which satellite is which; this may prove difficult in some cases, since your data gathering will be interrupted by a number of hours of daylight in between observing sessions. Observations of the relative brightnesses of the various satellites may aid you in this.

Attempt to determine the orbital period of one of the satellites and some indication of its distance from Jupiter (using the information that Jupiter is actually 143,000 kilometers in diameter). Insert this information into the Newtonian version of Kepler's third law and calculate the mass of Jupiter.

D. Photography of the Planets and Their Satellites

If the appropriate equipment is available and you have learned how to use it, any of the three activities above can be carried out photographically with greater accuracy. The first evening of photographing will not only give you your initial data but will also determine what the proper exposures should be to continue the activity. On the first night, take a large number of pictures at various exposures (following the instructions given for photographing the streetlight in unit 9) and, for subsequent data gathering, take a smaller number of pictures at the correct exposures.

E. Other Activities

If a shortage of time or disastrous weather should make the activities above unfeasible, several indoor activities involving aspects of the planets can be carried out. These activities do not involve the direct gathering of data and are therefore less of an observational experience; however, they lead to some important insights in understanding the study of the planets. Several of these activities, available from Sky Publishing Corporation, are listed in unit 22.

6 The Sun: Its Size and Daily Motion

BRITISH MUSEUM

Following the shadow of the sun throughout the day can give surprisingly precise information on its motion and insight into timekeeping.

OBJECTIVES

1. to use a gnomon to determine the altitude of the sun at various times during the day, beginning before noon and continuing to as near sunset as possible
2. to measure the azimuth of the sun at each altitude measurement
3. to determine the direction north from the observations
4. to determine the time of maximum altitude, and to explain why the sun has its maximum altitude at that particular time
5. to determine the azimuth when the sun's altitude was maximum, and to explain why the sun has that particular value at that time
6. to use the celestial globe to explain the variations in the altitude of the sun at different times of the year
7. to make a pinhole camera and measure the size of the sun's image
8. to determine the physical size of the sun

EQUIPMENT NEEDED

A gnomon, a large piece of paper, a protractor, a celestial globe, and a cardboard box.

The gnomon is one of the earliest astronomical instruments. By placing a stick vertically into a flat place on the ground, the Babylonians studied the daily motion of the sun in the sky by watching the stick's shadow. The tip of the gnomon's shadow marks the end of the straight line which connects the tip of the shadow, the top of the gnomon, and the sun. You will study how the length of the shadow is determined by the sun's altitude and its position by the sun's azimuth. The sundial is a relative of the gnomon. You will also build a pinhole camera and determine the size of the sun.

I. THE DAILY MOTION OF THE SUN

A gnomon is a straight stick or rod inserted vertically into the ground so as to cast a shadow which is easily measured. On a day clear enough to see shadows, find a location suitable for setting up a gnomon which will allow observations during most of the day. The site should have a relatively clear western horizon and be out of the traffic pattern of people and pets, since the paper cannot be moved during the series of observations. The ground should be level. A height of 7 to 15 centimeters (3 to 6 inches) is best. It is most important that the gnomon be vertical; an easy way to check that it is vertical is to hang a string with a weight from the top of the gnomon. the weight will hang straight down due to the pull of gravity. A protractor or T square may be used to check that the gnomon is vertical.

Put a large piece of paper or cardboard under the gnomon and mark a north-south line on the paper. Don't worry about being extremely accurate with the line (since the experiment itself will determine the line correctly); you simply want a rough estimate. You may obtain this from noting the direction of the polestar at night from your chosen location or perhaps from a map. A magnetic compass may differ considerably from celestial north, depending on your latitude and longitude.

At intervals during the day, mark the place on the paper where the end of the shadow extends to and write the time next to the mark. You should begin your observations before noon; how frequently you make them during the day can vary, but you should make an observation at least once per hour. When the sun is near its maximum altitude, you will get much better results if you measure every fifteen minutes. As explained below, continue your observations as close to sunset as possible.

For each measurement of the length of the shadow, you can easily calculate the altitude of the sun at that time. In figure 1, the stick and its shadow are represented as two sides of a triangle. If X designates the length of the stick and Y stands for the length of the shadow, then the ratio Y/X determines the angle A which is also the ALTITUDE of the sun. The relation of X, Y, and A forms part of an area of mathematics called trigonometry, but for our purposes in this unit a table is given at the end of this unit. For example, if X is 3 centimeters and Y is 1.5 centimeters, the ratio is 0.5 and the table shows that the angle A would be 63 degrees. Note that using an integer value for X makes the calculations easier.

Notice that you can calculate the altitude of the sun continuously, or you can wait until all your measurements are done. The former allows you to detect mistakes sooner! You may want to use the time between sightings to do the second activity in this unit.

On a piece of graph paper, plot your measurements of the altitude of the sun versus the time of day, and connect your data points by drawing a SMOOTH curve through them with a pencil. At what time was the sun at its maximum altitude? How accurately can this be determined from your data?

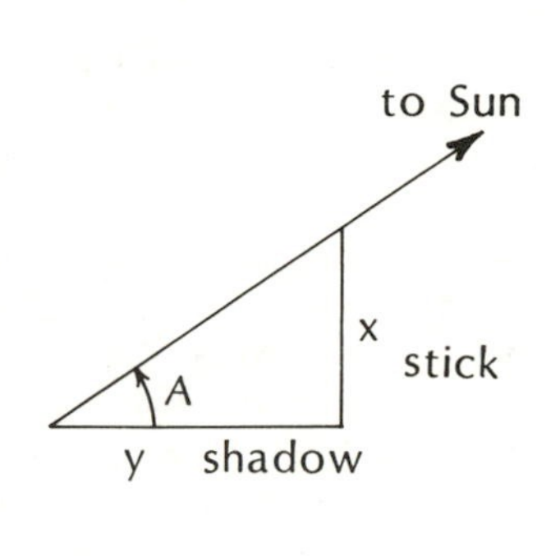

Figure 1

You can use your measurements to determine the direction of true north. A little reflection about how the sun and its shadow behave should allow you to determine the time when the shadow was pointing north. If you do not see how to do this, go to the celestial globe, set up the sun for the day of your observations, and see if it can help you determine when the shadow was pointing north. When you determine the north-south line, mark it on the paper with your observations.

Having a good north-south line, you are now in a position to determine the AZIMUTHS of the sun during the day. The azimuth is the angle, measured clockwise, from due south to the shadow, as indicated in figure 2. Use your shadow marks and a protractor to determine the azimuth of the sun for each of your markings, and plot the resulting values of angle versus time on graph paper. Can you determine from your observations the azimuth of the sun at sunset? (Note that this is easier and much more accurate if you continued your observations close to sunset.)

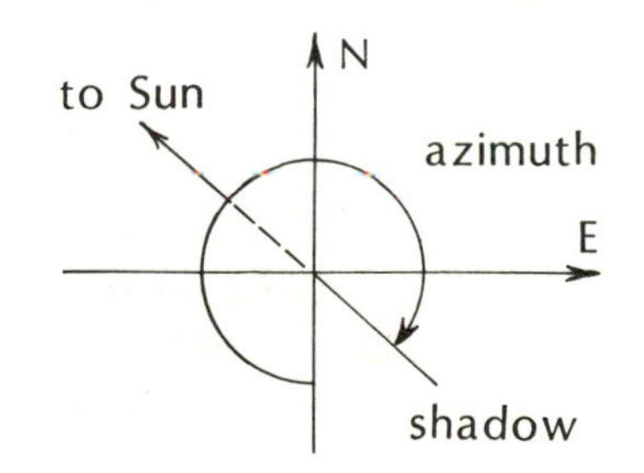

Figure 2

From your azimuth observations, determine the point on the horizon where the sun actually set (e.g., due west, 13 degrees south of west, 21 degrees north of west, etc.). Then use the celestial globe to predict the sunset point on the horizon for the day of your observations and compare the two values. Also use the globe to estimate the maximum altitude of the sun for your observation date to compare with your data. Which altitude do you think is more accurate -- the globe estimate or your measurement?

II. THE SIZE OF THE SUN

Take a cardboard box (the longer the better) and cut a small hole about 1 centimeter (1/2 inch) square out of one of the small ends. Tape aluminum foil or a piece of paper over the hole and punch a very small hole in the foil with a pin. If you point this end of the box toward the sun, the pinhole will form an image of the sun on the back surface. If your box has no top, you will see the image directly. If it has a top, cut a little flap on the top to be able to look inside and see the image. A box with a flap will be darker on the inside than a topless box; that is, the solar image will be easier to see. Figure 3 shows just the front and back ends of the box, to illustrate how the solar image is formed. A piece of film on the back surface could take a picture; such a device would properly be called a pinhole camera. Pinhole cameras are used in situations calling for a very wide angle photograph without distortion. Their disadvantage is obvious -- not much light gets through the pinhole.

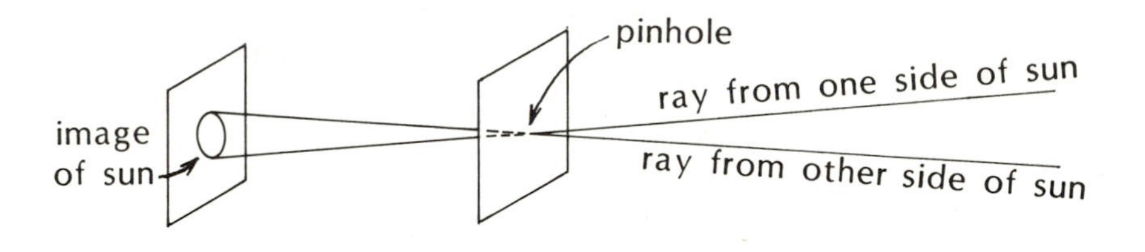

Figure 3

One possible refinement is to mount a piece of transparent ruled graph paper in a hole cut out of the back end of the box. This eliminates the need for a flap and makes it easier to measure the size of the image. Another variation is to block off all of a window which faces west, except for a pinhole. The entire room would be a pinhole camera at sunset, with the image on the far wall. This large image would be easy to measure.

If you examine the geometry of the situation (as shown in figure 4), you can see that, if the distance to the sun is known, we can use a similar triangles relationship to solve for the diameter of the sun. That is, we have

$$\frac{\text{diameter of the image}}{\text{distance from image to pinhole}} = \frac{\text{diameter of the sun}}{\text{distance from sun to earth}}$$

Assume that the distance from the earth to the sun is 150 million kilometers (about 93 million miles). Measure the quantities on the left hand side of the relation and determine their ratio; then solve for the diameter of the sun. As you measure the diameter of your image, try to estimate the amount of your measuring error before you do the calculation. Determine how uncertain your measurement of the diameter of the sun is, as a consequence of your measuring error. To aid in this error determination, do the following. Make a second pinhole next to the first one, only larger in size. Measure and compare the size of the solar image formed by the two pinholes; if there is a difference, explain why.

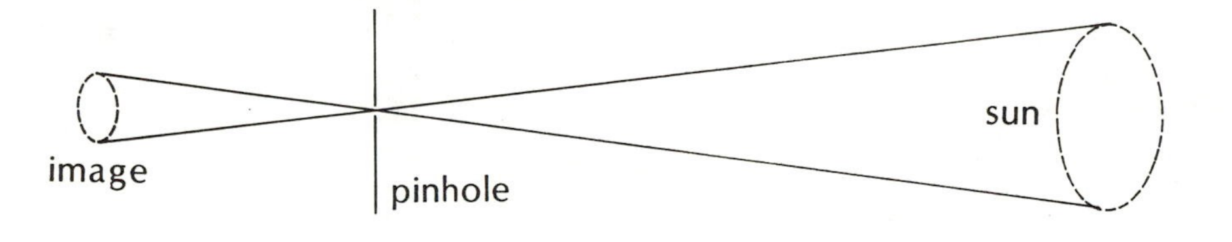

Figure 4

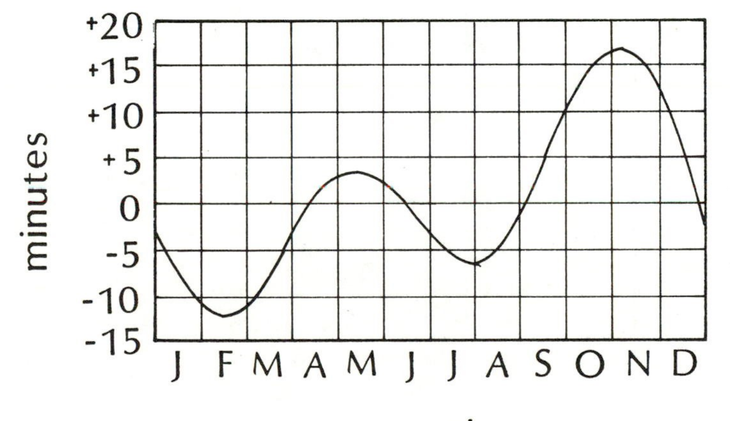

Figure 5

TABLE OF SOLAR ALTITUDE AS A FUNCTION OF THE RATIO Y/X

Ratio	Angle	Ratio	Angle	Ratio	Angle
0.00	90	0.58	60	1.73	30
0.02	89	0.60	59	1.80	29
0.03	88	0.62	58	1.88	28
0.05	87	0.65	57	1.96	27
0.07	86	0.67	56	2.05	26
0.09	85	0.70	55	2.14	25
0.11	84	0.73	54	2.24	24
0.12	83	0.75	53	2.36	23
0.14	82	0.78	52	2.48	22
0.16	81	0.81	51	2.61	21
0.18	80	0.84	50	2.75	20
0.19	79	0.87	49	2.90	19
0.21	78	0.90	48	3.08	18
0.23	77	0.93	47	3.27	17
0.25	76	0.97	46	3.49	16
0.27	75	1.00	45	3.73	15
0.29	74	1.04	44	4.01	14
0.31	73	1.07	43	4.33	13
0.32	72	1.11	42	4.70	12
0.34	71	1.15	41	5.14	11
0.36	70	1.19	40	5.67	10
0.38	69	1.23	39	6.31	9
0.40	68	1.28	38	7.12	8
0.42	67	1.33	37	8.14	7
0.45	66	1.38	36	9.51	6
0.47	65	1.43	35	11.43	5
0.49	64	1.48	34	14.30	4
0.51	63	1.54	33	19.08	3
0.53	62	1.60	32	28.64	2
0.55	61	1.66	31	57.29	1

THE TIME OF LOCAL NOON

Local Apparent Time (LAT) is the time that would be read by a gnomon or a sundial. The instant when the sun is crossing the local meridian and is therefore at its highest altitude would be local noon, or LAT = 12. Your observations of this unit have probably demonstrated that local noon need not be at "12" on the clock. There are several reasons for this.

Probably the most obvious is daylight savings time. Correct your clock time of local noon for DST, if it is in effect, by subtracting one hour.

Another correction arises from the existence of time zones. Adjacent geographical regions on the earth agree to set their clocks at the same time, for obvious reasons of convenience. All clocks within a time zone will read a common time, while a sundial would give different readings at any given instant for different longitudes within the zone. The only places in a time zone where the clock time will agree with solar apparent time will be located along the central meridian of the time zone. To convert your LAT to the center of the time zone, you must note that each degree of longitude you are away from the center of the time zone introduces four minutes of time difference. Correct your LAT to that of the central meridian of your time zone. If you are west of your central meridian add the correction, if you are east, subtract.

Zone	Central Meridian	Hours from Greenwich
Atlantic	60 W	4
Eastern	75	5
Central	90	6
Mountain	105	7
Pacific	120	8

One final effect that will cause your "corrected" LAT to disagree with clock time is due to the ellipticity of the earth's orbit. If the earth had a circular orbit, it would orbit the sun at a uniform rate. But the variable speed it has in an elliptical orbit introduces small variations in the length of the day (that is, the time interval between successive meridian passages of the sun) and hence small variations into the clock time at which apparent noon occurs. Astronomers have calculated the corrections due to this effect during the year, and they are summarized on figure 5, which is called the EQUATION OF TIME. By reading the correction from the graph for the appropriate time of year and adding it to the value you determined for the clock time of local apparent noon, you will finally obtain a quantity called the Local Mean Time. It should be 12. Any residual difference at this point will be indicative of observational errors in your determination of the time of highest altitude.

The Sun: Its Energy Output and Yearly Motion

7

The seasonal changes introduced on the earth by the yearly motions of the apparent sun have always figured prominently in determining the character of life, but recently the sun has gained increasing attention as the ultimate source of energy and power.

OBJECTIVES

1. to explain how the sun's apparent motion across the sky varies with time of year and latitude
2. to measure the amount of heat energy received by the earth each minute
3. to calculate the Solar Constant
4. to calculate how the Solar Constant is related to solar power

EQUIPMENT NEEDED

Gnomon (from unit 6) OR sextant, celestial globe, widemouthed bottle with leakproof screw lid and vertical sides, Celsius thermometer, BLACK plastic tape or BLACK polyethylene sheet, watch, ruler.

This unit has two separate but related parts. In the first part you will observe the sun's apparent ANNUAL motion by taking a set of observations near local noon. Changes in the sun's altitude will lead to an understanding of its long-term (yearly) apparent motion. In the second part you will measure the amount of heat energy received by the earth.

One important application of this information is for solar power. Much of our energy in the future may be derived directly from the sun. (Most of our energy comes indirectly from the sun. The world's present supply of energy is primarily obtained from fossil fuels, a form of solar energy produced many millions of years ago.) The most immediate applications will probably be in space heating and water heating for homes. Later there may be huge collectors, set up in regions receiving large amounts of sunshine, which will gather solar energy using photovoltaic systems to generate electricity. Most of the electricity will not be transmitted directly but will be used to manufacture hydrogen, methane (the principal component of natural gas), methyl alcohol (which can be used in internal-combustion engines), or similar fuels. These fuels can be transported and stored more cheaply than can equivalent amounts of electrical energy, at least with present technology. Their transportation via underground pipelines is less damaging to the environment and to natural beauty than high-tension wires -- and they are relatively nonpolluting when burned. This unit should make you more aware of how geographic location on earth affects the feasibility of solar collectors.

I. THE SUN'S APPARENT MOTION ACROSS THE SKY

Follow the instructions in EITHER A or B below.

A. Using the Gnomon

If at least one month has elapsed since the completion of unit 6, you may repeat the measurements of altitude and azimuth of the sun using a gnomon as you did before. Try to observe the position of the setting sun also, from exactly the same location as you did before. Using your data from both sets of observations, answer the following questions:

1. Is the maximum altitude of the sun greater or less than before?
2. Has the time of local noon changed?
3. Does the sun set farther north or south than before?
4. Is the sun above the horizon for a longer or a shorter time?

Proceed to C below.

B. Using the Sextant

An alternative way to accomplish this activity, even if unit 6 has not been done, involves using the sextant to observe the position of the sun in the sky very accurately. The use of the sextant is explained in appendix 2. Observe the sun during the middle of the day, measuring its altitude with the sextant every fifteen minutes. Start when the sun has not yet reached its maximum altitude, and continue observing frequently until the sun is well past its maximum altitude. Wait at least three or four days (more if your observations are taken near June 21 or December 21) and repeat this observation, determining the maximum altitude of the sun on both occasions. What does the change in the sun's maximum altitude reveal about its motion during the time between your observations? Answer the questions listed in A above.

Proceed to C.

C. Questions

Using your data from either A or B above, answer the following questions. Can you tell how the sun is moving in the sky relative to the stars? Can you estimate the date that the sun sets due west? By using the celestial globe, explain the answers to the questions and relate the seasonal motion of the sun to the cycle of the seasons (i.e., summer, fall, winter, spring). How does your latitude affect seasonal changes?

II. THE AMOUNT OF SOLAR ENERGY RECEIVED BY THE EARTH

The basic idea of this experiment is to set a container of water in the sun and observe the rise in its temperature as it absorbs solar energy. From the amount of the temperature rise, the amount of energy absorbed can be computed. By definition, it takes ONE CALORIE to raise the temperature of ONE CUBIC CENTIMETER of water by ONE DEGREE CELSIUS. Obviously, for this to work properly, the apparatus must be set up so as to ABSORB as much of the energy as possible, and the amount of energy leaking into the surrounding air must also be taken into account.

Cover the back half of the INSIDE of the jar with black tape or polyethylene (don't tape the neck portion). This black portion is to absorb the maximum amount of energy. It is important that it be on the inside of the jar, or it cannot transfer the heat to the water since glass is a poor conductor of heat. Stick the tape down tightly. (See figure 1.)

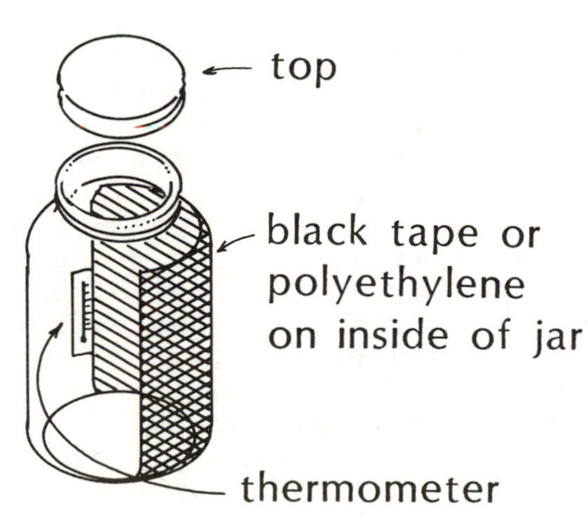

Figure 1

With SHORT pieces of tape, mount the thermometer at the EDGE of the black area, as shown in the drawing. (You will probably have to cut around the outside part of the thermometer if it is mounted on cardboard. Be sure not to cut off the scales.) The thermometer should be flat against the glass, so that it can be observed from the outside.

You will do the experiment alternately making measurements with the jar in the sun and in the shade. The measurements in the shade will be used to determine the amount of heat that leaks into or out of the jar.

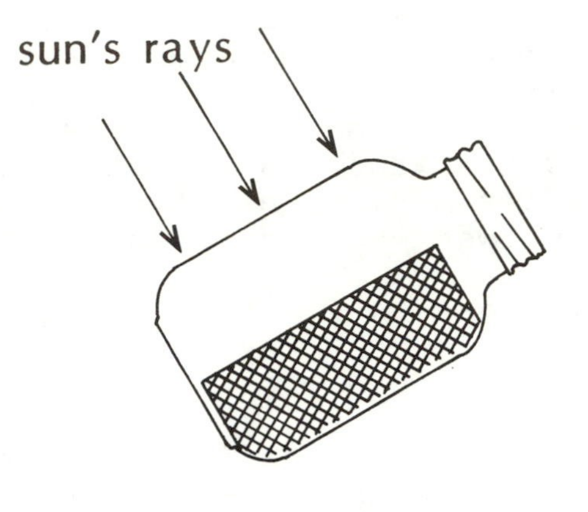

Figure 2

It is preferable to do this experiment on a VERY CLEAR DAY with no wind. For each measurement fill the jar with COLD tap water or ice water. The jar should be filled exactly to the top of the black area, leaving a small (e.g., one centimeter) air space at the top. It is very important that you use the same amount of water for each of the four measurements, so fill the jar carefully. Then place the jar in the sun, propping it up with rocks, books, or other such objects so that the transparent front of the jar is PERPENDICULAR to the direction of the sun's rays. (See figure 2.) This will maximize the amount of black area that is exposed to the sun. During the course of the experiment you may have to readjust the position of the jar to keep it pointing at the sun.

First make a measurement in the sun. Record the date on the chart. Then record the time and Celsius temperature to the nearest 0.1 degree. After EXACTLY fifteen minutes, record the time and temperature again. Next, make a measurement in the shade. Refill the jar with cold water and put it back in the same place as your first measurement. Place a sheet of cardboard or other barrier in front of the jar to keep the sun's rays from hitting it directly. Again, record the time and temperature and, after fifteen minutes, the time and temperature again.

Finally, repeat the above sequence by refilling the jar twice and making measurements with it once again in the sun and again in the shade. Figure 3 gives a model chart for record keeping.

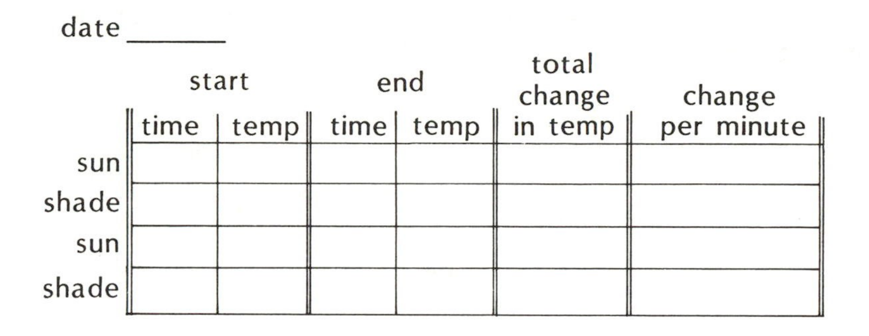

date ______

	start		end		total change	change
	time	temp	time	temp	in temp	per minute
sun						
shade						
sun						
shade						

Figure 3

Refer to your chart. Calculate the temperature change during each time interval (in the fifth column). (If the temperature goes down, it is a loss, or a negative number.) The readings taken in the sun should be about equal, as should the two readings taken in the shade. If they are not, THERE MAY BE SOMETHING WRONG WITH YOUR EXPERIMENT.

Divide the total change by the number of minutes (15) to get the change in temperature per minute. Then average the two numbers for the sun to get the AVERAGE CHANGE IN TEMPERATURE PER MINUTE in the sun; likewise, average the two readings taken in the shade to get the average change in temperature per minute in the shade. This average change in the sun has to be corrected for the loss (or gain) of heat to the atmosphere, which should not be counted as being directly due to the sun. If there was a loss of heat in the shade, then ADD the amount of the average change in the shade to the average change in the sun to get the NET TEMPERATURE RISE PER MINUTE. If there was a gain, then SUBTRACT.

Next, compute the VOLUME of water. First, express the depth of water in the jar and diameter of the jar in centimeters. The volume is approximately

(3.14/4) times diameter squared times (depth of water).

If we multiply this volume by the net temperature rise (or loss) per minute, we find the total number of calories absorbed (or radiated) per minute. Compute the COLLECTING AREA of the jar by multiplying the depth of the water in the jar by the jar's diameter. Finally, divide the CALORIES PER MINUTE by the COLLECTING AREA to find the number of calories absorbed (or radiated) per minute per square centimeter of collecting area. This we will call the PROVISIONAL SOLAR CONSTANT.

The Provisional Solar Constant depends on the sun's altitude, because at low altitudes the sun's rays have to pass through more of the earth's atmosphere than at high altitudes. See the figure, which shows that more energy is absorbed by the atmosphere when the sun is at certain altitudes. Ideally, this experiment should be done in a satellite above the atmosphere; however, it is possible to estimate the effect that the atmosphere has by measuring the Provisional Solar Constant several times during the day (when the sun is at different altitudes) and graphing the results. Do your measurements of solar heating show any evidence of this effect? (See figure 4.)

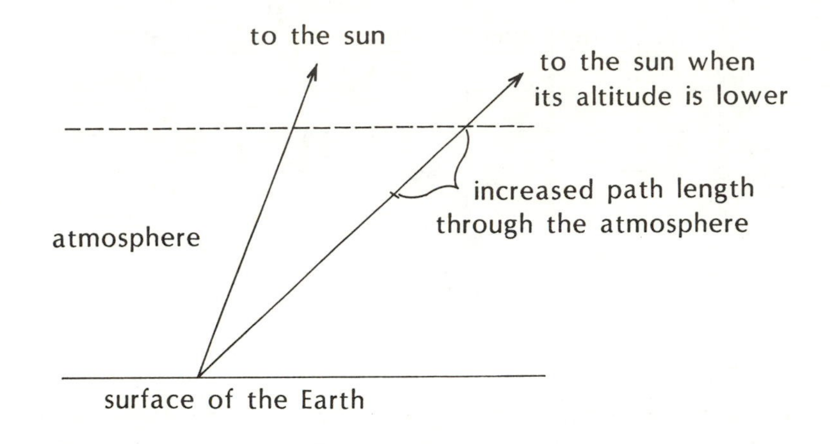

Figure 4

III. SOLAR ENERGY CALCULATION

Now you are in a position to estimate how big a collecting area (or "energy farm") would be needed to generate a large amount of power. To do this, use the relationship that one calorie per minute equals approximately 1/15 watt. How many watts fall on each square centimeter of the earth's surface? A square kilometer (about one-third of a square mile) contains ten to the tenth (10^{10}) square centimeters. How many watts fall on a square kilometer of the earth's surface? How many 100-watt bulbs would this much power keep burning? Do you think it would keep your town or city going? (Estimate the number of bulbs per person in your town or city.) If the world's yearly electricity consumption is 10^{13} kilowatt-hours and the efficiency of conversion from solar input to electricity was 10 percent, how much area would it take to supply the world's needs?

There are 3×10^{17} square kilometers of surface area in an imaginary sphere which has the radius of the sphere with the sun at the center and the earth at its surface. (See figure 5.) Through each of these square kilometers passes the amount of energy (in watts) you computed above. How many watts does the sun produce altogether?

Analyze the errors of this experiment. Can you think of additional sources of error not mentioned or not adequately treated?

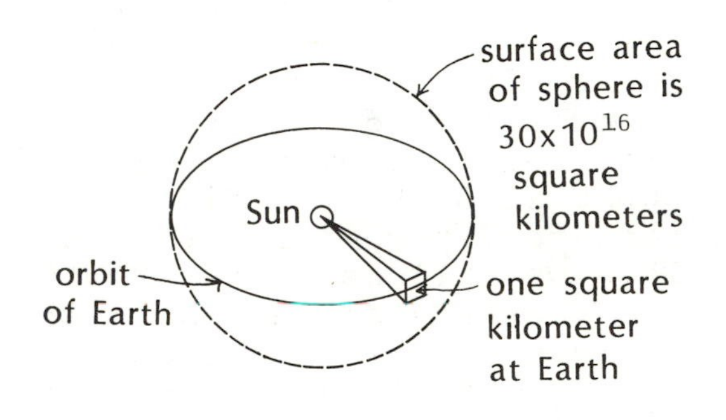

Figure 5

Properties of Lenses and Mirrors

8

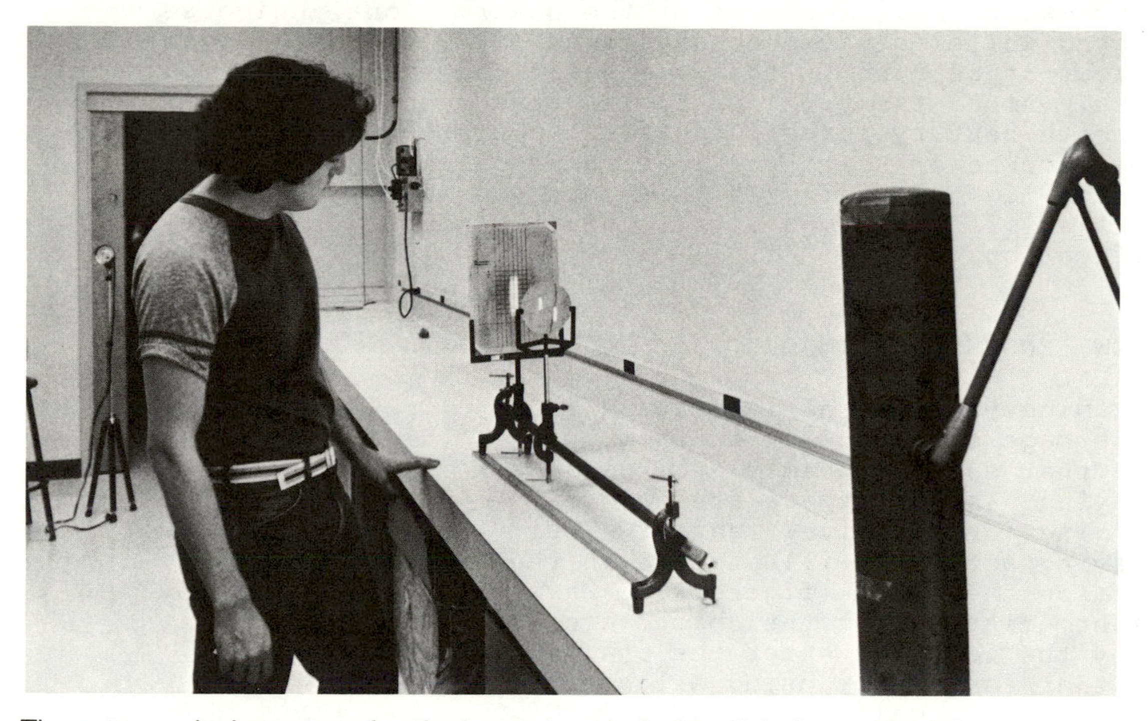

The astronomical process of gathering and analyzing the light from celestial sources begins with the imaging properties of simple lenses and mirrors.

OBJECTIVES

1. to form a real image with a lens or mirror and measure the focal length of the lens or mirror
2. to determine the relationship between the focal length of a lens and the size of the image it will form
3. to explain the difference between real and virtual images
4. to determine how far behind a lens an image will form, given the distance of the object and the focal length of the lens
5. to relate the size and distance of the image from the lens to the size and distance of the object from the lens
6. to determine the magnification that results when two lenses of differing focal lengths are used to form a telescope

EQUIPMENT NEEDED

An optical bench, a fluorescent bulb for a light source, a supply of lenses and a concave (astronomical) mirror, a ruler, a ray box and its accessories.

The invention of the telescope in 1609 changed our "world" view dramatically. The size of the observational universe increased while detailed observations of nearby objects changed the focus of astronomy from a mathematical science to a physical science. Among the problems facing seventeenth- and eighteenth-century scientists were the questions of how and why lenses focused or scattered light and the examination of the nature of light itself. In this unit you will discover some properties of lenses and mirrors, including the principle of magnification.

I. HOW LENSES FORM IMAGES

Examine your set of lenses and look at this page through each of them. Notice that some of them can be used as magnifying glasses; these are called converging (or positive) lenses. Their magnifying action is illustrated in the diagram on the right (figure 1). Rays of light emitted from the object are bent by the lens and enter the eye. The image that is seen by using a lens in this fashion is called a VIRTUAL IMAGE. The eye imagines that the light rays come in straight lines from the virtual image rather than from the object. The eye sees an apparently larger image, because it perceives only the direction the light rays come from and has no awareness of any bending that may have happened to the rays on their way from the object to the eye. Notice that a virtual image has no definite location in space; the apparent object (image) could be located anywhere along the line of sight of the eye.

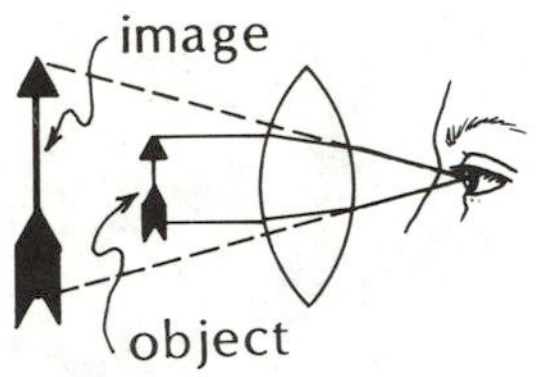

Figure 1

The converging action of a positive lens is illustrated in figure 2. It results from the fact that a light ray which passes from a less dense medium (in this case, air) into a denser medium (glass) is bent TOWARD THE PERPENDICULAR to the lens surface at the point of intersection. In the diagram, the ray is shown as bending toward the dashed line marked "first perpendicular," rather than proceeding in a straight line.

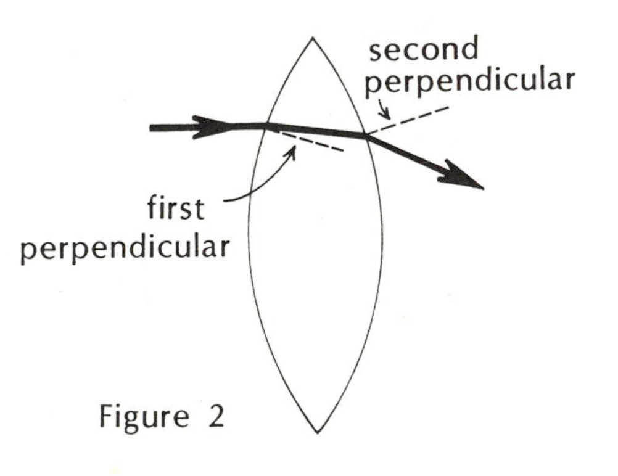

Figure 2

When the light leaves a dense medium and enters a rarer one, as on the other side of the lens, the light ray bends AWAY FROM THE PERPENDICULAR. In the diagram, notice how the light leaves the lens, not in a straight line path, but bending at the surface of the lens in a direction away from the dashed line marked "second perpendicular." Thus, both sides of the lens have served to deflect the ray of light in the same direction.

In figure 3, a ray of light is shown entering a prism-shaped piece of glass. Copy this figure into your notebook and indicate how the ray of light will continue on through the glass into the air on the far side by using the "bending" rules given above.

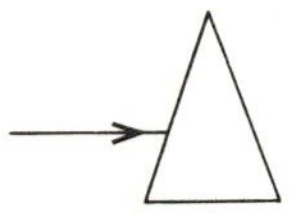

Figure 3

Figures 4 and 5 illustrate how the converging action of a positive lens allows the lens to form images. In the first, a bright pinpoint of light is shown being imaged into a point image. In the second, a lens is shown imaging two different sources of light, A and B. Since A and B could be two points on the surface of an extended source of light like a fluorescent bulb, this second diagram shows how the image of an extended source is formed.

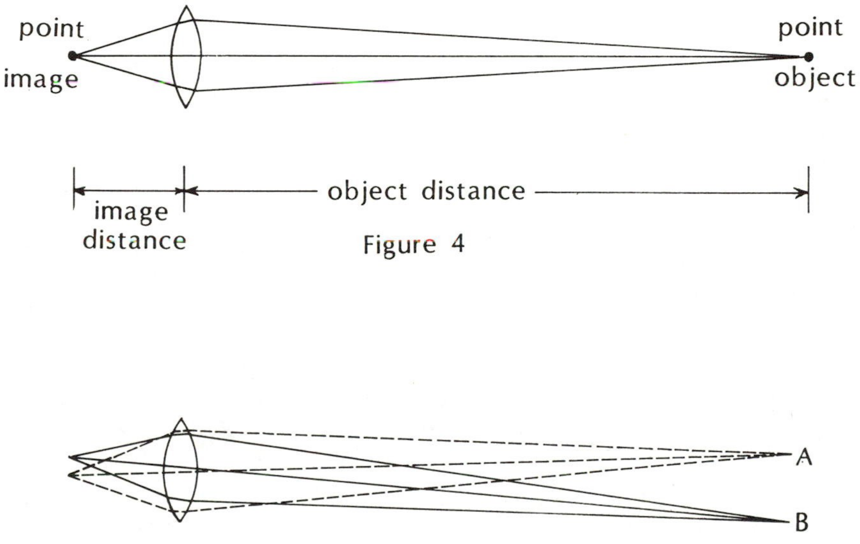

Figure 4

Figure 5

Using each of the lenses in the set in turn, stand a certain distance away from the light source (a fluorescent bulb will work better than an ordinary incandescent bulb) and form an image of the light source on a screen (or piece of paper or wall) behind the lens. Vary the distance from the lens to the screen in order to get the sharpest possible image. With each lens, note whether the image is erect or inverted. Is the image reversed right-to-left, as compared to the object? Are there any lenses which cannot be made to form images? Which ones? Examine those lenses and try to determine why they will not form images. Write all answers in your notebook.

The image cast on a screen by a lens is called a REAL IMAGE, to contrast it with the virtual image discussed earlier. Recall that a virtual image is formed by light rays entering the eye from a certain direction but that the image has no definite location in space. The real image, on the other hand, does have a definite location in space (e.g., on the screen), because it is formed by the INTERSECTION of rays of light from the object. Review the figures presented so far and note that there are indeed intersecting light rays when a real image is formed but not when a virtual image is seen. When you go to a movie, are you watching a real or a virtual image? When you look in a mirror, which are you seeing?

Take the largest lens that forms an image from the kit and form a sharp image with it. Now cover half the lens with a piece of paper. What happens to the image when you do this? Why? (Remember to answer all questions posed in this unit in your notebook, including the identification of each lens.)

II. THE RELATION BETWEEN OBJECT AND IMAGE DISTANCE

Take the largest lens and set it up on the optical bench as close to the fluorescent light as possible. Move it to the nearest point where an image will form. (This will be at a distance of about 1 meter.) Measure the OBJECT DISTANCE (the distance from the light to the lens), the IMAGE DISTANCE (the distance between the lens and whatever surface the sharpest possible real image has been formed on), and the SIZE OF THE IMAGE. Then move the lens farther away from the object, forming a sharp image, and then measure the object distance, image distance, and image size until you have moved as far away from the source of light as you can. Keep a table of your measurements.

Now examine the data you have collected and answer the following questions. How does the image size change as the object distance increases? How does the image distance change as the object distance increases? To see the answer to this latter question most clearly, it is best to GRAPH the numbers you have measured; make a plot of image distance versus object distance. When are the changes in image distance the greatest -- when the source of light is close or when it is far away?

III. THE FOCAL LENGTH OF A LENS

When the source of light is an infinite distance away from the lens, light rays move in parallel lines coming into the lens and are focused at a characteristic distance behind the lens called the FOCAL LENGTH. In figure 6, rays of light from a distant star are shown entering the lens and being brought to focus in a point image at a distance behind the lens equal to one focal length. In the following diagram, light rays coming from two different stars, or two different parts of a distant object like the moon, are shown being focused at two different points in the FOCAL PLANE of the lens.

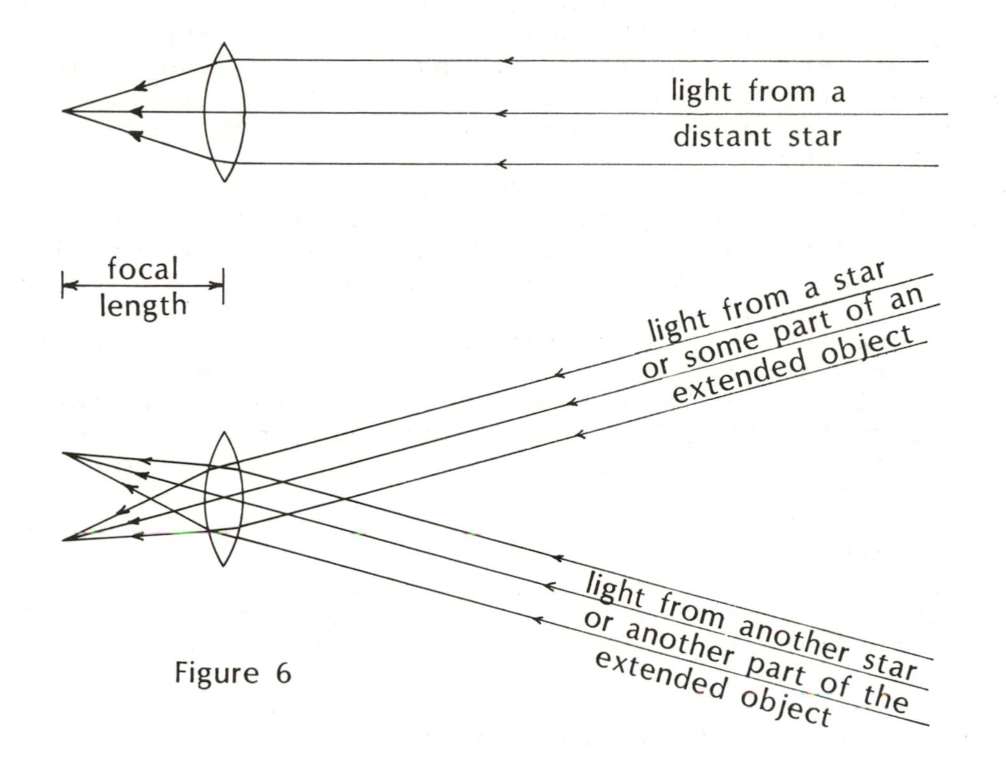

Figure 6

In the confines of the lab it is not possible to move the light source an infinite distance away from the lens, but in practice, as long as the object distance is VERY LARGE compared to the focal length of the lens, the light rays are entering the lens almost parallel and the image distance is almost equal to the focal length. Recall your graph of image distance as a function of object distance from the previous section; after the object distance became fairly large, the image distance no longer changed very much.

Assume that, when you are across the room from the fluorescent lamp, it is effectively an "infinite" distance from the lens. Measure the focal length of the lens by finding a sharp focus and determine the image distance.

You can now determine a more accurate value for the focal length of the lens by using the formula that relates object distance, image distance, and focal length as follows:

$$\frac{1}{\text{object distance}} + \frac{1}{\text{image distance}} = \frac{1}{\text{focal length}}$$

On the left hand side of this formula, substitute the numbers you measured for the image distance and object distance of the large lens in part II, at various distances from the fluorescent lamp. Then solve the equation for the focal length of the lens. You will have several different determinations of the focal length, one for each set of measurements you obtained. The variation between them will be due to observational errors. Reduce these errors by averaging all the numbers together. Compare this value of the focal length you determined by measuring the image distance when the source of light was across the room. How much error resulted in assuming that "across the room" equals an infinite distance?

Notice also that in this formula, if you had a lens of known focal length, you could predict the image distance for any object distance or, conversely, solve for the object distance given the image distance. More generally, given any two of the quantities in the formula, you can solve for the third.

This equation is an important one for cameras. When the object to be photographed is at various distances, the distance between the lens and the film obviously has to change for sharp focus. Since the film is fixed in position, cameras are structured to allow the lens to move in and out for sharpest image on the film. Fixed-focus cameras have no such adjustment and simply set the lens-film distance for the image distance of an average subject.

Use the formula above to solve the following problems:

1. A lens 25 centimeters from a source of light forms a sharp image 20 centimeters behind the lens. What is the focal length of the lens?
2. If the source of light is moved 15 meters away from this same lens, what will the image distance be?

IV. IMAGE SIZE

You should also use your table of measurements to verify that the following relationship is true (by substituting your data and verifying that both sides of the equation agree).

$$\frac{\text{size of the light bulb}}{\text{size of the image of the bulb}} = \frac{\text{object distance}}{\text{image distance}}$$

This relationship is based on the geometry of similar triangles that is set up by the imaging process, as illustrated in figure 7.

When you substitute your data, which measurements agree most closely? Can you think of a reason why this is so?

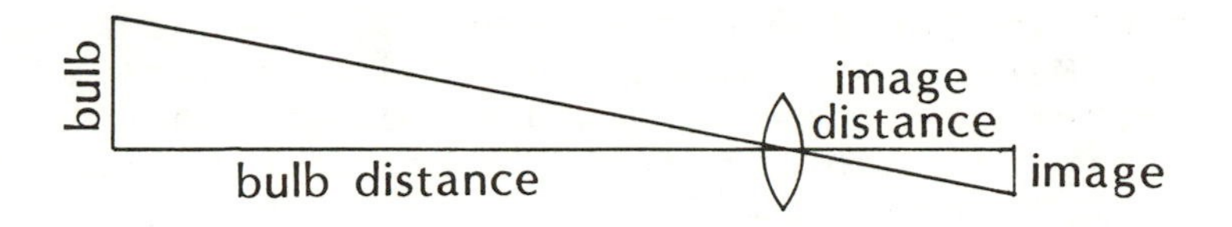

Figure 7

V. RELATIONSHIP BETWEEN FOCAL LENGTH AND IMAGE SIZE

You have now measured the approximate focal length of the first lens and also the image size that it forms when the source of light is very distant (across the room). Measure the diameter of this lens and start a table of focal length, image size, and lens diameter. Measure the focal length, image size, and lens diameter of all the other lenses in the kit which form an image. Make sure that each of the lenses is the same large distance away from the source of light when the focal length and image size are measured. Place the lenses at the "across the room" location which you utilized in part III.

You should also form an image with the concave MIRROR in the kit. The mirror forms an image of a distant source of light by reflection, as illustrated in the figure 8. Note that a concave mirror has a clearly defined focal length. Measure also the size of the image it forms, as well as the mirror diameter.

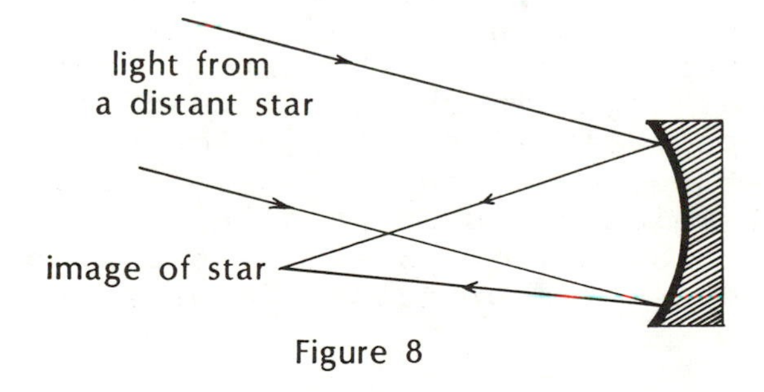

Figure 8

The mirror in the kit is silvered on its front surface; handle it with care, since it can be scratched. Reflecting telescopes are the most commonly constructed systems today, since mirrors are easier to support rigidly against flexing, while a lens can be held only around its perimeter since the light must pass through it. Also, mirrors focus all colors of light at the same point. Lenses do not bend and focus all colors of light the same way; this defect -- called chromatic aberration -- is one of the main disadvantages of optical systems using lenses.

Having measured the diameters, focal lengths, and image sizes for a number of lenses and one mirror, plot your measurements and examine them. Make two graphs, one of image size versus focal length, the other of image size versus lens diameter. What is the relationship between the focal length of a lens and the size of the image it forms? How does the size of the image formed depend upon the diameter of the lens?

VI. RAY TRACING

Given an object to be imaged, and a lens of known focal length, there is a simple graphical procedure to find out approximately where the image will be formed. In figure 9, the arrow on the left is imagined to be glowing, and we are interested in determining where the image of the arrow will be formed by the lens on the right.

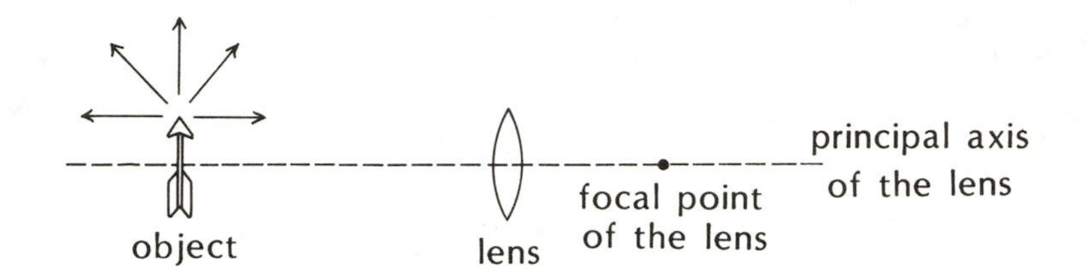

Figure 9

Light rays stream out from the glowing arrow in all directions; in the preceding figure a number of rays have been drawn radiating out from the tip of the arrow. We will consider how the lens images the tip of the arrow. The lens can focus only those rays of light which come toward it. Certain rays, however, have especially simple trajectories through the lens and can assist us in seeing how the image is formed. In particular, any ray of light which comes in parallel to the principal axis of the lens passes through (for positive lenses) or diverges from (for negative lenses) the principal focus of the lens. (See figure 10.)

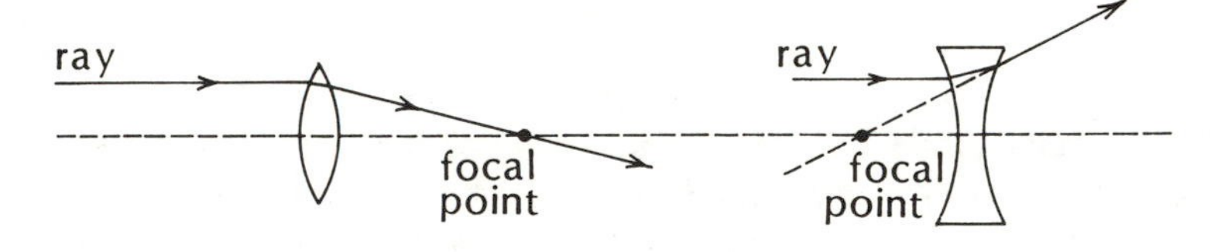

Figure 10

Any light ray that passes through the very center of the lens is not bent at all but passes straight on. (See figure 11.)

Figure 11

If the tip of the arrow is our object, consider two particular light rays leaving it and entering the lens -- one going through the center of the lens and one moving parallel to the principal axis until it intersects the lens. (See figure 12.)

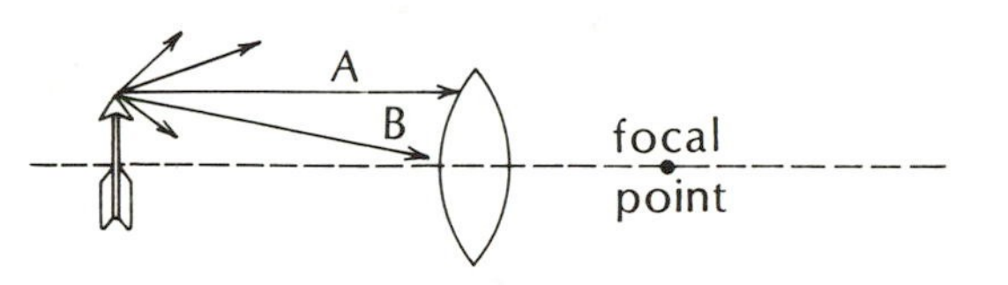

Figure 12

Ray A must pass through the focal point F. Ray B goes through undeviated. Since a real image is formed by the intersection of light rays, the image of the tip of the arrow must form where the rays A and B intersect. All the other rays from the tip of the arrow will also pass through this intersection point, but only for rays A and B is the point easy to locate. (See figure 13.)

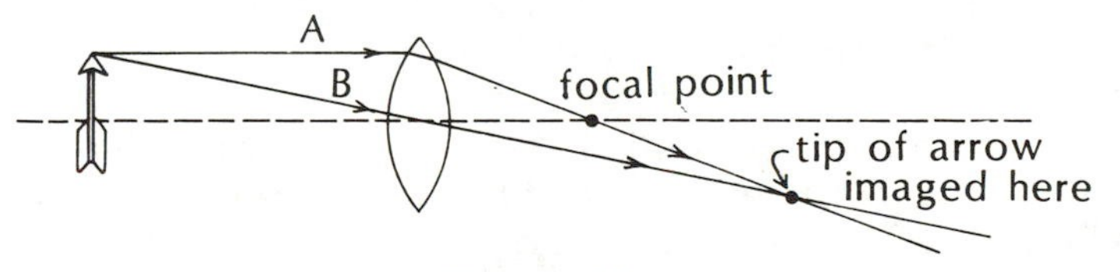

Figure 13

In your notebook, do the same thing for the tail of the arrow and demonstrate that the image is indeed inverted.

Check out the ray box and its special set of lenses, and verify these principles of ray focusing to your satisfaction.

VII. THE ASTRONOMICAL TELESCOPE

As shown in figure 14, the real image formed by a lens can be examined by a short-focus converging lens near the eye (called an eyepiece). This combination forms a so-called Keplerian, or astronomical, telescope. On the left, parallel rays of light from an infinitely distant source of light are shown entering the objective lens of the telescope system.

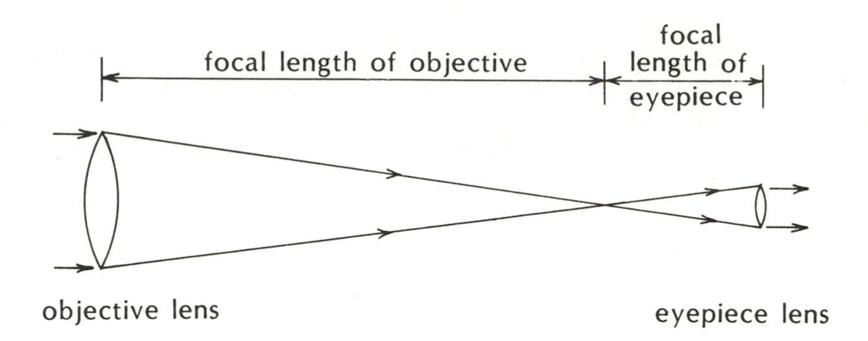

Figure 14

Make a telescope on the optical bench by mounting a long-focus objective lens and a short-focus eyepiece as illustrated in the diagram. Examine something with it and adjust the lenses until you get the sharpest possible image. What is the orientation of the image? Is the field of view large or small?

Estimate the magnification of this telescope by an comparison of the object with the image of the object. A good procedure for this is as follows. Draw on a blackboard a set of parallel lines, about 15 centimeters or more long and about 5 centimeters apart. Place the "telescope" you have assembled on a table so you can sight the lines you have drawn. Keeping both eyes open, look through the lenses at the lines with one eye while looking directly at the lines (i.e., not through the lenses) with your other eye. With a little practice, you will be able to see the lines and the image of the lines superimposed in your field of view, and you can determine the magnification directly by seeing how many lines in the pattern fit between two lines in the magnified pattern.

Compare your estimate of the magnification with the number predicted by this formula: magnification of a telescope equals focal length of objective divided by focal length of eyepiece.

Cameras and Photography

9

Time exposures allow astronomers to look more deeply into space and time. In spite of the mystique surrounding them, darkroom procedures are very straightforward.

OBJECTIVES:

1. to identify the parts of the camera and describe their functions, and state what settings are used when photographing stars
2. given any two values -- focal length, lens diameter, or f-ratio -- to calculate the third
3. given an f-stop and shutter speed, to determine an equivalent exposure with a different f-stop and shutter speed
4. to describe the chemistry involved in exposing and developing film
5. to take and develop at least one roll of properly exposed film following the guidelines in this unit

EQUIPMENT NEEDED

Single lens reflex camera, separate camera lens, tripod, cable release, roll of black and white film (Plus-X), negative darkroom and supplies.

In this unit you will learn to use a camera, take and develop film. Most astronomical observations are obtained photographically, rather than visually. A camera is superior to the eye in being able to add together information on the photographic emulsion as long as the exposure is continued; this is why it is possible to photograph objects fainter than the eye can see. Although the eye is extremely sensitive, it cannot take time exposures. Further, the large amount of detailed information captured on one photograph can be retained for as long as desired and studied at leisure in a variety of ways. Most of the body of twentieth century astrophysical knowledge has been acquired through photographic studies and even with the advent of sensitive new photoelectric detectors in astronomy, the photographic emulsion (on a plate or film) will always remain an important tool for research.

I. THE CAMERA

A. Introduction

In its simplest form, the camera is simply a box with a piece of film in one end and a hole in the other. An example of this is the pinhole camera (as studied in unit 6). Rays reflected off an image and through the pinhole strike a chemically sensitized surface, and form a picture. In figure 1 notice that the rays from the top of the image pass through the hole and fall near the bottom of the film. Only that spot on the film registers an image from the top of the object. Similarly, rays travel from each point on the object to particular points on the film. Thus an image is formed reversed and upside down. The problem with a pinhole camera is that very little light gets through. By enlarging the opening and placing a converging lens in it, a sharper image is obtained. The lens also emits enough light to take pictures in fractions of a second. Every camera works this way, only differing in how well and how easily each gets light onto film in order to form an image.

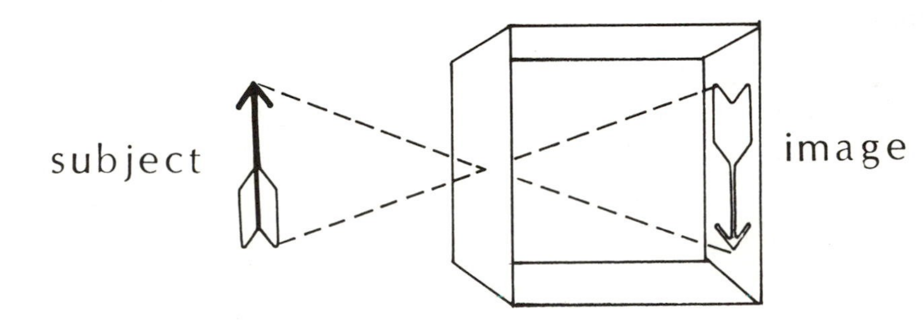

Figure 1

Pinhole Camera

Most cameras, except the very cheapest, have certain features which help them to do their job properly. First, there must be a viewing system which shows the scene the picture will cover. In a RANGEFINDER camera, one looks through a separate lens aligned with the main lens. The best way to see what the camera sees is to look through the camera lens itself. By using a mirror and prism, the SINGLE-LENS REFLEX camera can do this. The recommended camera for this unit is a single-lens reflex (SLR). Both types, rangefinder and SLR, are shown in figure 2. Examine your camera as you read the next sections.

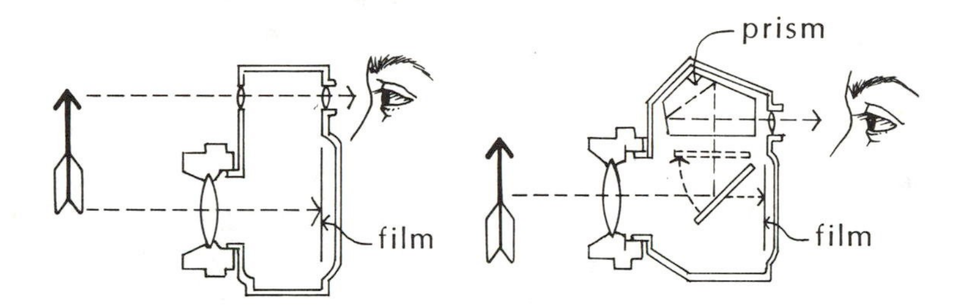

Figure 2

B. Focusing Control

The need for a focusing control is obvious. Recall unit 8 when you compared object distances with image distances. As the light source moved closer or further away, the distance from the lens to the image changed. Because the film is always threaded through the same place in a camera, the focusing control moves the LENS back and forth to create a sharp image on the film. Rotate the focus adjustment on your camera to the infinity (∞) setting and then back in the opposite direction. Which way does the lens move? Can you explain this? Write the answers to all questions in your notebook.

Focusing for most types of photography is putting your most important object in sharp focus and allowing closer and farther objects to be less sharply defined. For celestial photography, the focusing control is simply set at infinity. Consider the definition of focal length and remember that at the infinity setting, the film is exactly one focal length behind the lens.

C. Shutter

One of the greatest advantages of adjustable cameras is their ability to take pictures in a variety of lighting conditions from bright sunlight to semi-darkness. There are two light controls which make this possible: the SHUTTER and the DIAPHRAGM. The shutter is a movable, protective shield that opens and closes to permit light to strike the film for a measured length of time.

Shutter speeds -- the amounts of time which the shutter remains open -- range from 1/2000 to one second on some cameras. Most SLRs also provide for exposures of several seconds, minutes, or even hours if the photographer desires. These long exposures are referred to as "time exposures," and are used predominantly for astronomical photography. For terrestial photography, there are suggestions for shutter speeds under different lighting conditions given in the "Film Instruction Package". In celestial photography, most objects of interest are remote and the light we receive from them is very faint. In order to build up the photographic image, long time exposures are made. To determine the correct exposure times for astronomical objects, several exposures of different lengths are made. Then, after developing the film, the best exposure times can be determined.

D. Diaphragm

The aperture control, another light-modifying device, is made of over-lapping metal leaves which form an adjustable hole. The diaphragm can be opened to let more light in or partially closed -- "stopped down"-- to restrict the passage of light. The individual graduations are generally called f-stops, and the differences between any two is one "stop".

Get out the separate lens with a variable aperture control, or use the camera lens, and set it up approximately eight meters from a light source. Set the aperture control on its lowest numerical setting and focus the image on a screen. Measure the distance from the lens to the screen. Now set the diaphragm on an intermediate setting and then on the highest setting, and make the same measurements. What effect does the aperture size have on the focal length? How does the size of the image change? How does the light intensity change? Can you explain what happens?

Stopping down a camera lens may be done when the scene is well-lit or special effects are desired. A small opening (e.g., f/16) gives a greater depth of field. Depth of field is simply the range of distances which will be in focus at a particular aperture setting. In astronomical photography, the lens is set at full aperture in order to gather as much light as possible since there cannot possibly be a depth of field problem.

E. Light Meter

On the assumption that most photographers place their main subject matter at or near the center of the scene, and want that area exposed for optimum detail, many manufacturers have designed SLRs whose meters make measurements mainly on light from the center area. Their cameras have what is called a "center weighted metering system". Using a light meter which has been set for the speed of the film (the ASA rating), the light available can be measured and a diaphragm-shutter speed combination can be selected.

F. ASA Setting

The usual way of describing film is by its sensitivity or speed; this is indicated by its ASA rating -- a numerical system, devised by the American Standards Association, that grades film according to the amount of light needed to form a normal image. Higher numbers mean a greater sensitivity to light, so that a photographer can get pictures with shorter exposures or under conditions of lower illumination. Photographic terminology often refers to films as being SLOW (ASA in the 20 to 50 range), medium (100 to 200), or FAST (400 and above). The appendix discusses the emulsion characteristics of several common films.

G. The F-ratio And Exposure Time -- READ CAREFULLY

In unit 8 you learned that the size of the image formed depends upon the focal length of the lens. Doubling the focal length doubles the scale of the image; that is, the image is twice as long and twice as wide so that the area of the image increases by four. In general, the area of the image depends upon the SQUARE of the focal length. Now recall the observations you made in part D above with the variable aperture lens. You should have noticed that as the lens was stopped down, the size of the image did not change but its brightness did. When you make the lens smaller, the amount of light it gathers is also less. The light gathered by a lens is proportional to its area. For a circular lens, we have the formula

$$\text{Area} = 3.14\ R^2 \qquad (R = \text{radius} = \text{diameter}/2)$$

Stopping down the aperture decreases the light collected by the lens.

You now have enough information to understand how exposure times are determined in a camera. The amount of light gathered by the lens depends upon the SQUARE of the DIAMETER, but the area on the film over which this light is spread (which is the size of the image) depends upon the SQUARE of the FOCAL LENGTH. Consider as an example, two lenses as given below:

lens A: diameter = 5 cm   focal length = 5 cm
lens B: diameter = 5 cm   focal length = 10 cm

Both lenses have the same diameter so they collect the same amount of light. But lens B forms an image which is four times the area of the image formed by lens A. This means that the light is four times less concentrated on the film, and exposure times will have to be four times longer. As a reward for the longer exposure, however, you do get a larger image.

Consider as a second example the following two lenses:

lens C: diameter = 5 cm   focal length = 10 cm
lens D: diameter = 10 cm   focal length = 10 cm

Answer the following questions:

1. Which lens gathers the most light?
2. By what factor?
3. Which lens has the largest image?
4. For a given amount of light, which lens will take the fastest picture?

The important factors determining your exposure are thus the diameter of the lens squared and the focal length squared. The two factors are frequently combined together into a single number (called the f-number or f-ratio) which is an indication of the speed of the lens system. By definition,

$$f = \frac{\text{focal length}}{\text{aperture}}$$

For example, a lens with a diameter of 50 mm and a focal length of 200 mm would have an f-ratio of four (this is written f/4). From the discussion above, it should be clear that the exposure time depends on the SQUARE OF THE F-RATIO. Making the f-ratio twice as large makes the exposure four times as long. If we compare two camera settings, one at f/16 and the other at f/2, since the difference in f-ratio is a factor of 16/2 = 8, we can say that the f/2 setting will take a picture 8 squared = 64 times faster than the f/16 setting.

On a given camera, such as the Pentax, the focal length is of course constant. In order to change the f-ratio, and hence the speed of the camera, the effective diameter of the lens is changed by using the variable aperture diaphragm. As the lens diameter is decreased, the f-ratio becomes larger and the exposure times must be made longer. Set the f-stop ring to 2 on the camera and observe that the aperture control diaphragm is as wide as it will go. Since the focal length of the lens is 55 mm and the largest diaphragm setting is f/2, compute the diameter of the lens in the camera.

If we set the Pentax camera at f/16, how much longer will an exposure take?

Different lenses can be attached to the Pentax camera body which allow for special effects. For a 135-mm focal length lens whose f-stop range if f/2.5 to f/22.0, what are the minimum and maximum apertures?

The idea that short exposures with large diaphragm openings are equivalent to long exposures with small openings is referred to as the Law of Reciprocity. This law holds for exposure times between 1/1000 of a second and 20 seconds, but not for extremely long or short exposure times. See figure 3.

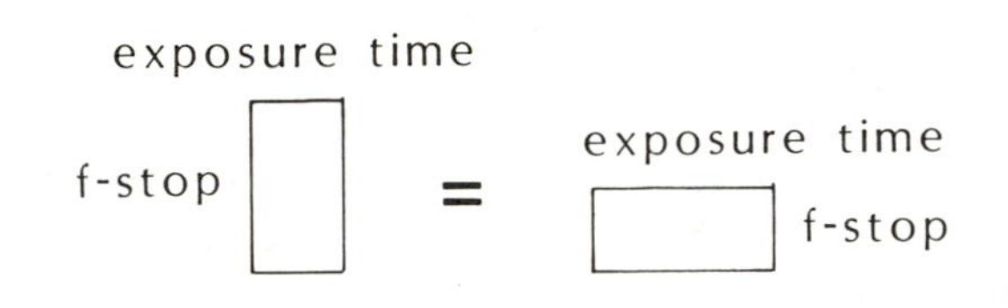

Figure 3

Law of Reciprocity

II. TAKING PICTURES

A. Your Photographic Log

If you do not have a camera of your own, check one out from your instructor and read the operating manual carefully. Then before loading the film, prepare a Photographic Log. You may design your own log, but a sample is illustrated. It is especially important to keep a log for astronomical photographs, since you will usually take a sequence at various exposure times, and then examine the results for the best exposure. When you find it, you will want to know the exposure time you used, and the log will have all the information you need. Keep a record of all the photographs you take.

Photography Log

Name ____________ Film Type ______

Date ______ Camera ______ Roll No. ______

Frame No.	F-stop	Exposure	Subject	Remarks

Figure 4

Now load the camera according to the directions in the operating manual and take a roll of 20 exposures following the guidelines listed below. Each time a picture is taken, be sure to check the following:

1. The lens cap is off.
2. Film is in the camera.
3. F-stop setting is correct and recorded.
4. Field of view is in focus.
5. Exposure time setting is correct and recorded.
6. For shutter speeds slower than 1/30 second, use a tripod and cable release.

B. Guidelines For Your Test Roll

The main objective is to take a wide variety of pictures of different types under different exposure conditions. It is suggested that you first take some ordinary pictures of your choice, getting used to the exposure controls on the camera. Include situations in which the behavior of light is captured in some interesting manner -- reflection, refraction, etc. For example, a stick in a glass of water illustrating refraction, or any interesting sky phenomena could be taken. Then, do a series of observations that relate more directly to astronomy. A distant street light can simulate a bright star for us, but since it is so bright exposures will be shorter. The basic procedures will be the same as for stellar photography in which we take a series of photographs in order to ascertain the correct exposure. We do not know in advance what the correct exposure will be, and a star in general is too faint for the exposure meter. To determine the proper exposure one uses a procedure of trial-and-error called "bracketing". That is, start out with an exposure that is probably too fast (like 1/500 second, or as close as your camera allows to that time), then do a series of exposures increasing the time interval until you are sure you are overexposing the film. For example, on a streetlight which is far enough away to appear as a point source, try exposures of 1/500, 1/50, 1/5, 2, 20, 200 seconds and 10 minutes. (Except for the last, these changes are factors of 10.) When you develop your film, you can judge which exposure was the best one.

Take some star trail photographs. Point the camera towards the sky and open the shutter for a while. Remember to always record your exposure time and camera setting. The setting for star-trails should be obvious. Use a cable release, as it will help stop "camera jiggle". The suggested exposures are two with the camera pointed toward the pole star (Polaris) of 5 minutes and 20 minutes, and two pointed in another part of the sky (East, West, South, or overhead -- wherever the sky seems darker), of 5 minutes and 20 minutes. A tripod is very helpful in setting up your camera for these photographs, but not essential.

III. THE PHOTOGRAPHIC PROCESS

The process that creates a picture on a piece of film involves a reaction between light and a layer of chemicals embedded in the film. Film consists of a plastic base material for support (professional astronomers often use glass plates since they are more rigid), with a thin layer of a transparent gelatin called an emulsion spread on it. Suspended in the emulsion are microscopic crystals of compounds called silver halides (silver bromide with a trace of silver iodide) and these halides are the light sensitive agents. When struck by light, these halide crystals undergo small modifications in their electrical structure as a consequence; at this point they are referred to as "sensitized" and the pattern of sensitized halide crystals on the film is referred to as the "latent image", even though at this stage no actual image is visible. The DEVELOPER is a solution of several chemicals which makes the latent image visible by breaking down the silver halide molecules and leaving atomic silver deposited on the film. It is this silver that forms the image you see. The developer acts on all the silver halide molecules in the film, but acts more rapidly on those that have been sensitized by exposure to light. Therefore, the developing action should go on only for a finite length of time, sufficient to break down most of the sensitized molecules but not long enough to break down most of the non-sensitized molecules. The net result of the developing action is to leave grains of metallic silver on the film at those places where it was originally exposed to light. This metallic silver blocks the passage of light through the film and hence appears dark, so the image on the film at this point is a negative -- the film will be dark where the original scene was light. A STOP BATH is used as a chemical brake to halt the action of the developer rapidly. If the emulsion is then placed in a liquid called a FIXER, the silver halides that were never sensitized by light in the first place are dissolved and washed away (or else they would become sensitized and ruin your image as soon as you took the film out of the developing tank into the light), leaving just the grains of metallic silver as the final image. Finally, a running water RINSE is used to wash away chemical traces that might injure the image.

IV. PROCESSING THE FILM

1. OPEN THE FILM CASSETTE. In total darkness, open the film container by prying off the flat end of the cassette with a can opener. If a reuseable film cassette has been employed, a sharp blow on the counter will pop it open.
2. CUT THE FILM. Still in the dark, the tongue at the end of the film needs to be squared off with scissors. Note that the film is also taped to the spool at the other end, and this needs to be cut off squarely when the film is unwound to that point.
3. GETTING THE FILM INTO THE DEVELOPING TANK. There are several different types of reels to hold the film before inserting into the tank. With the metal (Honeywell-Nikor) reels, bow the film slightly with the thumb and forefinger and hook it to the center of the reel. Rotating the reel will draw the film naturally onto the reel. For the (plastic) Patterson reels, the film is inserted at the outer edge in a pair of grooves and then pulled part way onto the reel. Then grasp the reel in both hands and with a back and forth motion, "walk" the film onto the reel. For the Kodak film blanket reels (long lasagna-like plastic strips), the film is inserted in one end and cradled by the strips as they are wrapped up and inserted into the tank.

   For all these film tanks, a demonstration will probably be needed. Be sure that the film in whatever reel you use does not touch itself anywhere, or development will be inhibited at that point. Put the reel into the tank and cover it. The lights may now be turned on.
4. DEVELOPING IN THE TANK. Pour sufficient developer into the tank to fill it and start the timer. For most black-and-white films, use either D-76 or Microdol developer, whichever is supplied in the darkroom. The length of time the film should develop is given on the white sheet that comes with the roll

of film (or posted on the wall of the darkroom, if you did not buy the film). The development time depends upon the concentration of the developer and its temperature. Measure the temperature of the developer before starting with a thermometer, or during development if you use a tank with a built-in thermometer.

5. TO REMEMBER WHILE DEVELOPING: The tank needs to be agitated while development is going on, to dislodge air bubbles that might stick to the film and prevent development at those points. The agitation (shaking) of the tank need not be continuous, but it should be frequent. If the tank has a cap on it, be sure to use it to prevent the spilling of developer.
6. STOP DEVELOPMENT. When the proper time has elapsed, pour the developer out. In general, developer can be re-used. Ask your instructor if you should save the developer or pour it down the drain. Fill the tank immediately with STOP BATH, dislodge air bubbles and agitate as before for about 30 seconds. Stop bath can be re-used many times; when the time is up, pour the stop bath into a beaker to later return to the storage container.
7. FIXING THE EMULSION. After pouring out the stop bath, pour in the FIXER and agitate for the proper length of time (posted on the darkroom wall). Save the fixer to recycle, just as you did with the stop bath.
8. WASHING THE FILM. After fixing, the tank can be opened. Remove the top and direct a gentle stream of cool water into the center of the tank to flush away all remained chemicals. Empty the tank after a minute or so, and then continue flushing with the stream of water for another 20 minutes. The fixer contains dissolved sulfur compounds which will tarnish the silver image if they remain on the developed film.
9. DRY THE NEGATIVES. Hang the film to dry with clips on the bottom to prevent curling. Gently wipe it with a sponge or rubber squeegee to avoid streaking or spotting, or dip the film in "photo-flo", a fluid designed to prevent spotting, if it is available. Let the film dry for 30 minutes to one hour. It is easiest to store your film by cutting it into envelope sized portions.
10. CLEAN UP THE DARKROOM. Clean up after your darkroom session very thoroughly. Wash off and put away everything. Pay particular attention to counters, tubs, and containers that chemicals may have touched. This is not just a matter of courtesy to the next user; a dirty darkroom can contaminate the entire film processing sequence.
11. PRINTING. Printing is not a required part of this project. It is included as part of unit 18, Celestial Photography. If you intend to do unit 18, you might save any negatives of special interest and print them then. You could also take your negatives to a film shop and pay for their printing.

V. COMPLETING THE UNIT

After your darkroom session, check your developed negatives to verify that your photographic effort has satisfied the unit objectives. In particular, examine the series of bracketed exposures of the distant streetlight with a magnifier and determine for yourself what is the best exposure. Can you formulate a good criterion for "best exposure"? Examine your star trail photographs with the magnifier also. Which part of the sky showed the longest trails for a given exposure time? Why is this so?

APPENDIX: FILM EMULSION CHARACTERISTICS

A. Black and White Films

1. Tri-X (ASA 400)
   This film is good for indoor and some outdoor work. However, it may be too "fast" for direct sunlight. It is good for astronomical work because of its speed, but it is somewhat grainy.
2. Plus-X (ASA 125)
   This film must be used outdoors or with a flash. Though its speed is slow (requiring a longer exposure time), it has a very fine grain. Thus it is suited for astronomical photography of bright objects, as prints may be greatly enlarged without appreciable graininess.
3. 103a-F (no ASA)
   This film, used primarily for astronomical work, has a very low reciprocity failure (thus no ASA rating) and can therefore be used for long time-exposures.* It also has the advantage of sensitivity over a large range of wavelengths. However, this film is <u>very</u> grainy.
4. 103a-O
   This film is similar to 103a-F, but is sensitive to a bluer color range.

*Reciprocity failure refers to the fact that the response of most films is usually not linear with time for exposures of over a few seconds duration. The fact that 103a-F is a low reciprocity failure film can mean that in order to photograph stars one magnitude fainter, the exposure time must be increased by a factor of three. However, for high reciprocity failure films (such as Tri-X or color film), increasing the exposure time by a factor of three or four may allow you to record stars only half a magnitude fainter.

B. Color Film

1. High Speed Ektachrome (ASA 160)
2. Kodachrome 64 (ASA 64)
   These two types of film are similar in that they are both good general purpose films. However, the Kodachrome 64 is better suited for astronomical work (especially for planets) as it has a fine grain.

Using a Small Telescope

10

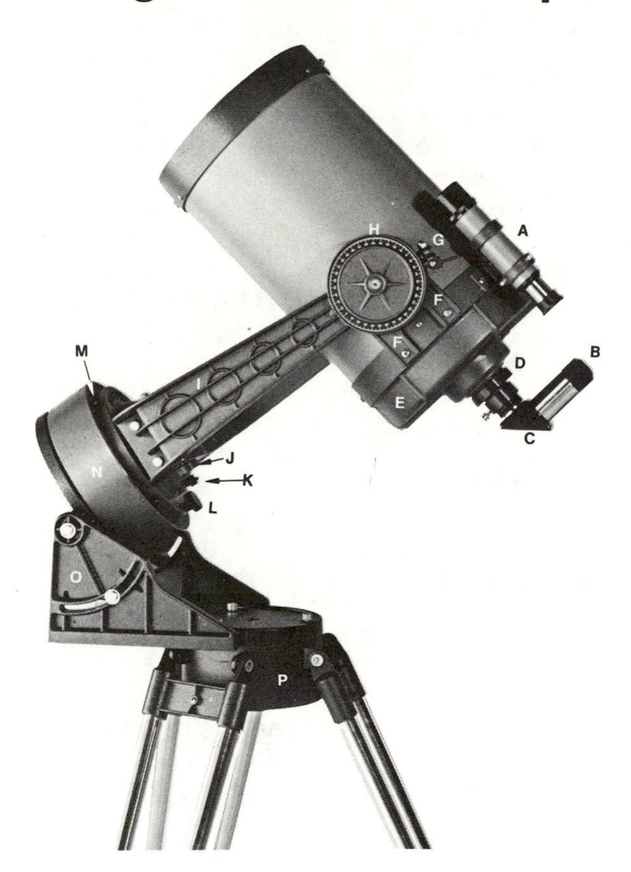

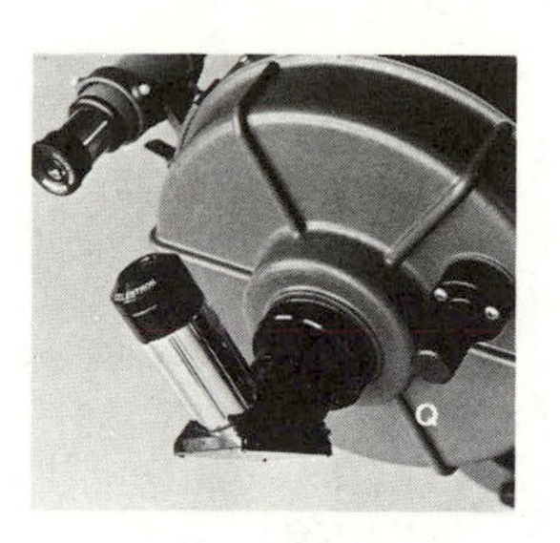

The Celestron Telescope
(A) Finderscope (B) Ocular (C) Star diagonal (D) Visual back (E) Rear cell (F) Tube saddle (G) Declination clamp (H) Declination setting circle (I) Fork tine (J) Declination slow-motion knob (K) Right ascension clamp (L) Manual right ascension control knob (M) Right ascension setting circle (N) Drive base (O) Wedge (P) Tripod (Q) Focusing knob

Even a small telescope, such as the Celestron 8 shown here, gathers 1,000 times more light than the eye. This corresponds to looking 30 times farther into space and an increase of almost 30,000 in the volume of space sampled.

OBJECTIVES

1. to describe the functions of the telescope
2. to describe and illustrate the differences between the two major types of telescope -- reflector and refractor
3. to explain the different types of foci for reflecting telescopes and their purposes
4. to equatorially mount a telescope and set it up on a particular object, given the coordinates of the object
5. to determine the field of view of an eyepiece-objective combination by observation
6. to determine from observations of sunspots the rate of rotation of the sun
7. to describe and sketch several different types of celestial objects, observing their size, color, and other discernible characteristics (illustrations of nonstellar objects should emphasize differences between objects of different types).

EQUIPMENT NEEDED

Small telescope or binoculars, larger telescope (to be mounted equatorially, a "Norton's Star Atlas."

Astronomers made steady but unspectacular progress in unraveling the mysteries of the fixed and wandering stars for the several millennia leading up to an event which was to have a monumental influence upon the future progress of the science -- the invention of the telescope by a Dutch spectacle maker, Hans Lippershey, in 1608. Upon hearing a description of the instrument and its properties, Galileo constructed his own. His first telescope had a magnification of three, but he built many others of increasingly greater size and power during his life. Galileo was the first man to turn a telescope toward the heavens, and as a consequence he accumulated a list of astronomical firsts which will probably never be matched. Among his discoveries were:

1. Many nebulous blurs (e.g., the Praesepe in Cancer) were actually clusters of stars.
2. The Milky Way itself was a myriad of faint, unresolved stars.
3. Jupiter had four satellites which revolved about it. This discovery was significant because of the philosophical debates of the era; many anti-Copernican scholars had argued that the earth could not revolve around the sun, for if it did the moon would be left behind.
4. Venus has phases. Venus goes through the same phases shown by the moon (crescent, quarter, gibbous, and full). This was the first observational disproof of the Ptolemaic theory that the earth was the center of the universe and that everything else, including the sun, revolved about it. The phases of Venus demonstrated that at least Venus must revolve about the sun.
5. The surface of the moon has mountains, "seas," and other structures.
6. The sun is blemished by sunspots which allow the sun's rotation to be calculated.
7. The moon has certain irregularities in its motion, called librations.

I. TELESCOPES -- GENERAL INFORMATION

Most telescopes have two optical parts which are connected by a tube or framework. The OBJECTIVE is a large light-gathering device such as a lens or a mirror. Light entering the telescope from astronomical objects arrives in parallel rays. The objective, therefore, focuses a real image of the source of light onto a FOCAL PLANE exactly one focal length from the lens. The EYE-PIECE, usually a lens, is used to view and enlarge the real image formed by the objective. See figure 1.

A good astronomical telescope must have several capabilities. The major one is its LIGHT-GATHERING POWER which depends upon the area of the lens or mirror. Faint objects cannot be seen with the naked eye because the pupil of the eye is limited in its ability to enlarge and, therefore, in its light-gathering ability. The large lenses and mirrors in telescopes gather larger quantities of light. The amount of light collected depends on the area of the objective. For example, the 200-inch telescope at Mt. Palomar can gather about 640,000 times more light than the 1/4-inch lens of a human eye. How much more light does a 25 millimeter camera lens gather than the eye? (Write all your answers in your notebook.)

Another important capability of a telescope is its RESOLVING POWER. This ability to resolve detail depends on the diameter of the objective and the quality of both the eyepiece and the objective. When sources of light are close together, they may appear to blend into one; stars which appear as one star to the naked eye may actually be double or triple stars. The resolving power of a telescope is the ability to distinguish objects which are too close for the eye to separate, and this is given by the following formula (for visible light):

RP(arcsec) = 4.56 arcsec / diameter of objective (inches)

A simpler formula to remember is:

RP (radians) = wavelength/diameter
where wavelength and diameter have the same units.

This formula works at all wavelengths. For optical telescopes the value in radians is usually a tiny fraction that must be converted to seconds of arc (arcsec). But the formula is especially useful when applied to other types of telescopes, such as radio telescopes. Working at a wavelength of one foot, what is the resolving power of the 300-foot dish at the National Radio Astronomy Observatory in West Virginia?

A third -- probably overemphasized -- capability of a telescope is MAGNIFICATION. Recall from unit 6 that magnification is given by the ratio of the focal length of the objective to the focal length of the eyepiece of the telescope. The magnification is in practice set by the stability of the atmosphere. (See section IV below.)

A good telescope should be ACHROMATIC -- free from chromatic aberration -- and have good definition. Definition is the power of a lens to show an object in good, sharp outlines and applies to extended images of objects, such as the moon or planets.

II. TYPES OF TELESCOPES

Telescopes which have a lens as the objective are called REFRACTORS. They converge light by refracting it. Figure 1 shows such a telescope. The image is formed in a focal plane and examined with the eyepiece. A refracting telescope is very rigid and quite suitable for precise measures of small angles and observing fine details of planetary surfaces, but there is a limit to the size of the objective and, therefore, a limit to the light-gathering power. Objective lenses cannot be made extremely large because they would bend under their own weight and distort the image. For this reason, the largest refractor is only 40 inches in diameter. The large modern telescopes, such as the

200-inch at Mt. Palomar, are REFLECTORS. The objective which gathers and converges the light is a mirror with a parabolic surface. This mirror is usually made of either Pyrex glass or fused quartz (so as not to be susceptible to flexing during temperature changes) with a thin coating of highly reflective aluminum on the surface. A further advantage of reflectors is that all incoming light rays are reflected in the same manner regardless of color. Thus, a reflector is achromatic; it has no chromatic aberration.

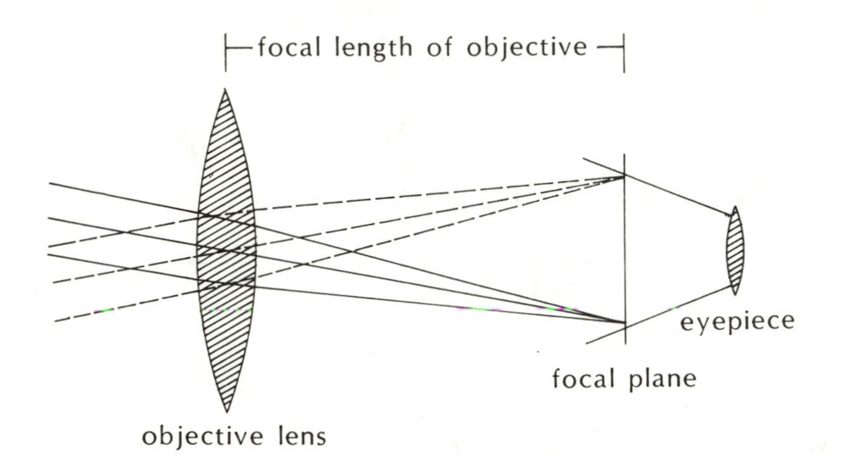

Figure 1

In a reflector, as in figure 2, parallel light rays strike a parabolic surface and are reflected to converge at a point called the PRIME FOCUS. Large telescopes can actually put an observer at prime focus in a hanging cage, but in smaller telescopes such usage would block too much light from the incoming beam and the converging light must be conducted out of the tube by means of mirrors or lenses. The various methods by which light rays are conducted out of the tube result in different focusing systems (see figure 4). Large reflectors, being expensive, are usually designed to be used at several different focal points depending upon what the observer requires.

Recall from unit 8 that image size is related to focal length; the different foci have different focal lengths and different image sizes. Prime focus has the smallest image, cassegrain focus larger, and coude focus the largest image. A large image shows more detail, but there is a penalty -- INCREASED IMAGE SIZE means less light concentration and hence SLOWER EXPOSURES. Choosing a focus for the telescope means determining whether image size or speed is the most important consideration.

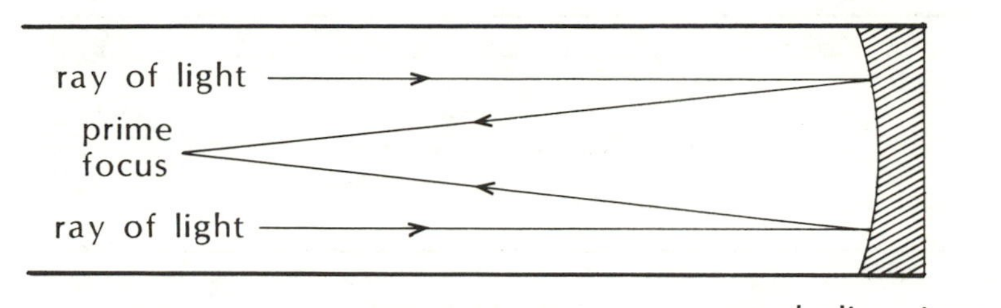

Figure 2

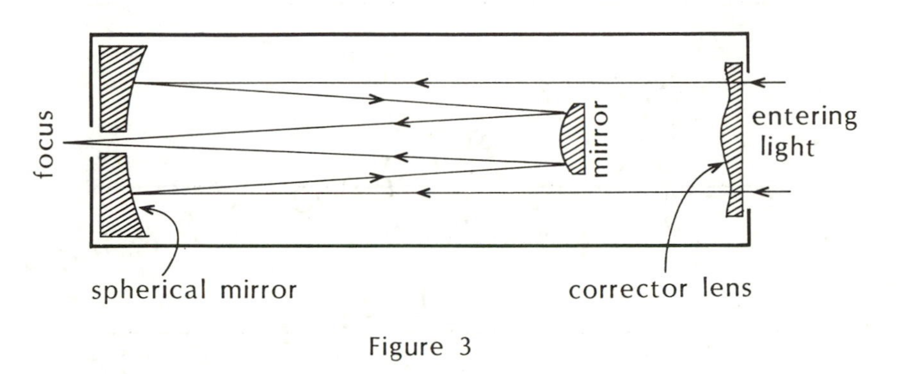

Figure 3

Schmidt Optical System

A telescope used at prime focus gives small images, but the light is concentrated on the film and the exposures are faster. At the other extreme, the coude focus of a telescope gives large images, but this means that the collected light is spread over a larger area of film in the focal plane. At the coude focus exposures are long and slow.

Another type of reflector is the Schmidt telescope. This system uses a spherical mirror (which is easier to grind than a parabolic mirror and gives a wider field of view) combined with a correcting lens to reduce the spherical aberration (see figure 3). The advantage of this system is its compactness and high quality imaging over a wide angular field. The Celestron telescopes are of this type. Note that, since the optical path doubles back on itself, it is possible to have a longer effective focal length in a relatively short tube.

III. TELESCOPE MOUNTS

The most practical telescope mount for serious observing is an equatorial mount. This type of mount allows the telescope to be pointed at any part of the sky and to follow the slow daily motion of the stars caused by the earth's rotation. The equatorial mount consists of two axes about which the telescope can be moved. The POLAR AXIS is pointed toward the celestial pole, parallel to the earth's axis of rotation. Rotation of the telescope about this axis enables it to follow the westward rotation of the sky. The DECLINATION AXIS is perpendicular to

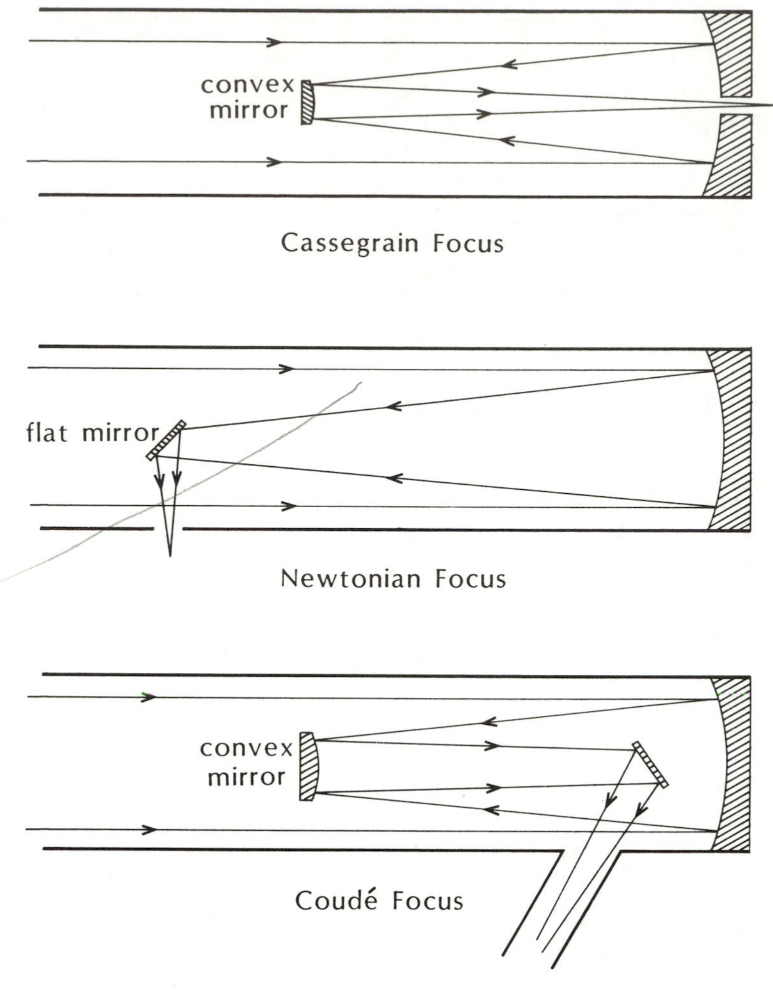

Figure 4

TYPES OF FOCI

the polar axis and is used to set the angular distance of the star from the celestial equator.

In figure 5 the polar axis of a telescope is illustrated. On the left, the telescope is shown pointing perpendicular to the north celestial pole -- this would be in the direction of the celestial equator, and is designated by the greek letter delta (δ). On the right, the telescope is pointed to an object which is north of the celestial equator. The diagram illustrates the telescope aimed at a star approximately 45 degrees north of the celestial equator; such a star is said to have a DECLINATION of 45 degrees. The declination of a star is measured in degrees north (+) or south (-) from the celestial equator.

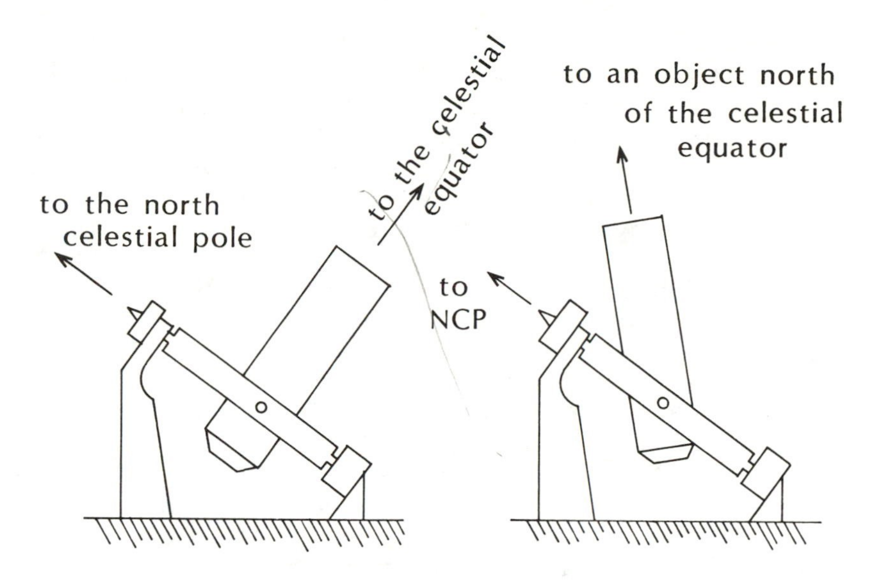

Figure 5

The telescope can track a star in its westward motion across the sky by swinging about the polar axis. The east-west position of a star on the plane of the sky is called its RIGHT ASCENSION, designated by α, the Greek letter alpha. The zero point for α is the intersection of the celestial equator with the ecliptic, and the coordinate increases eastward around the sky. It is measured in units of time, with twenty four hours in right ascension equaling one full sweep around the sky.

These celestial coordinates should already be familiar to you from unit 2, but if you are rusty you might study the celestial globe briefly to refresh your memory of how the positions of stars are designated by two angular coordinates.

IV. FACTORS AFFECTING OBSERVATION

In addition to the type and quality of your telescope, several other factors affect the quality of your observations. A telescope placed in orbit about the earth, or on the moon, is limited only by its own quality and methods of data transmission. Terrestial telescopes observe celestial objects through the murky window we call the atmosphere.

1. "Seeing"

Seeing refers to the stability of the atmosphere. An unstable atmosphere causes the pinpoint images of stars to be blurred into disks of finite size in the eyepiece of the telescope. (A seeing disk of one arcsec is considered a very good evening. For what aperture telescope would this be the resolving power?)

Magnification is also limited by seeing. If the seeing on any particular night is poor, increased magnification causes a greater magnification of the distortion and blurring.

2. Sky Brightness

Moonlit nights are not satisfactory for many types of viewing because too much light is scattered in the upper atmosphere and the sky itself becomes bright. The best time for viewing is when the moon is absent or low in the sky (either younger than first quarter or older than third quarter); the dark sky then gives good contrast. The brightness of city lights also reduces the visibility of faint stars. For some localities this effect may be worse than that from the moon.

3. Atmospheric Interference

Absorption by atmospheric gases and obscuration by dust and other pollutants in the atmosphere also contribute to the reduction of visibility. When objects being observed are close to the horizon, they are being observed through approximately SIX times more dust and gas than objects at the zenith. Consequently, objects near the horizon are usually more difficult to see and often appear redder in color. This reddening is more noticeable when observing the sun or moon, but it also affects starlight.

V. OBSERVING PROCEDURES

Before taking any telescope out to use for observations, familiarize yourself with its parts and controls by reading the instruction manual. Celestron or Dynamax 5 or 8 inch telescopes are commonly used for this project; the opening figure of this unit shows a Celestron 5 with the important parts noted. Any equatorially mounted telescope with an objective of this size (or larger) is suitable.

1. Finder

Before using the telescopes outside for observations, adjust the finderscope. To do this, center an object in the field of view of the telescope. (This can be done in the daytime.) Being careful not to move the telescope, adjust the finder until the object is in the center. Once the finder has been adjusted do not touch it or use it to move the telescope, as this will move it out of alignment.

2. Mount

Larger telescopes are already mounted. For medium-size telescopes (such as the Celestrons) a permanent mount may be available which automatically aligns the polar axis properly. If a permanent mount is not available, a tripod can be used. If you are using a tripod, check the wedge setting for your latitude. Then arrange the tripod so that the polar axis points north. For our purposes, finding Polaris will be sufficient.

3. Setting Circles

To observe on any night, you must first set the right ascension circle of the telescope. To do this, you should set visually upon an object whose celestial coordinates you know.

Obvious candidates are the stars Vega and Sirius, the brightest stars of the summer and winter sky. Their coordinates are:

Vega R.A.= 18 h 35 m Dec = 38 deg 44 min
Sirius R.A.= 6 h 43 m Dec =-16 deg 38 min (Epoch 1950)

Center the star in the telescope and set the movable right ascension circle to read the value of the right ascension of the star. Now that the setting circle has been properly set, other objects can then be located by setting the telescope to their celestial coordinates, as given in published catalogs and other astronomical references.

4. Determining the Field of View

When you go out to begin your observations, take a small telescope with you, in order to make comparisons. Determine the field of view of both the small telescope and the main telescope with each eyepiece available. To determine the field of view, locate a star on or near the celestial equator and line it up on the edge of the field of view. Use your watch or a clock to time how long it takes the star to drift from one side of the field to the other. Since 24 hours = 360 degrees, 1 hour = 15 degrees and 4 minutes = 1 degree. When you have computed the field of view, you will be able to estimate the angular size of things you see in the telescope, such as the angular separation of double stars, by estimating what fraction of the field of view they occupy. Note that the field of view depends on the magnification; thus for each eyepiece there is a different field of view. Answer the following two questions in your notebook: why have we specified that you observe a star near the celestial equator to determine the field of view? If the moon is up, how could you use it to check your field of measurement?

5. Keeping a Logbook

Keep a detailed log of your observations. For each observation that you make, sketch what you see, note the time and date, and indicate which telescope and eyepiece you were using. Record any comments on color, brightness, ease of locating, etc. Knowing your field of view, estimate the size of any star groupings, nebulae, or planets which you see. Try to locate at least one of each type of the celestial objects discussed in the following section. Orient each object with respect to the direction in the sky. For example, your oriented sketch of a binary star might look like figure 6.

The low-power eyepiece is always the first to be used. Never search for anything with the high-power eyepiece. When changing eyepieces or checking a star chart, you may need a light. As in unit 2, a light with a red filter is recommended, since red light has the least effect on your dark adaption. Note that, when you change eyepieces, you must refocus the telescope.

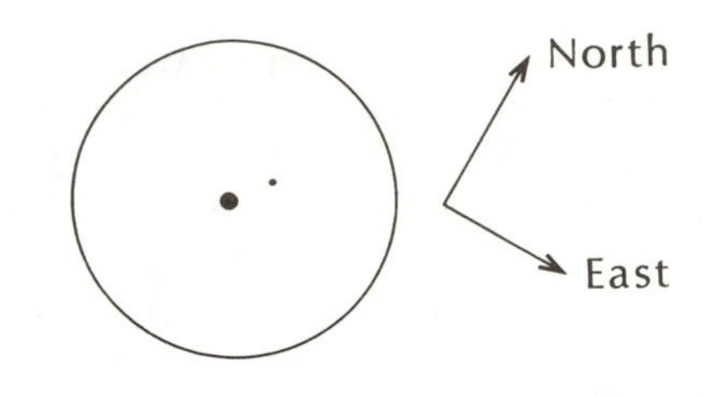

View through the eyepiece

Figure 6

VI. FINDING OBJECTS TO OBSERVE

Coordinates of celestial objects can be found from star maps, atlas, or catalogs. Several of these are listed at the end of the unit under "References." Objects which change their positions, for example, planets, are found in periodic publications, also listed under "References." To find out what star maps you will need, set up the celestial globe for the time at which you want to observe. Check the right ascension at both the eastern and the western horizons. Then, using these right ascensions, locate the appropriate maps. Remember that stars are often listed in constellations by Greek letters, alpha being the brightest. In "Norton's Star Atlas" each map specifically points out interesting nebulae, star clusters, galaxies, and binary stars in that area of the sky. Burnham's "Celestial Handbook" lists the objects by constellation. An additional source of nonstellar objects is the Messier catalog, assembled by Charles Messier. All 103 M-objects can be seen with small telescopes, although city lights cause difficulties with many of them.

1. Sun

The sun can be observed by projecting the image -- NEVER by looking through the telescope. PERMANENT EYE DAMAGE AND/OR BLINDNESS CAN RESULT FROM DIRECT VIEWING OF THE SUN THROUGH ANY TELESCOPE!

Sunspots appear as dark spots which may appear and disappear, or may last several weeks and travel across the face of the sun. Project the image of the sun on a screen behind the eyepiece of the telescope and record any sunspots by tracing the sun and its spots on a piece of paper. If you are using a telescope with a high-power eyepiece, again trace the spots, concentrating on the detail of their pattern. Keep track of the sunspots for several days and use your observations to determine the rate of rotation of the sun. In which direction does it turn?

2. Moon

The moon is probably the first astronomical object most amateurs observe. Look at the moon through the telescope and see how much detail you can sketch. Estimate the angular size of the smallest feature you can observe.

3. Planets

Planets can be distinguished from stars because they shine with a steady light while stars seem to change rapidly ("twinkle") in both color and brightness. Planets also appear as disks in a telescope while stars remain as points. Over a period of time, planets move among the background of fixed stars. (Their name comes from a Greek word meaning "wanderer.") Try to locate and identify a planet among the stars. Draw its position with respect to the background stars and, by observing it twice, show that it moves. The coordinates for the planets can be obtained for each day of the year from "The Astronomical Almanac," which is available for reference in many libraries. Approximate positions for planets are illustrated monthly in "Sky and Telescope" magazine.

4. Double Stars

Stars which appear single to the naked eye but which can be split into more than one star under telescopic magnification are called DOUBLE or BINARY stars. There are two kinds of double stars: OPTICAL DOUBLES, which are far apart in space but line up by chance, and PHYSICAL BINARIES, which revolve about each other with a common center of gravity. The brighter of a binary combination is called the primary and the fainter is called the secondary or companion. Look up the coordinates for several binaries or doubles and see if you can locate them in the sky. Determine their angular separation and orientation in the sky, and compare your value with that of your references. Note the colors of the stars in these systems. In general, visual binary systems have periods of many years, so you will not have the opportunity to actually observe the orbital motion.

5. Globular Clusters

A globular cluster is a closely packed, ball-shaped group of thousands of stars. To the naked eye, these clusters appear as fuzzy patches, but some of them can be resolved into individual stars with a telescope on a good night. Locate one and determine its angular size. How might you estimate the total brightness of the cluster?

6. Open Clusters

An open cluster is a loosely arranged group of associated stars found in the plane of our galaxy. The typical galactic cluster has from ten to several hundred stars, all moving in a common direction. Like the stars of a globular cluster, the stars in an open cluster originally formed out of one large interstellar gas cloud. Locate at least one open cluster and estimate its angular size. Look for differences in magnitude, color, and orientation in each type of cluster. How many stars can you see in your cluster?

7. Nebulae

This word is used to refer to a variety of types of nonstellar objects, which are studied in units 13 and 14.

a. Emission (gaseous) nebulae absorb ultraviolet radiation from nearby "hot" stars and reradiate it at visual wavelengths.

b. Reflection nebulae are interstellar clouds of dust which reflect light from nearby bright stars.

c. Planetary nebulae (which are round and green in appearance, like planets) are composed of gases which have been ejected from and are illuminated by radiation from an aging central star.

d. Extragalactic "nebulae" appear as fuzzy patches of light in all but the very largest telescopes. These "nebulae" are not gaseous but are aggregates of billions of stars -- galaxies like our own Milky Way.

Locate several nebulae and see if you can identify the type of each. Use the descriptions in your references to select and observe the brightest nebulae available at your time of year. Draw each as best as you can, determine the size and shape, and list any other properties you can isolate.

References

Norton's Star Atlas and Reference Handbook, Sky Publishing Company, 1978.

Celestial Handbook (in 3 volumes) by Robert Burnham Dover Publications, 1978.

Field Guide to the Stars and Planets by D. Menzel Houghton Mifflin, 1964.

The Astronomical Almanac United States Government Printing Office, annual publication.

Observer's Handbook of the Royal Astronomical Society of Canada, annual publication.

"Sky and Telescope", monthly magazine.

Introduction to Spectroscopy

11

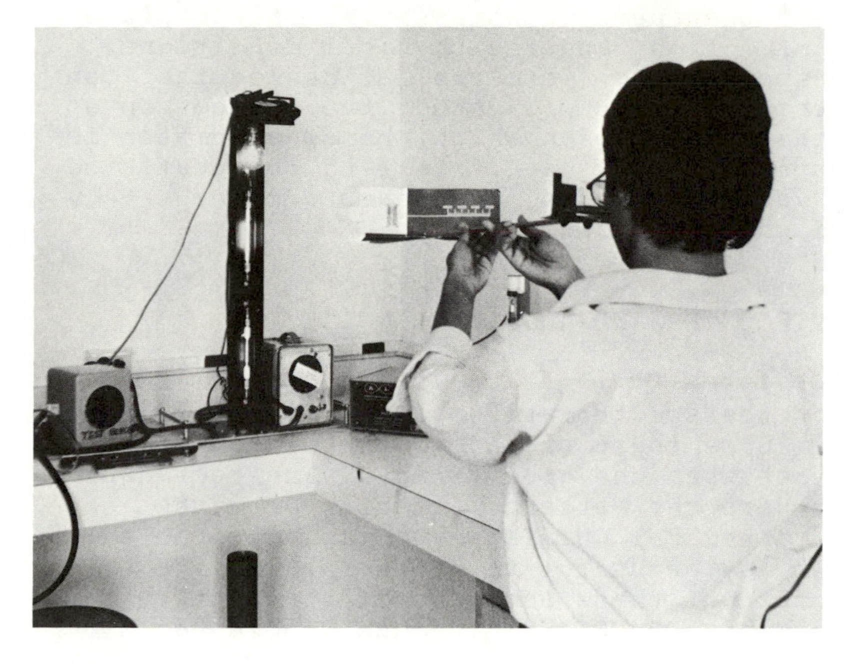

A simple spectrometer with an inexpensive grating can measure wavelengths of light to an accuracy of 20 Angstroms. This is one five-millionth of a centimeter!

OBJECTIVES

1. to construct a spectrometer with an adjustable slit
2. to view the diffraction of light and to determine how the diffraction of light depends upon its color
3. to examine the continuous spectrum emitted by an incandescent bulb and measure the central wavelength of the various colors exhibited
4. to measure the wavelengths of the emission features in the spectra of discharge tubes such as mercury, hydrogen, and neon
5. to use the spectrometer to examine various lights, such as streetlights, "neon" signs, or fluorescent tubes, in order to determine the type of spectrum they emit and identify the gases (if any) in the light sources
6. to be able to predict the type of spectrum that will be emitted by various types of light sources
7. to observe absorption features in the solar spectrum
8. to distinguish between systematic and random errors

EQUIPMENT NEEDED

Spectrometer pattern (on insert), a NEW double-edged razor blade, medium-weight cardboard, rubber cement (or glue), diffraction grating, meterstick, scissors, and stapler.

One of the astronomer's most basic tools is the spectrograph. With it can be determined the chemical composition of the planets and stars, their state of motion, and the physical conditions prevailing in them. A spectrometer is a similar device that can be used visually. In this unit you will study the phenomenon of diffraction (which is the principle on which the spectrometer is based), make a simple spectrometer, use it to find out what kind of light is emitted by various sources, and measure some properties of light and unknown light sources.

I. ASSEMBLING THE SPECTROMETER AND ADJUSTABLE SLIT

The black page which forms part of the insert in this book contains the pattern for the spectrometer scale and a piece to be used as a brace for it. Glue these pieces to the medium-weight cardboard and cut them out when the glue is dry. The rectangular area marked CUT OUT is where the pattern will slip onto the meterstick; the square marked CUT OUT is where the adjustable slit is constructed. It may be useful to use a razor blade instead of scissors to cut out these sections of the pattern in order to have neat edges, but DO NOT USE the NEW double-edged blade for this purpose.

The spectrometer scale folds toward you along the dotted line and mounts on the meterstick at the 53.3-centimeter (21 inch) mark, as indicated in figure 1. The brace fits on the backside to give the scale rigidity.

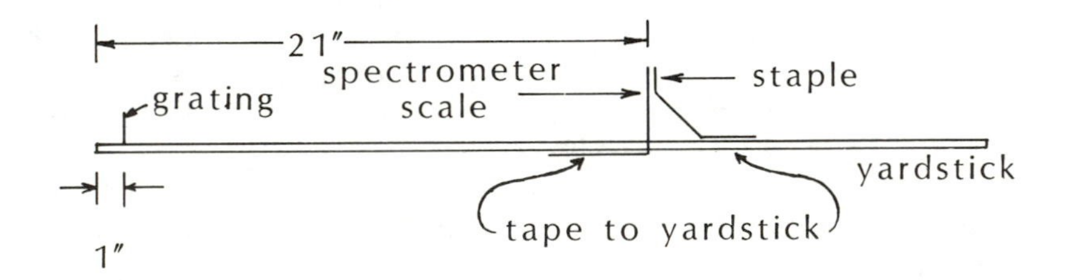

Figure 1

To construct the adjustable slit, break a new double-edged razor blade in half lengthwise (careful!). Mount one-half of it with tape to the spectrometer scale, along the line marked "align slit here," with the razor edge facing left, as illustrated in figure 2. Note: if you cannot obtain a razor blade, a piece of thick plastic or exposed film, cut with scissors, also makes an adequate slit.

For the other half of the slit, cut out from a piece of cardboard a rectangular piece approximately 2.5 by 6 centimeters and tape the other half of the razor blade to it as illustrated in figure 3. This piece will be the movable half of the slit.

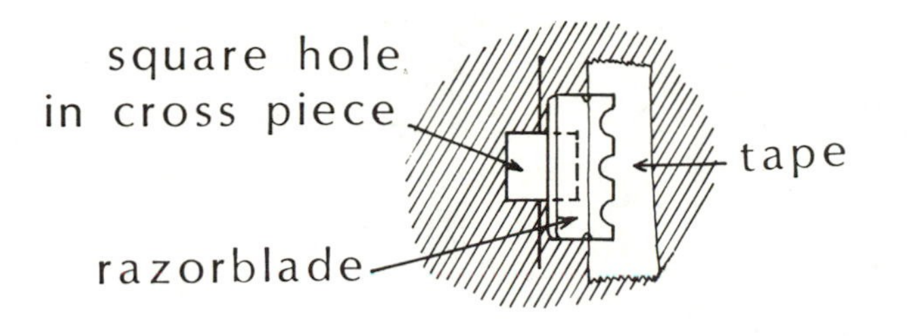

Figure 2

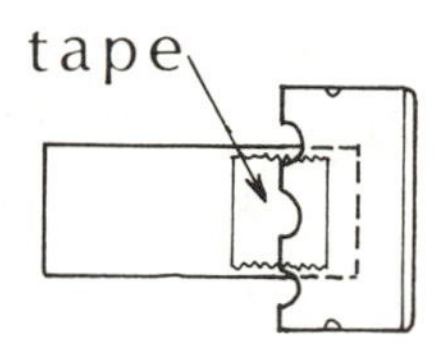

Figure 3

With the movable half in place on the spectrometer, the total slit assembly will look like the picture in figure 4. The loose, movable piece can be held in place with a strip of paper or cardboard stapled crosswise on the crosspiece, as illustrated.

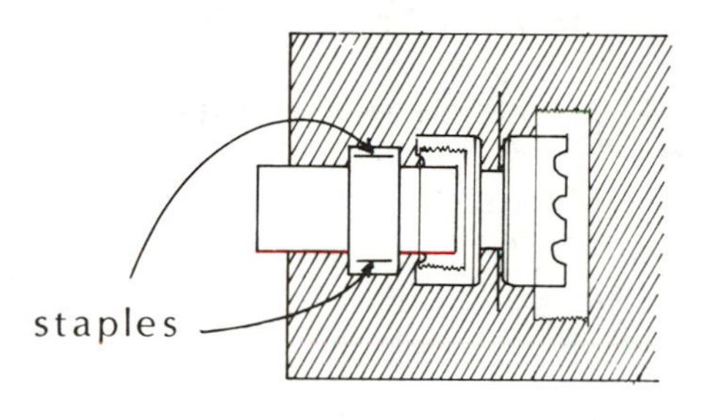

Figure 4

Mount the grating (which is mounted in a 2-by-2 inch cardboard slide) 2.5 centimeters (1 inch) from the eye end of the meterstick. The simplest way to do this is to use the sliding crosspiece from the CROSS-STAFF to hold the grating and to use paperclips to fasten it to the cross-piece.

II. OBSERVING DIFFRACTION WITH A SLIT

Light is composed of WAVES, as was first demonstrated about 150 years ago by Thomas Young. Because of the wave nature of light, it is DIFFRACTED, or SPREAD OUT, as it passes through a narrow slit. The waves can "bend around" the edges of the slit. It is the same phenomenon that makes it possible to hear someone talking around a corner. The only difference is that because the wavelength of light is much shorter the effect is much smaller.

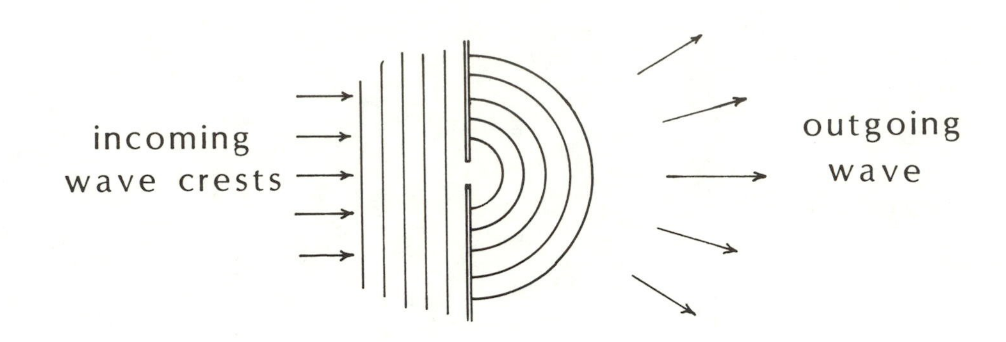

Figure 5

In the first observation of this unit, you will examine various sources of light through the adjustable slit ONLY, without the grating. Hold the crosspiece of the spectrometer up very close to your eye, so that the slit is as close to your eye as possible. It will be possible to easily view the diffraction of light.

Perform the following tasks with the slit:

1. Hold the slit up close to your eye in front of a bright, evenly lit surface. SLOWLY close the slit until it is very narrow (this takes a little practice). Hold the slide firmly between thumb and forefinger for greater control. Describe in your notebook what you see.

2. Find a distant, bright, and SMALL source of light (examples: the sun reflected off a distant window, a streetlight). Observe it as you SLOWLY close the slit. Can you see it spread out? In what direction (relative to the slit) does this happen? Are there any noticeable color effects? Describe and draw the appearance of what you see when the slit is narrow.

3. In a darkened room arrange a set of gas discharge tubes (if they are available), or the lights you plan to examine. Mercury vapor tubes glow blue and neon tubes glow red. These vapor tubes contain rarefied gases, excited to emit radiation by passing a high voltage through the tube. Care is required in turning on these tubes since the voltages involved are very high (several thousand volts) and QUITE DANGEROUS. Observe these laboratory spectrum tubes through the slit, held parallel to the tubes. Describe and draw in your notebook what you see. Does the amount of the diffraction effect change as you vary the slit width? Is it the same for a red tube as for a blue tube? Which color is most affected?

III. OBSERVING AN ORDINARY LIGHT THROUGH THE GRATING

A grating serves the same purpose as does a prism, namely, to break the light up into its component wavelengths. Your grating is actually a very fine set of parallel lines etched on plastic (too fine to see with the naked eye). These lines behave like many small parallel slits and spread out the light in the same way your single slit did in the observations above. Different colors are spread by differing amounts. If the light contains a mixture of colors, the fact that each color is diffracted differently serves to separate the various colors in the source and display the "spectrum" of the light source. Separating the colors in this fashion is called DISPERSION.

Light of different colors is actually electromagnetic vibration at different frequencies and wavelengths. Astronomers generally characterize light by its wavelength and avoid the ambiguities of describing colors. The unit for measuring the wavelength of light is the Angstrom, abbreviated Å . In metric units, the Angstrom is one hundred-millionth of a centimeter, that is, ten to the minus eight centimeters. The eye is sensitive to light having wavelengths between about 4000 and 7000 Angstroms. Notice that the scale on your spectrometer has been calibrated to read wavelengths over just such a range.

With the grating in place, observe an ordinary light bulb (also called an incandescent bulb) by looking at it THROUGH THE GRATING and also THROUGH THE SLIT of the spectrometer. Note that a clear bulb (one in which you can see the filament) works best. Narrow the slit until it is about 1.5 millimeters in width. When you look at the bulb in this manner, you should be able to see a spectrum spread out to the right of the slit along the wavelength scale. By a spectrum we mean a continuous band of color running from blue through red. If you do not immediately see the spectrum, move the slit back and forth across the light source until you do. You will soon find the proper position. If instead of the spectrum you see a streak of light above and below the slit, turn your grating 90 degrees in its holder and you will then see the spectrum.

Figure 6 illustrates how to use the spectrometer to see the spectrum of a source of light.

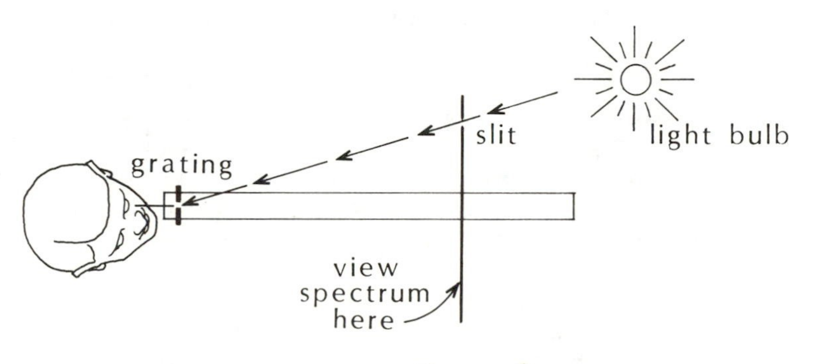

Figure 6

Notice that each color in the spectrum falls on a different portion of the wavelength scale. Write down the colors you see and measure their wavelengths. Since one color spreads over a range of wavelengths, you can measure either the center of a color band or its edges, but be sure to specify which. The light bulb is clearly emitting energy at all visible wavelengths. Such a spectrum is called CONTINUOUS. It has been found by experiment that a continuous spectrum is given off by HOT SOLIDS, HOT LIQUIDS, and HOT DENSE GASES. An incandescent bulb is a hot solid, since inside the glass there is a tungsten filament which is heated to glowing by passing a current through it.

If the light bulb is attached to a rheostat control, you can vary the current passing through the filament. Lots of current means a high temperature for the filament, so when you set the rheostat control on maximum, the temperature of the filament is as high as it can be. As you turn the rheostat control down, the filament temperature decreases, and the decrease in brightness of the bulb is obvious. The total energy emitted actually depends upon the fourth power of the temperature; if you have done unit 12, recall that the discussion on surface brightnesses of stars used a temperature relationship of a similar format. Start with the rheostat control on high and turn it down to lower settings until the filament is just barely visible. What happens to the color of the bulb? Now repeat this observation using the spectrometer; that is, start with a high setting and turn the rheostat slowly down to a very low setting, looking at the changes in the SPECTRUM that take place. What changes in the spectrum do you see, and how do they explain the changes in the color of the bulb that you saw without the spectrum? Put all your observations in your notebook.

IV. OBSERVING DISCHARGE TUBES WITH THE SPECTROMETER

Return to the blue and red discharge tubes you studied earlier and observe them through the slit and grating. For these light sources, you will not see a continuous band of light; instead you will see only a few discrete features -- thin lines of light at certain specific wavelengths. This type of spectrum is called an emission line or "bright line" spectrum. Measure and write down the wavelength of each discrete feature you see in the spectrum. Begin with the mercury vapor (blue) tube.

1. The blue tube contains mercury vapor which is excited by the electrical current passing through it. It is apparent that the gas emits energy only at certain specific wavelengths -- in particular, you should see a feature in the blue, one in the green, and one in the yellow part of the spectrum. Measure and write down the wavelengths of these features.

2. When you observe the red tube (neon), notice that its spectrum is somewhat more crowded, with numerous emission features close together. To be able to see as much detail as possible, rest the spectrometer on something rigid so that the shakiness from your hand does not hinder the measurements, and

also narrow down the slit as much as possible (consistent with your still being able to see the spectrum). It should be clear from your observations that the spectrum consists of a series of "lines" only because the light comes from a long thin tube. Even if it didn't, passing the light source through a long thin slit would cause a line spectrum to appear. The spectrum is actually a series of rectangular images of the slit in the various wavelengths where the source is emitting energy. By changing the width of the slit, the width of the lines in the spectrum can be changed. If a gas emits energy at two or more wavelengths which are very close together, we would want to use a very narrow slit to be able to see those features as separate entities. A wide slit produces wide lines, which would tend to overlap and hence not be separately resolved. Thus, we increase the RESOLUTION of the spectrometer by narrowing the slit, but we pay a price since the amount of light that comes through is also reduced.

Try varying the slit width and noticing its effect on the crowded neon spectrum. With the slit as narrow as practical, observe and describe the neon spectrum, writing down the wavelength of EVERY feature you can see as a distinct feature.

A bright line spectrum is characteristic of HOT RAREFIED GASES (as opposed to hot, dense gases which give off a continuous spectrum). As you have observed, mercury gas has a certain pattern of emission and neon gas has another characteristic pattern. Each element in the chemical table has its own characteristic pattern of emission features at certain wavelengths. By recognizing these patterns, we can identify the various elements in the spectra of celestial objects. The various patterns shown by the chemical elements are related to their differences in atomic structure.

Examine the neon spectrum again and answer the following question: why does the neon gas glow red when you look at it without the grating? Why does the mercury vapor look blue to the eye?

3. Turn on the discharge tube which contains hydrogen gas. Write down the wavelengths of the spectral features that you see. (Note that there may be a spurious feature in the yellow due to impurities in the tube and due to evaporation of the elements in the glass itself. Ignore the yellow spectral line in your measurements. It is actually caused by sodium emission.) What are the colors of the observable features of the hydrogen spectrum?

4. Observe the spectrum of a fluorescent lamp. Describe the spectrum and write down the wavelengths of any discrete features you see. Comparing your measurements to the earlier measurements you made of the gases in the discharge tubes, can you identify the gas inside the fluorescent tube? Give reasons supporting your conclusion.

V. CALIBRATION AND ERRORS

The most important lines of hydrogen are at 6563 A, 4861 A, and 4340 A. There is also one at 4101 A, but it is hard to observe. Mercury has strong lines at 4358 A, 5461 A, and 5780 A.

For each of these lines which you have observed, calculate the error (observed value minus true value). Some of your errors may be negative. Do not take the absolute value. Plot the error versus the observed wavelength, and draw the straight line which best represents the errors. This line can be used to correct all your other wavelength measurements. It represents the SYSTEMATIC ERROR of the instrument, which may be due to misalignment, misuse, etc. Drawing this plot is a way of calibrating the instrument in order to correct and eliminate the systematic error. As long as you do not change your spectrometer construction or the way in which you use it, this will be your error curve. If you borrow someone else's instrument, it will have different errors associated with it.

The scatter of the individual points around the straight line is the RANDOM ERROR of your observations. This allows you to estimate the basic accuracy of the observations. To estimate the amount of scatter, measure the separation of each point from the straight line and average their absolute values. Repeated measurements will often reduce the scatter, but it cannot be eliminated.

VI. OTHER OBSERVATIONS

There are many other observations that you can examine with your spectrometer. Several are listed below.

1. Observe an incandescent bulb whose light has passed through a piece of blue plastic before entering the spectrometer. Describe and measure the spectrum you see. What effect did the plastic have on the light passing through it, especially on its color? Make the same measurements using a piece of red plastic.

2. Examine some streetlights. Look for several of different color. Is the spectrum continuous, bright line, or dark line? (See section VII below for an explanation of a dark line spectrum.) Write down the wavelengths of any discrete features. If any bright line features are present, can you identify the gases responsible for them?

3. Look for a bright red orange sign. Can you recognize the pattern of features?

Observing hint: for all the observations, getting close to the source will make the spectrum brighter and easier to examine. When the light is large, narrow down the slit in order to resolve the lines. Use the rest of the spectrometer scale to block out stray light and unwanted light sources.

VII. OBSERVATIONS OF THE SUN

There is a third type of spectrum which a light source can exhibit. If a source of continuous radiation (such as an ordinary incandescent bulb) passes through a cooler, rarefied gas, the cool gas ABSORBS energy from the continuous spectrum. See figure 7. Through the spectrometer we will see a continuous distribution of light but with DARK LINES at certain specific wavelengths where the cool gas is absorbing. The wavelengths absorbed are the same ones the cool gas would EMIT if it were heated up and viewed by itself as a bright line emitter.

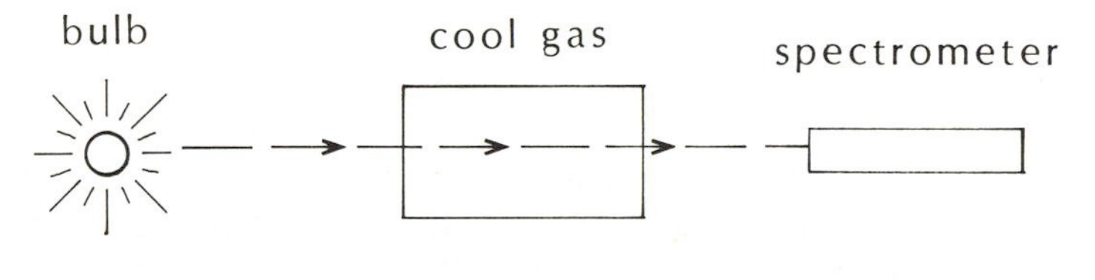

Figure 7

The easiest way to see a dark line spectrum is to locate a pink orange high-pressure sodium streetlamp. If you cannot find one of these, then observe the sun. The sun is a large ball of hot dense gas which gives off a continuous spectrum. It is surrounded by a cooler outer atmosphere which absorbs energy at certain specific wavelengths, and these dark line features can be viewed with your spectrometer.

The observation of the sun is the most delicate observation of the unit, so read the following instructions through carefully before proceeding. Find a place where there is a relatively clear view of the horizon, and observe the sun very late in the afternoon, near sunset when it is close to the horizon in the west, or very early in the morning, near sunrise when it is very close to the horizon in the east. The spectrometer can be held level, and the sun's brightness is considerably attenuated at these times by passing through so much of the earth's atmosphere. To further protect your eyes from irritation, narrow down the slit of the spectrometer until it is very thin and then observe the sun through the slit ONLY, taking care to keep the spectrometer scale between your eye and the sun. You will then see the solar spectrum on the scale. Note: since there are no lenses in the spectrometer, there is no danger of concentrating the sun's light. These instructions are given in order to minimize the marginal discomfort that might be experienced from looking directly at the rising or setting sun. As you know, when the sun is high in the sky, it is quite bright and you should not look directly at it. But when it is close to the horizon, and when the spectrometer scale masks most of it, there is no danger in this observation.

Your observation of the sun will probably not succeed if you try to hand-hold the spectrometer, since the dark line features are not as obvious as the bright lines in the discharge tube spectra. Arrange for the instrument to rest on something while it points at the sun. You could tape it to a tripod, lay it on a windowsill or table, or rest it on a tree branch.

When you observe the sun, you will have no difficulty observing its continuous spectrum. But, narrowing the slit down as much as is practical without losing the spectrum, try to notice the thin, vertically oriented DARK lines in the spectrum. Write down the wavelengths of any dark lines that you see. For example, do you see a dark line in the yellow part of the spectrum? Such a line should be among the most visible in the solar spectrum; it is caused by sodium gas in the atmosphere of the sun. Do you see any features which indicate the presence of hydrogen gas in the solar atmosphere?

VIII. ADDITIONAL ACTIVITIES (OPTIONAL)

The spectra of all stars exhibit dark lines of this type, and it is by identifying their wavelengths and measuring their strengths that astronomers can determine the chemical compositions of the stars. Unit 12 shows how the spectra of stars are related to their temperatures, masses, and radii. Unit 17 compares the detailed spectra of Beta Draconis with that of the sun. In Unit 16, many spectral features are observed with the solar telescope. Unit 14 discusses how spectrum lines may be used to determine the motion of an object.

If you have already done the photography unit (unit 9), you can make a SPECTROGRAPH by simply taping a grating in front of a camera lens and photographing various light sources. By projecting the 35-millimeter slide taken of the discharge tubes onto the wall, you can measure the wavelengths with a meterstick. After calibrating this procedure with mercury and hydrogen tubes (since they have known wavelengths), you could then photograph unknown sources.

12

Distances and Fundamental Properties of Stars

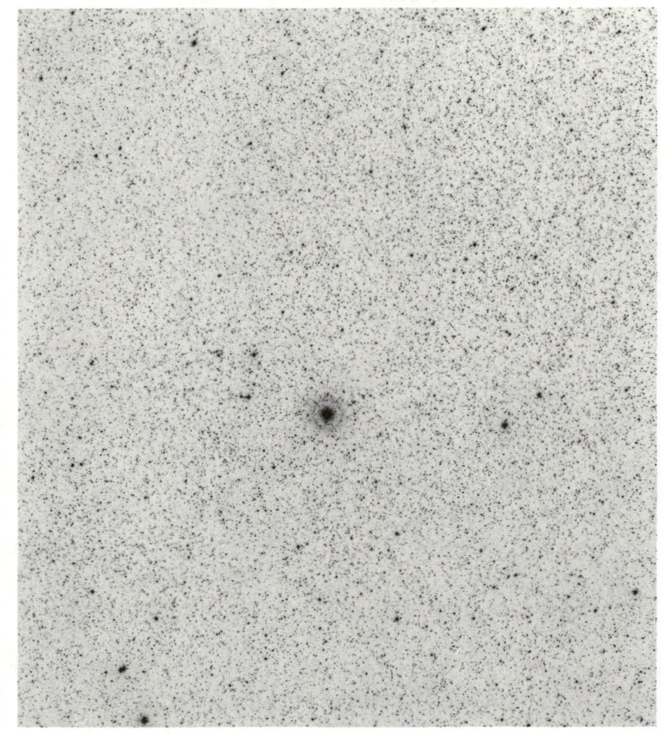

NATIONAL GEOGRAPHIC, PALOMAR SKY SURVEY

In this photograph of the Cygnus region, every speck is a star. The astronomer's problem is—which ones are the interesting ones?

OBJECTIVES

1. to use simple equipment to measure the distance of an object approximately one mile away, by a triangulation process similar to that used to determine distances to nearby stars
2. to plot data relating the brightness and physical temperatures of stars, and to use the plots to infer other characteristics of stars such as physical size and mass
3. to be able to explain why the "main sequence" of stars exists
4. to be able to explain how the size and surface temperature of a star combine to determine its total luminosity
5. to be able to describe the process of obtaining a "spectroscopic parallax" for a star
6. to be able to convert magnitudes to brightness and vice-versa

EQUIPMENT NEEDED

Cross-staff, sextant.

12-2 Distances and Properties of Stars

In this unit we examine some of the intrinsic properties of stars (their luminosities, temperatures, and radii) and their distances. These quantities are among the most interesting and the most difficult for an astronomer to study. The sun is the only star near enough to us for detailed study. The topics in this unit form the basis for all stellar studies and for galactic structure investigations.

I. PARALLAX

Relatively nearby stars may yield a measurement of their distance by making use of the fact that the earth orbits the sun and thereby changes our perspective when we view them at different times. Consider figure 1, with the sun at S and the earth's orbit shown. When the earth is at E-1, the nearby star is seen at a certain place on the celestial sphere with respect to the more distant (background) stars. Six months later, when the earth has orbited to position E-2, the nearby star will seem to have shifted its apparent position with respect to the background stars. The background stars are chosen to be sufficiently distant that the motion of the earth does not affect their apparent position to any measurable extent. Measuring the apparent change in the position of the nearby star gives us the angle E-1→star→E-2, which is shown in the diagram. Since the distance from the earth to the sun is known, we have a triangular relationship which can be solved for the distance to the star.

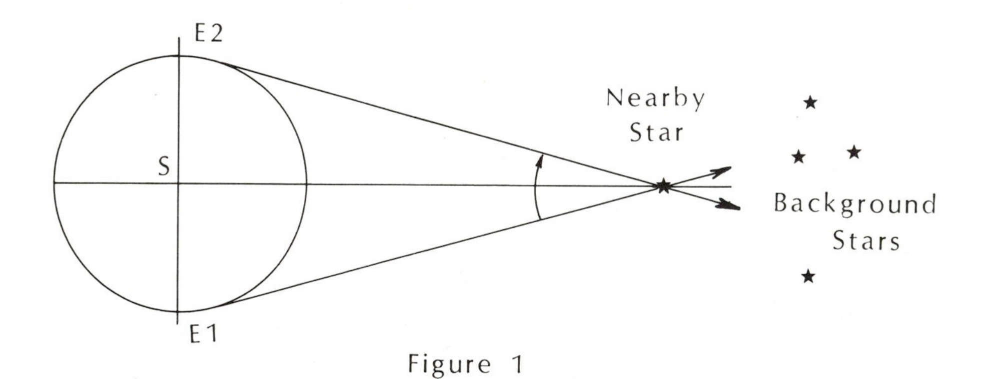

Figure 1

Consider the right triangle formed by the earth, sun, and nearby star, as illustrated in figure 2. The side labeled A is the distance from the earth to the sun; this distance is also called one astronomical unit (abbreviated 1 AU) and is equal to 93,000,000 miles or 150,000,000 kilometers. The angle θ (theta) is one-half of the angle marked in figure 1; it is called the PARALLAX of the star.

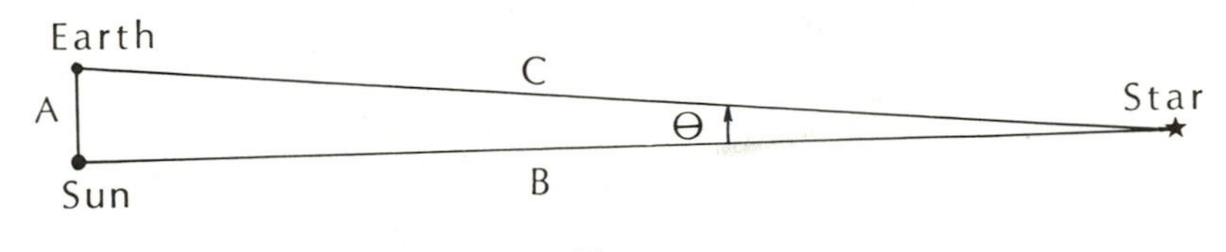

Figure 2

Trigonometry tells us that the tangent of the angle θ equals side A divided by side B. That is

$$\text{tangent } \theta = A/B$$

If θ has been measured, we can look up the value of its tangent in a book of tables (or use a calculator). Knowing A, we could then solve for B, the distance of the star. The distance A is often referred to as the BASELINE DISTANCE. This process is further simplified by the fact that, when θ becomes small (as usually occurs in astronomical observations), the tangent of the angle is approximately equal to the angle itself measured in radians. In other words, if θ is smaller than a few degrees,

$$\text{the tangent of } \theta = \theta \quad \text{(in radians)}$$

Thus the relationship of the quantities in figure 2 is now:

$$\theta = A/B \quad (\theta \text{ in radians})$$

We will use this IMPORTANT FORMULA in the following example. The nearest star is Alpha Centauri, approximately 300,000 AU from the sun. If its parallax were measured, the value would be:

$$\theta = 1 \text{ AU}/300{,}000 \text{ AU}$$

Convert this value of θ, which is in radians, into degrees. Then convert the value into seconds of arc (arcsec). You should obtain 0.69 arcsec. This is a very small angle. These small angles are very difficult to measure, and obtaining the parallax of a star often takes several years and extreme care. Can parallax measurements be made for very distant stars? Explain.

II. MEASURING THE PARALLAX OF AN OBJECT

You can apply this method of triangulation to the measurement of an object approximately one mile away. To measure its parallax angle, you will use the cross-staff and nomograph from unit 1. Take these to the roof of a building or another high place. Pick a nearby building or object to serve as the object whose distance is to be measured. A very distant object is needed to serve as a "background star." Figure 3 illustrates how the nearby object should be within a few degrees of the line of sight to the distant object. The figure is not drawn to scale, since the distant reference object must be VERY distant to serve as an adequate reference point. Go to the extreme corner of the roof and measure the angle between the nearby and distant objects. Repeat this measurement at least twice to obtain consistency.

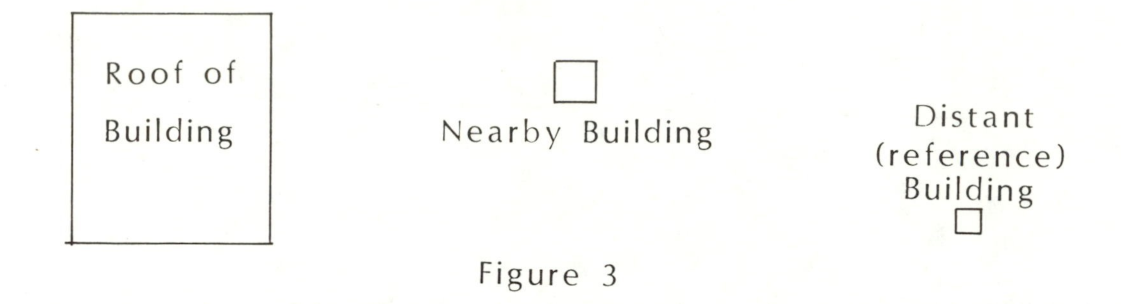

Figure 3

Next measure the angle between the nearby and distant objects from the other extreme corner of the roof. These measurements are illustrated in figures 4 and 5. Take care to measure the angle between the SAME points on each building. Repeat this measurement at least twice also.

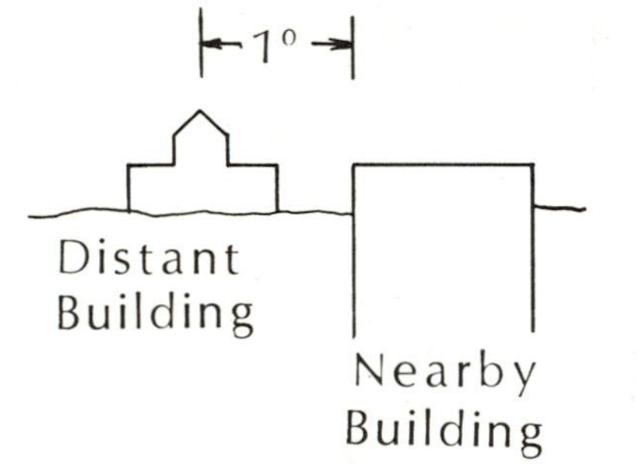

Figure 4

The First Measurement

Total Parallactic Shift

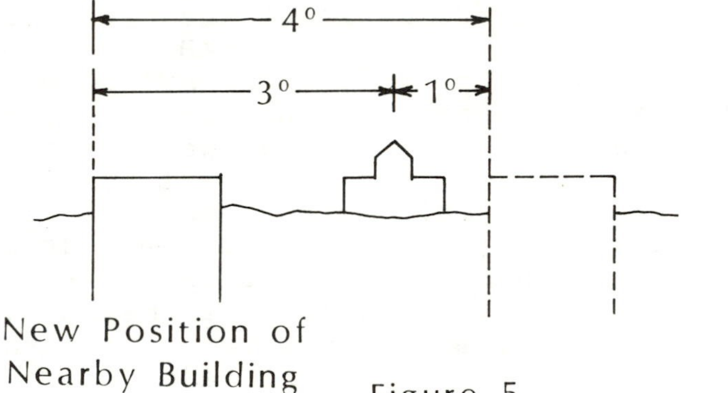

Figure 5

The second measurement compared with the first

Refer to figure 6, which shows the geometry of the situation you have measured. The distance reference object is so distant that your lines of sight to it from both corners of the roof are effectively parallel. The two angles you have measured are α (alpha) and β (beta) in figure 6.

From plane geometry comes a theorem that the angle γ (gamma) in figure 6 is actually equal to the sum of α and β. Therefore, γ is actually the total parallactic shift discussed above. However, the quantity "parallax," used by astronomers, is one-half the parallactic shift. The parallax of the nearby object is

$$\theta = 1/2\ \gamma = 1/2\ (\alpha + \beta)$$

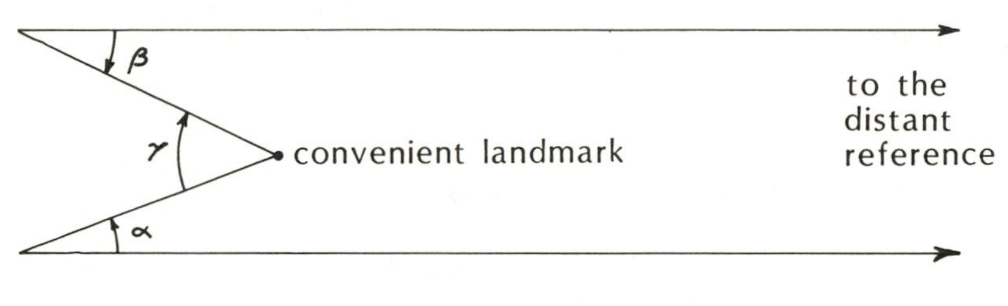

Figure 6

Knowing θ, return to the basic formula and calculate the distance to the nearby object. A is one-half the total distance of the reference baseline across the roof. Pace off the distance, or use some other method of measuring it. With θ and side A, solve for side B, the distance to the object, since distance equals half the baseline divided by the parallax angle (in radians), or A/θ.

Note that your geometry could be similar to that in figure 7. This is no problem, since the geometry provides a simple solution in this case:

$$\theta = 1/2\ (\alpha - \beta)$$

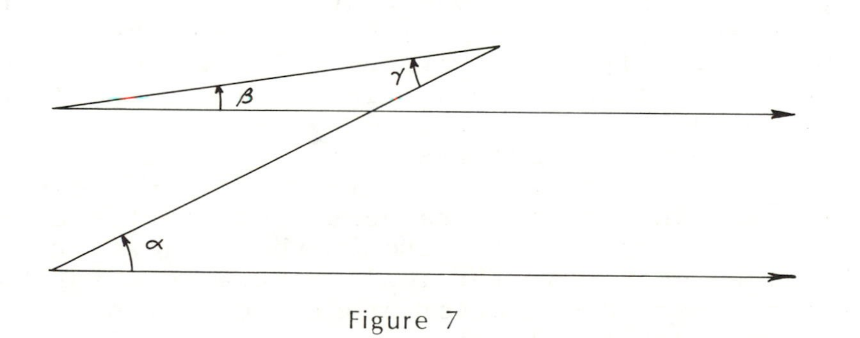

Figure 7

III. MEASURING PARALLAX WITH A SEXTANT

In this section, you will use a sextant to remeasure, with greater accuracy, the same angles you measured in section II.

Refer to appendix 2 (The Sextant) for instructions. Stand at each end of the roof, as you did before, and measure the angle between your distant reference object and nearby object. Make several measurements and average your results. After computing the distance, compare it with your previous determination of distance.

Astronomers find the units of meters, feet, or even miles too small to be practical. For our planetary system, the "astronomical unit" is commonly used. Stellar distances are often measured in units of "light years" (abbreviated LY). One light year is the DISTANCE light travels in one year. One parsec (abbreviated pc) is the distance of an object whose parallax is one second of arc. 1 pc = 3.26 LY.

IV. THE HERTZSPRUNG-RUSSELL DIAGRAM

Early in this century, a Danish astronomer, Ejnar Hertzsprung, and an American astronomer, Henry Norris Russell, independently proposed a method of comparing stars by graphing their luminosities against their surface temperatures. Such graphs are so generally useful and informative that astronomers continue to employ them today to illustrate a wide variety of concepts.

The LUMINOSITY of a star is its total energy output per second. It can be measured in absolute units; that is, in the centimeter-gram-second system of units, the luminosity of the sun is 4 times 10 to the 33 power ergs per second. This is written 4×10^{33}. It can also be measured in relative units, comparing the energy output of a star to that of some standard reference like the sun. For example, the star Arcturus is approximately 100 times more luminous than the sun. If the luminosity of the sun is designated by $1\ L_\odot$, the luminosity of Arcturus would be $100 L_\odot$.

To determine the luminosities of stars, it is necessary to know their distances from us. Only then can an APPARENT BRIGHTNESS be converted into ABSOLUTE PHYSICAL units of luminosity. A bright star may appear bright because it is close to the sun in space, or it may be bright even when quite distant if it is intrinsically very luminous. The distance must be known to distinguish these situations. In the beginning of this unit you saw how the distance to relatively nearby stars may be measured by geometrical parallax methods. If a star is too far away, the angle θ becomes too small to measure, even with the best telescopes, and astronomers must adopt indirect methods for determining the distance. All such methods, however, are ultimately based on the geometrical measurements for the very nearest stars. Section VI describes the use of statistical parallax, a nongeometrical method.

Values for the brightness and surface temperature for the thirty-seven nearest stars are listed in table 2. Notice that some of the "stars" are multiple systems so a total of forty-nine separate stars are actually listed in the table. The units of brightness and temperature in the table need some discussion for the benefit of those who have had no previous astronomy or physics instruction.

A. The Magnitude Scale

As explained in unit 2, the apparent brightnesses of stars as seen by the eye are measured on a scale of APPARENT VISUAL MAGNITUDES in a system which began with the Greeks. With telescopic assistance, stars with apparent magnitudes of 25 have been photographed. The apparent magnitude of a star is usually designated by lowercase "m"; a third magnitude star would have m = 3.

When precise methods of measuring stellar brightnesses were developed, it was found that a first magnitude star was about 100 times brighter than a sixth magnitude star, and so this factor of 100 was adopted as defining the ratio of brightnesses between different magnitudes. Mathematically, if two stars differ in brightness by one magnitude, they differ in brightness by a factor of 2.512. Thus a first magnitude star is 2.512 times brighter than a second magnitude star, whereas a star with m = 2 is 2.512 times brighter than a star with m = 3. Measurements also showed that some stars were actually brighter than first magnitude (these were called zero magnitude stars), while four stars are so bright that they actually need negative numbers to express their magnitudes (a star of magnitude -1 is 2.512 times brighter than a star with m = 0).

A first magnitude star is 2.512 x 2.512 = 6.31 times brighter than a third magnitude star. In general, if two stars differ in brightness by N magnitudes, they differ in brightness by a factor of 2.512 to the N power, that is 2.512 multiplied by itself N times. We can construct a table relating a difference in magnitude to a brightness ratio (see table 1).

Table 1. Brightness Table

Difference of Magnitude	Brightness Ratio
0.5	1.6
1	2.51
2	6.31
3	15.85
4	39.81
5	100.0
6	251
7	631
8	1,585
9	3,981
10	10,000

Using the above table, you should find that a third magnitude star is 100 times brighter than an eighth magnitude star. Answer the following questions in your notebook. (1) How much brighter is a second magnitude star than a fifth magnitude star? (2) How much brighter is a fourth magnitude star than a seventh magnitude star? (3) How much brighter is a fourth magnitude star than a ninth magnitude star?

B. Absolute Magnitudes

So far we have discussed only the APPARENT magnitude of stars, that is, how bright they APPEAR to be. But, if we knew their distances, we could convert these apparent magnitudes into quantities proportional to their intrinsic luminosities. This quantity, called an ABSOLUTE MAGNITUDE, is usually designated by a capital M. The absolute magnitude of a star is the apparent magnitude it would have if it were located at a standard reference distance of 32.6 light years away from the sun (this curious distance is chosen because at this distance a star would have a parallax of 0.1 arcsec). As an example, consider Alpha Centauri, the nearest star. Its apparent magnitude is zero; it is very bright -- about 2.5 times brighter than a first magnitude star. But it is bright primarily because it is close to the sun. If it were magically transported to a standard distance of 32.6 light years from the sun, it would have a brightness of 4.4 magnitudes (its ABSOLUTE magnitude is $M = 4.4$). The sun itself has an absolute magnitude of 4.8, which means that its intrinsic brightness is very close to that of Alpha Centauri. The sun's apparent brightness is $m = -26.7$, since it is so close to the earth.

In table 2 the quantities M are the absolute visual magnitudes of the stars, quantities which are proportional to the total luminosities of the stars. As a further example, consider the star Sirius A in the table. It has $M = 1.4$; thus it is 3 magnitudes more luminous than the sun and gives off about 16 times more energy per second.

C. Temperatures of Stars

That the color of a star depends upon its surface temperature will come as no surprise to anyone who has ever watched an object heating up in a furnace. It becomes first red-hot (corresponding to a temperature of around 3,000 degrees Celsius), then white-hot (at a temperature from 4,000 to 6,000), and finally blue-hot, if the furnace is hot enough. A similar behavior is seen in stars; for example, Sirius appears bluish because its surface temperature is about 10,000 degrees. As a result of this high temperature, Sirius emits more light at blue wavelengths than it does in the red. On the other hand, the star Betelgeuse in Orion has a surface temperature more like 3,000 degrees and appears quite red. By measuring the color of a star (i.e., comparing the energy it gives off in various wavelength bands), astronomers can determine the surface temperatures of stars, even if they are quite distant. Column 3 of table 2 gives temperatures for the nearest stars. Note that astronomers specify temperatures in degrees Kelvin, which are simply degrees Celsius plus 273. The Kelvin scale is a metric temperature scale in which there are no negative temperatures. Water freezes at 273 degrees Kelvin and boils at 373 degrees Kelvin. The techniques explained in unit 11 are used by astronomers to obtain temperatures of stars and other celestial objects.

D. Plotting the Nearest Stars

Using the graph paper of figure 9, plot the data in table 2: The Nearest Stars. For each star, make a mark on the graph paper corresponding to its value of M and temperature.

E. Plotting the Brightest Stars

In the second page of table 2 are indicated the twenty BRIGHTEST stars in the sky. Using a different-colored pencil, locate these stars on the same piece of graph paper which you used in the previous plot. (The data in table 2 were taken from Allen's "Astrophysical Quantities", with updated values for some stars supplied by Dr. Harry Shipman of the University of Deleware.)

V. INTERPRETING THE HERTZSPRUNG-RUSSELL DIAGRAM

There is a wealth of information in the diagram you have plotted. First we will examine the differences between the nearest and the brightest stars. This will be followed by a discussion of the intrinsic differences in the basic physical characteristics of stars. These quantities are among the most interesting to astronomers and are also very difficult to obtain.

A. Differences between Nearest and Brightest Stars

Note on your plot that the two groups of stars are of rather different types; only three objects (Sirius, Procyon, Alpha Centauri) overlap between the two lists. The brightest stars in the sky are not generally the nearby stars; they are instead usually extremely luminous objects, so intrinsically bright that even when they are not nearby they remain the most prominent objects of the night sky. It appears that there is an enormous range in the intrinsic luminosities of stars. The luminosity difference between the most luminous and the least luminous star on the plot is 24 magnitudes. Calculate what ratio of brightness this represents.

Since the brightest objects in the sky are superluminous objects at various distances, it should not be difficult to convince yourself that the diagram of the nearest stars is more representative of the AVERAGE population of stars. In plotting the nearest stars we see the statistics for ALL the stars within a certain volume of space -- a sphere of radius about 15 light years, centered on the sun. To make an analogy, you could probably get a better idea of the characteristics of citizens by surveying everyone in your hometown than you would by surveying only the people whose names appeared in the newspaper headlines. People whose names appear in the headlines are usually atypical, being rich, or famous, or notorious, etc. Similarly, we expect that the nearest star diagram would tell us more about the average star than would the brightest star diagram.

The argument of the previous paragraph breaks down if the solar neighborhood just happens to be an extremely atypical part of our galaxy. In our analogy above, surveying everyone in sight would give you a misleading notion of citizens if you did it in a nursing home, or, on a college campus. We note that modern research indicates that the solar neighborhood IS representative of stellar populations in the disk of our galaxy.

Besides demonstrating the range of intrinsic luminosity of stars, the diagram also gives information concerning their relative numbers. It should be apparent that the superluminous objects are relatively rare and that the majority of the stars fall in the lower (less luminous) part of the diagram. The fainter the star, the more numerous it seems to be (within the limits of this diagram -- we will examine this point more carefully below). From the diagram, what kinds of stars appear to be the most common? Write the answer in your notebook. In the plot of the nearest stars, how many are more luminous than the sun, and how many are less luminous?

B. The Main Sequence, Giants, and Supergiants

In the diagram you have plotted, it should be obvious that some parts of the graph are empty, while others are crowded with stars. Perhaps the most obvious feature is the band of stars running across the diagram from the upper left to the lower right -- this grouping is called the MAIN SEQUENCE. Among the nearby stars, most fall in this region. Since the nearby stars are more representative of the average stellar population, this means that most stars are indeed found on the main sequence. Below we shall see that there is a sound physical reason for this and that the main sequence is where most stars spend the majority of their lifetimes.

When we consider the brightest stars, we find that many are not main sequence stars. For example, we note that, of stars having a surface temperature of 4,000 degrees, there are some with M = 7 and also some with M = 0. The second star is about 600 times intrinsically brighter than the first. How might this be explained? We know that both stars have the same surface temperature, and this tells us that a square centimeter on the surface of each star is radiating the same amount of energy. Thus, the more luminous star must have more surface area -- it must be larger. The stars above the main sequence with absolute magnitudes around zero are called GIANT stars. Aldebaran is a giant star, significantly larger than the more common main sequence stars. Aldebaran has 600 times the surface area of our sun, or a radius about 25 times larger than our sun.

This same argument implies the existence of even larger stars on the diagram. Consider stars with a surface temperature of 3,300 degrees Kelvin. Note that we have such stars with M = 10 approximately and also with M = -5. This is a factor of 1,000,000 in luminosity, so the brighter star must have that much more surface area, since the surface temperatures are the same. The more luminous star is called a SUPERGIANT. Antares, in Scorpio, is a good example of a supergiant star; if were placed where the sun is, its surface would stretch out to the orbit of Mars.

Using the relation we deduced above, let us further compare stars on the main sequence. Consider the stars Wolf 359 and Sirius. The surface temperatures are 2,400 and 9,500 degrees K respectively. So, if the two stars were the same size, their surface temperatures would give a factor of (10,000/3000) to the fourth power, or 3.3 X 3.3 X 3.3 X 3.3, which is 123. We note from the table that the two stars differ in luminosity by about 1,000,000. Therefore, Sirius is not only hotter than Wolf 359 but also much larger. Wolf 359 is an example of a class of stars called white dwarfs. A white dwarf has a mass similar to that of our sun, but a radius not much larger than that of our earth.

The relation between the luminosity, temperature, and size of a star may be expressed very simply: the luminosity is the total energy per second given off by the star. We may write this as the product of "the energy radiated by a square centimeter of the star's surface" times "the surface area of the star in square centimeters." The first term in this expression is given by Stefan's Law: the energy radiated by one square centimeter of a hot surface is given by the fourth power of the temperature times a constant (which depends upon the system of units used). Thus we may write the first term in the luminosity of a star as

$$\sigma T^4$$

where σ is the constant. If we use centimeters, grams, and seconds for our units,

$$\sigma = 5.67 \times 10^{-7}$$

For a spherical star, the surface area is very simply expressed as 4π times the square of the radius of the star. If R is the radius, the second term in our luminosity expression is then

$$4\pi R^2$$

Thus the luminosity of a star may be expressed as

$$L = \sigma T^4 \; 4\pi R^2$$

Answer the following questions:

1. If two stars have the same temperature but differ in radius by a factor of two, how much brighter will the larger star be?
2. If two stars have the same radius but differ in temperature by a factor of two, how much brighter will the hotter star be?
3. From the table, compare the stars Antares A and BD +43 44 B. How much larger is Antares than the dwarf star?

You must consider both size and surface temperature when deducing the luminosity of a star. You have already seen in the diagram that there are stars 100,000 times more luminous than the sun and stars 10,000 times less luminous. Some of this range is due to differences in surface temperature, since normal stars range from about 30,000 degrees K down to about 3,000 degrees K, a factor of about 10. Since temperature is raised to the fourth power, this can account for a factor of 10,000 in luminosity alone. The rest of the luminosity differences are accounted for by differences in size, and when we compare stars, we find that they range from approximately 1000 times larger than the sun to about 1000 times smaller. The sun is often referred to as a typical star, and these comparisons show that there is some justification for this terminology; but we must not lose sight of the fact that the range of stellar properties is really quite large. Whenever the word "average" or "typical" is used, it actually has little meaning until we are also told the range of the property being discussed.

C. Explanation of the Existence of the Main Sequence

Our limited plots indicate that most stars are found on the main sequence, and modern astronomical research confirms that conclusion. The fainter on the diagram you go, the more stars there are; even on the main sequence, the more luminous stars are rare. Less luminous stars are more common. There is a real physical reason for the existence of the main sequence. Observational studies of masses of stars have demonstrated that the luminosity of normal stars depends upon the MASS of the star; the more massive the star, the more luminous it is. The luminosity of a star is determined by nuclear fusion reactions going on at the core of the star. The energy generated by these reactions depends upon the central temperature of the star. The central temperature is in turn determined by the inward pressure exerted by the great mass of the star overlying the central core. Thus, the more massive the star, the higher its central pressure and temperature, the faster the nuclear reactions proceed, and the more energy is generated. The figure shows the relationship; note that it is a logarithmic plot and that the increase of luminosity with mass is quite rapid. An increase of a factor of 2 in the mass increases the luminosity by a factor of approximately 10.

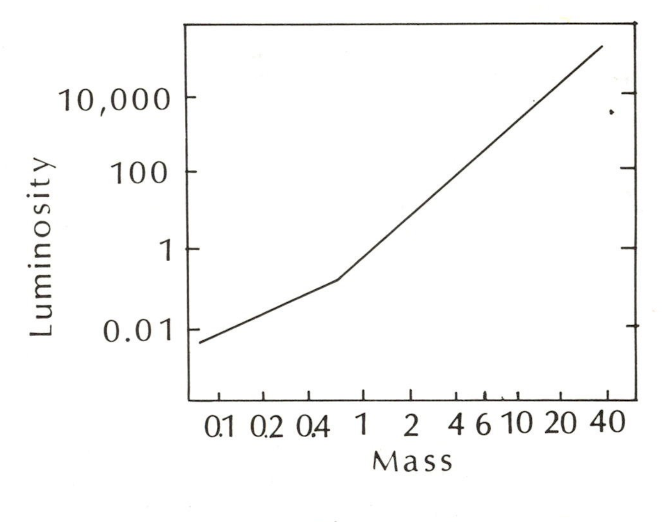

Figure 8

The main sequence is thus a MASS sequence for normal stars, with the most massive stars being the luminous objects at the top (upper left) and the least massive being the low luminosity objects at the bottom (lower right). The most massive stars appear to have masses of about 100 solar masses. The least massive stars have about 1/50 of a solar mass.

In summary, stars at the top of the main sequence are more luminous, hotter, and bigger than stars at the bottom of the main sequence. It is apparent from the diagrams that the less massive and less luminous stars are the most numerous. We have explained the existence of the stars along the main sequence but not the giants, supergiants, or white dwarfs. It is beyond the scope of this unit to discuss the entire life cycles of stars (unit 13 deals with some aspects of stellar evolution), so we simply note at this point that, after a star spends the overwhelming majority of its lifetime in a stable main sequence configuration, the later and final stages of its life are marked by dramatic and rapid changes in physical characteristics. Because of internal changes brought about by eons of nuclear fusion reactions in the stellar core, a star begins to enlarge and grow in luminosity toward the end of its life. This evolution produces the giant and supergiant phases; these phases are of short duration, which is why only a small fraction of all the stars are seen in these stages at any moment. Most stars then go through a period of instability and mass-loss, eventually ending up as burned-up remnants collapsing under their own gravitational forces -- the white dwarfs. The future of a white dwarf is dull and consists of a long period of slowly cooling and shrinking. Research shows that the majority of stars are not single but are instead found in some sort of association with one or more other stars. If you examine the entries in table 2, our most representative sample of stars, it can be seen that only twentyseven out of fortynine entries are single stars.

These conclusions concerning the masses and sizes of stars, as illustrated from their positions on H-R diagrams, have also been confirmed by other measurements of stars. In particular, the close gravitational interaction of stars in double star systems allows astronomers to determine masses for these stars, and, for certain types of double systems, the radii of the stars may be determined as well.

VI. SPECTROSCOPIC PARALLAX

Astronomers can obtain a star's temperature not only by examining its color but also by analyzing the exact distribution of its energy at each wavelength emitted. This second method is called spectroscopy and may be studied in more detail in unit 11. The spectrum of a star reveals its temperature and the relative density of its atmosphere, for example, a supergiant star has a thin, rare atmosphere, a main sequence star has a denser atmosphere. The star can be placed on an H-R diagram which was constructed using stars for which geometric parallaxes were known. When this placement is done, we then know the APPROXIMATE ABSOLUTE MAGNITUDE of the object. Comparing the absolute magnitude of the star with its observed apparent magnitude, we can calculate the distance of the star. A distance calculated in this fashion is called a SPECTROSCOPIC PARALLAX. As a simplified example, suppose we had examined a star's spectrum and found that it was a giant with a surface temperature of 3,300 degrees Kelvin. By examining your H-R diagrams, you can see that such an object has a characteristic absolute magnitude of about M = 0. The apparent and absolute magnitudes are related through the distance of the object (by the definition of absolute magnitude). If this star had an apparent magnitude of m = 5, it would mean that it appears 5 magnitudes fainter than it would if it were located 32.6 light years from the sun. Five magnitudes fainter means a factor of 100, so the star is 100 times fainter than if it were 32.6 light years away. Thus it must be farther away than 32.6 light years, and, since brightness varies as the square of the distance of an object, we know that the star is 10 times farther away than 32.6 light years, or 326 light years. (Note that 10 times farther away makes a star 10 squared = 100 times fainter.)

While it is not required for this unit, some students are interested in the relation between m, M, and the distance of a star. It is:

$$m - M = 5 \log(D) - 5$$

where D is measured in parsecs. A small (m - M) means the object is nearby, while a large value of (m - M) means the object is distant.

Using the above formula, the astronomer can find the distance to any object for which an absolute magnitude can be determined. (The apparent magnitude is easy to obtain.) Spectroscopic parallax is one example of obtaining distances for objects which are too distant for direct trigonometric measurement. Note that the word "parallax" is being used loosely here as synonymous with "distance." The spectroscopic parallax method uses the known properties of stars to estimate their distances, and no angle measurements are involved. Trigonometric methods are good out to a distance of 300 LY at best. Since the galaxy is 100,000 LY in extent, distances for the overwhelming majority of stars must be estimated by some indirect method.

Table 2: The Nearest Stars

Star	Absolute Visual Magnitude	Surface Temperature
α Centauri A	4.4	5800 K
α Centauri B	5.8	4000
α Centauri C	15	2600
BD +50 1725	8.3	3300
Barnard's Star	13.2	2600
Wolf 359	16.8	2400
Lalande 21185	10.5	3100
CD -37 15492	10.3	2900
Sirius A	1.4	9500
Sirius B	11.5	8100
Luyten 726-8 A	15.4	2500
Luyten 726-8 B	15.8	2400
CD -46 11540	11.3	2700
Ross 154	13.3	2650
Ross 248	14.7	2500
ε Eridani	6.1	4500
Ross 128	13.5	2600
Luyten 789-6	14.9	2500
61 Cygni A	7.5	4000
61 Cygni B	8.3	3700

BD +20 2465	11.1	2700
Procyon A	2.7	6500
Procyon B	13.0	7000
ε Indi	7.0	4000
Σ 2398 A	11.1	2700
Σ 2398 B	11.9	2600
BD+43 44 A	10.3	2950
BD+43 44 B	13.2	2700
CD-44 11909	12.8	2600
τ Ceti	5.7	5000
CD-36 15693	9.6	3100
BD+5 1668	11.9	2700
CD-39 14192	8.7	3300
Kruger 60 A	11.8	2900
Kruger 60 B	13.4	2650
CD-49 13515	11.0	2900
Kapteyn's Star	10.8	3300
Ross 614	13.1	2650
BD-12 4523	12.0	2650
von Maanen's Star	14.3	5800
AOe 17415-6	10.7	2900
Ross 780	11.8	2600
Lalande 25372	10.2	3100
CC 658	12.5	9000
Wolf 424 A	14.4	2500
Wolf 424 B	14.4	2400
40 Eridani A	6.0	4700
40 Eridani B	9.4	17000
40 Eridani C	12.5	2600

Table 2: The Brightest Stars

Star	Absolute Visual Magnitude	Surface Temperature
Sirius A	1.4	9500 K
Sirius B	8.4	28000
Canopus	-3.1	6400
α Centauri A	4.4	5800
α Centauri B	5.8	4000
α Centauri C	15.8	2600
Arcturus	-0.3	3900
Vega	0.5	9700
Capella A	-0.7	5000
Capella B	9.5	3200
Capella C	13.0	2600
Rigel A	-6.8	11000
Rigel B	-0.4	10000
Procyon A	2.7	6500
Procyon B	13.0	7000
Betelgeuse	-5.5	2700
Achernar	-1.0	13500
β Centauri	-4.1	20000
Altair	2.2	7700
α Crusis A	-4.0	19500
α Crusis B	-3.5	16500
Aldebaran A	-0.2	3500
Aldebaran B	12.0	3100
Spica	-3.6	19500
Antares A	-4.5	2700
Antares B	-0.3	15000
Pollux	0.8	4100
Fomalhaut A	2.0	8900
Fomalhaut B	7.3	4200
Deneb	-6.9	9400
β Crusis	-4.6	20500

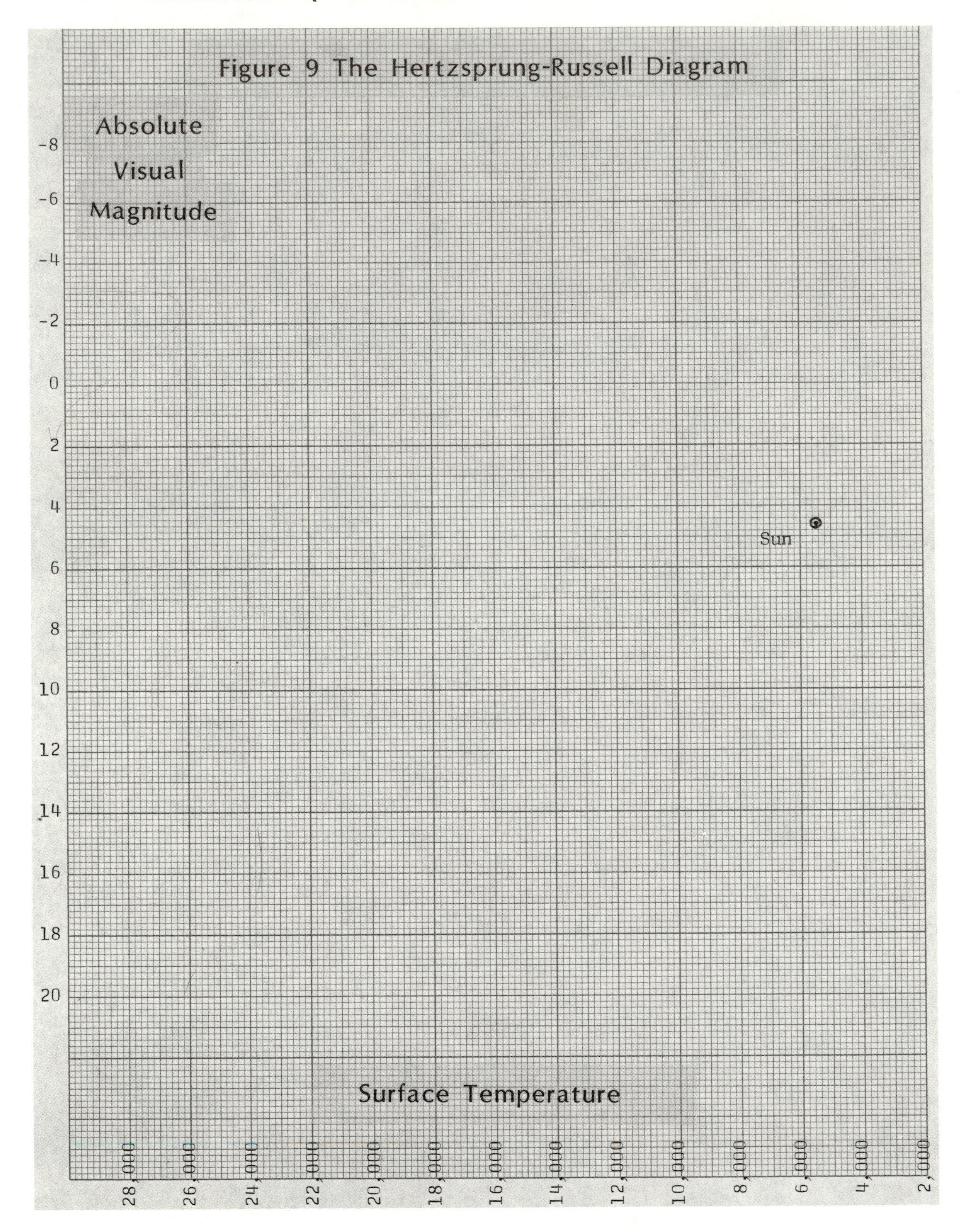
Figure 9 The Hertzsprung-Russell Diagram
Absolute Visual Magnitude
-8
-6
-4
-2
0
2
4
6
8
10
12
14
16
18
20
Sun
Surface Temperature
28,000
26,000
24,000
22,000
20,000
18,000
16,000
14,000
12,000
10,000
8,000
6,000
4,000
2,000

13

Components of the Milky Way: Dust, Gas, and Stars

MIKE SCHOLTES

One wide-angle photograph of the Milky Way in Sagittarius contains enough diverse objects to keep several astronomers busy for a lifetime. The diversity and interactions of stars, star clusters, nebulae, gas, and dust are the subject of this unit.

OBJECTIVES

1. to locate the following features on Sky Survey prints of the Cygnus region: H II regions, filaments, a planetary nebula, reflection nebulae, dust clouds, globules, and prominent stars
2. to graphically determine the relationship between stellar brightness and image diameter, and between a star's "color" and its surface temperature
3. to describe how H II regions are formed
4. to state the differences between H II and H I regions
5. to list two sources which could be responsible for the ionization of filaments
6. to explain the physical differences between emission and reflection nebulae and their causes
7. to briefly describe the appearance of a dust cloud, a planetary nebula, and a globule
8. to relate the presence of gas and dust to stages in the life history of a star
9. to explain how scattering by dust affects starlight

EQUIPMENT NEEDED

NGS-POSS prints E-754, O-754, E-1099, and O-1099; magnifier, wedge scale or reticle.

13-2 Components of the Milky Way

The Milky Way is a spiral galaxy approximately 100,000 light years in diameter. The disk of our galaxy contains a mixture of stars, gas, and dust, the interstellar material spread out in an irregular distribution with concentrations appearing in the spiral arms of the galaxy. Clouds of gas and dust are called by their Latin name: NEBULAE. Even dense nebulae have so few particles that they would be considered a vacuum on earth. In this unit we will consider several methods by which astronomers can observe various components of the Milky Way and try to relate these observations to events in the life histories of stars.

I. THE SKY SURVEY PRINTS

The National Geographic Society - Palomar Observatory Sky Survey was the first project to employ the 48-inch Schmidt telescope-camera, built in 1949. Seven years later, the project was completed, having accumulated 1,758 prints covering the entire sky as visible from the Mount Palomar Observatory in southern California. A single exposure with the Schmidt telescope covers an area of about 6 degrees by 6 degrees with unusually good definition -- this is approximately the area of the sky covered by your fist held at arm's length. According to Ira Bowen, former director at Mount Palomar, it would take the 200-inch Hale telescope about 10,000 years to record the same area, due to its one-quarter degree field of view. The survey prints show stars to a distance of about 600 million light years. Red plates record stars to a limit of magnitude 20 while the limit for blue plates is 21. (To put this in perspective, these stars would be about 100 million times fainter than Vega.) Exposure times were about 10-15 minutes for blue (O) plates and 40-60 minutes for red (E) plates. As the plates were developed nightly, new discoveries were made -- one of which you will replicate in part VI.

The survey prints which you will use for this unit were originally made on glass instead of film. Why do you think this was done? Each print is identified in the upper left corner in a small rectangle which gives information about its location and the date of observation. The print number is preceded by an O- or E-. These refer to the light sensitivity of the photographic emulsion: E- plates are red-sensitive while O- plates are blue-sensitive. These are negative prints which appear as a photographic negative does, with light and dark reversed. Stars and other luminous objects will appear black, and dark areas of the sky will be similarly reversed. Negative prints are used because greater detail can be preserved. Our contact prints are the same size as the glass photographic plates exposed at the telescope.

II. ORIENTATION

Spread out the four photographs and take a moment to orient yourself to their correct relationship. When you place the E-prints together correctly, you will notice that there is an overlap of about one inch. The same overlap is seen for the O-prints. What reason can you devise for this overlap? Now that you have the prints correctly oriented, make a LARGE sketch of them in your notebook. First, draw an outline of the print borders, including the overlap, and place several of the most prominent stars in their approximate locations. When this is complete, refer to the star chart at the end of this unit. The plate coordinates will help you determine which part of this chart to use. (Hint: The bright star in the upper left of print 754 and lower left of print 1099 is Alpha Cygni, also known as Deneb.) Name as many of the stars you have drawn as possible. Refer to the coordinates of the center of each plate and draw the lines of right ascension and declination on the map which you are making.

This activity should make you aware of the prints' location on the celestial sphere and realize the relatively large area of the sky which they cover. You may have noticed several unusual features on the prints. An example is the photographic "ghost" in the lower right of print O-754. This is caused by reflections within the telescopic system and is placed symmetrically across the optical axis from the original light source. Ghosts appear only from the very brightest images. Many brighter stars also have halos and spikes. These are caused by the diffraction of light within the telescope. Halos and spikes occur only for point sources, such as stars, never for extended objects. Occasionally a photographic defect occurs, due either to a flaw in the plate or to dust or lint during developing. How would you determine if a feature on a plate was spurious or not?

III. STARS

Beginning with prints O- and E-1099, compare the appearance of stars on the prints. Find some examples of stars which are intrinsically blue or intrinsically red. Record their positions or names and your reasons for choosing them. For the majority of the stars, the brightness is indicated by the size of the image. For extended objects, such as galaxies and nebulae, and for the very faintest stars, brightness is indicated by the intensity of the image.

The apparent brightnesses of stars as measured in units of apparent magnitude was discussed in unit 2. To review briefly, the Greeks called the brightest visible stars "first magnitude" and the faintest stars which the eye could detect "sixth magnitude." With the invention of the telescope in 1609 even fainter stars could be seen, and these were named with even larger numbers. Thus, a small number for magnitude means a bright star, while a large number means a faint star. Quantitatively, a 5 magnitude difference between two stars is a factor of 100 in their ratio of brightness. If two stars differ by one magnitude, their ratio of brightness is 2.512; a difference of N magnitudes equals a difference in brightness of 2.512 to the N th power.

In this unit you will consider the apparent brightness of stars in two wavelength regions (colors), red and blue, since the photographic plates used were sensitive to either reddish light (peak wavelength at about 6,500 Angstroms) or bluish light (peak wavelength at about 4,000 Angstroms). Figure 1 is a diagram (at the same scale as the prints) of the faint stars near the center of plate 1099. Table 1 lists their number or name, apparent visual magnitudes, and temperatures. Measure the size of the image for each star on both the E- and O- prints using either the wedge scale or reticle. The reticle has a finely graduated scale that can be seen through the eyepiece. The pattern for the wedge scale appears as part of this unit. Use this pattern with a thermofax or duplicating machine which accepts transparent plastic pages to make a clear impression of the scale. Slide the scale along until the star image touches each side, then record the corresponding number. Record the values from the reticle or wedge scale in your notebook.

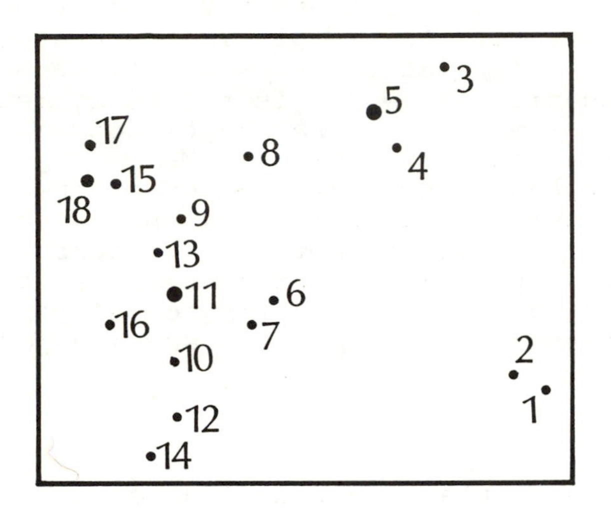

Figure 1

Table 1

NAME	MAG	TEMP
1	7.9	12,800 K
5=43 Cyg	5.7	7,600
6	6.6	4,760
7	7.8	6,000
8	7.3	5,520
11=45 Cyg	4.9	20,300
13	8.3	5,520
15	6.5	9,700
18= ω^2 Cyg	5.6	3,500
51 Cyg	5.4	20,300
Deneb	0.1	9,700

Construct a graph on which you PLOT the diameter of the image against the apparent visual magnitude for the red print and the blue print. Draw smooth curves through your data on each graph. On which print does the diameter-magnitude ratio seem most valid? Why do you think this occurs? Would you get better results if you could also plot the diameters of the stars versus the photographic (blue) magnitudes? These prints record stars that are as faint as magnitude 21. Can you determine the diameter of a magnitude 21 star by extrapolating from the data on your graphs? Record in your notebook the expected diameter. Measure some of the smallest star images you can locate. Does their diameter agree with your calculations? Describe some possible sources of errors in this activity.

For each of the stars in the list, calculate:

(red diameter minus blue diameter)/average diameter.

Plot this quantity against temperature. How does this quantity vary with temperature? The quantity (red diameter minus blue diameter)/average diameter is called a "color index" by astronomers, because it gives a method of attaching a precise number to the otherwise rather subjective concept of color. What color are the hotter stars? The cooler stars? You may want to refer to unit 11 to compare this relationship with that observed for the incandescent bulb in the variable rheostat settings.

IV. H I AND H II REGIONS

The interstellar gases, like the stars themselves, are composed primarily of hydrogen; about 9 out of every 10 atoms in the universe are H atoms. Most of this gas is cold and nonluminous, with temperatures on the order of 100 degrees Kelvin (-273 degrees Celsius). However, if a gas cloud is near a hot star, the cloud can be lit up by the star and shine brightly. The basic operating mechanism is as follows:

hydrogen atom + UV radiation ⟶ proton + electron

The hydrogen atom can absorb ultraviolet radiation from the star and as a consequence be split up into a proton and an electron. The process of removing an electron from an atom is called ionization, because an ion (a positively charged particle) is created. Thus, the reaction indicated above is the ionization of hydrogen.

The radiation from the star ionizes the hydrogen gas to a gas composed of protons and electrons. These protons and electrons, however, are moving at high speeds, and they frequently encounter each other. When they do, RECOMBINATION can occur as follows:

proton + electron ⟶ hydrogen atom + radiation

Some of the radiation that is created in the recombination process is emitted in the visible part of the spectrum, whereas originally all of it was in the unobservable ultraviolet part of the spectrum. As you recall from unit 11, hydrogen emits a great deal of energy at 6,563 A, an emission line in the red part of the spectrum. In fact, a glowing mass of hydrogen gas usually appears red to the eye because it emits so much energy in this particular transition in the red part of the spectrum. However, as you noted in unit 11, hydrogen gas also emits energy at 4,861 A (a turquoise feature in the spectrum) and at 4,340 A, in the far blue end of the spectrum.

The mechanism causing the gas clouds to glow is illustrated in figure 2 -- ultraviolet radiation from the star is absorbed by the hydrogen gas, creating protons and electrons, and, as the protons and electrons recombine, visible radiation is given off. Gas clouds seen under these conditions are called EMISSION NEBULA.

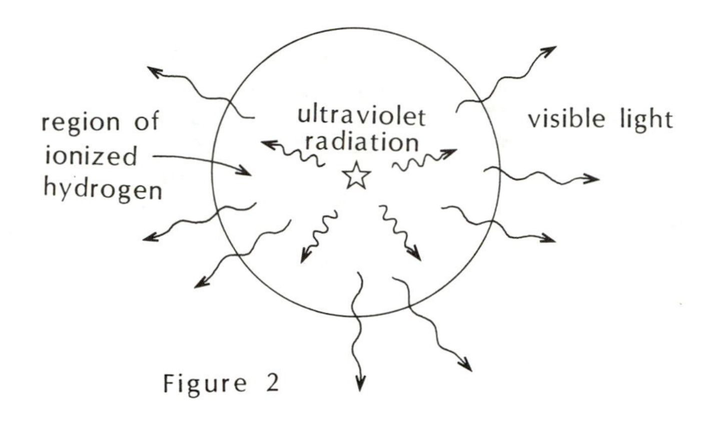

Figure 2

Neutral hydrogen gas is designated spectroscopically by the notation "H I," whereas ionized hydrogen is denoted "H II". Therefore, the regions of ionized H around hot stars are called H II regions. The famous Orion Nebula (M 42) is the most prominent H II region in the sky.

On the prints, the luminous H II regions will appear black. The degree of blackness is related to the actual brightness of the nebula. Since the bright H-alpha (6,563 A line) is in the red region of the spectrum, on which prints would you expect to see H II regions most clearly?

The physical size of an H II region is directly related to the temperature of the hot star that creates it; the hotter the star, the more ultraviolet radiation it emits, and the larger the ionized region that it can create. The stars that create H II regions are typically hotter than 30,000 degrees Kelvin, and theoretical stellar models indicate that such stars are more massive and younger than the sun. Astronomers find these hot young stars associated with gas clouds much more frequently than chance would predict -- leading to the conclusion that these young stars have FORMED FROM THE CLOUDS, by some type of condensation and collapse of regions of the gas. Draw the H II regions you locate in your sketch. If the gas were uniformly distributed, the regions should be spherical -- why is this so?

The term "H I" region refers to clouds of neutral gas that are not excited by a nearby star. As a consequence, they are cool (near 100 degrees Kelvin). Such regions of gas are not visible directly by either reflected or emitted light; they can be detected optically by the ABSORPTION of starlight passing through them. In unit 11 it was explained that, if light from a source of continuous radiation (such as a star) passes through a cooler gas, the gas will absorb radiation at certain specific wavelengths. This was the mechanism by which the dark absorption features in the solar spectrum are formed -- absorption by the cool gases in the sun's outer atmosphere. Similarly, when starlight passes through a cool interstellar gas cloud, discrete absorption features in the spectrum due to the cloud can be seen. (See figure 3.) These absorption lines formed by the cool cloud will be considerably sharper and narrower than the lines formed by the star's own atmosphere. Also, since the star's atmosphere is hotter than the gases in the cool cloud, different absorption features are produced by the cloud. Since they are awkward to observe in the visible part of the spectrum, most of our knowledge concerning H I regions come from observations at radio and millimeter wavelengths.

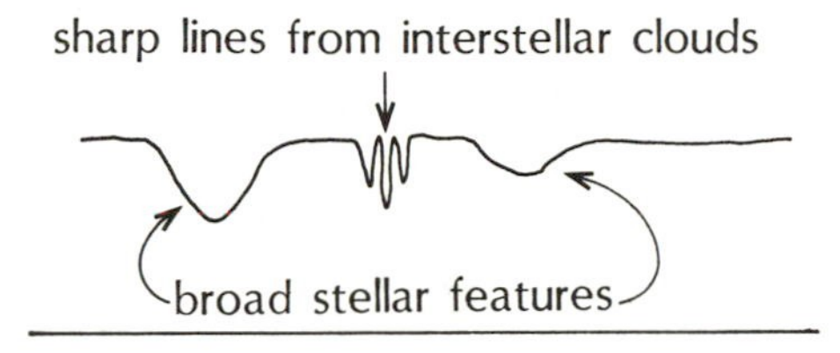

Figure 3

V. FILAMENTS

The delicate structures that are prominent on E-1099 are another type of emission nebula called filaments. Since these appear dark on the prints, we may assume that they are emitting light. Their distinct appearance on the red-sensitive print is due to the H-alpha emission line at 6,563 Angstroms, which also illuminates the H II regions. What makes these filaments appear different from the H II regions is the absence of stars nearby to supply the radiation necessary to ionize the hydrogen gas. Some theories have been suggested to explain the origin of the filaments, one being that the gas in the filaments is moving rapidly and being ionized by violent, high-speed collisions with stationary H I clouds. Another is that the gas was ionized by a source which we cannot see. A supernova could provide ample radiation for the initial ionization. While the gas is still partially ionized from the burst of initial radiation, the supernova would have had time to fade.

A supernova is the cataclysmic event produced by a large star from 4 to 60 solar masses at the end of its life cycle. Theory suggests that a star of 4 to 10 solar masses probably leaves no corpse behind but throws its material off in all directions when it explodes. A larger star of about 10 to 20 solar masses produces a supernova explosion which leaves the core of the star as a remnant. This surviving core would most probably be in the form of a neutron star, a collapsed, incredibly dense object which could contain a solar mass or more of material, yet be only ten to twenty kilometers in diameter. Radio astronomers in the late 1960s discovered strange objects which they named "pulsars." Later, theoretical astrophysicists showed that pulsars could be explained as clouds of gas excited by spinning neutron stars inside them. Thus, the pulsars that radio astronomers can detect with relative ease can lead us to the positions of ancient supernovae. If these positions relate properly to nearby filaments, the original power source for the nebulae we see today may be discovered. Even larger stars of 20 to 60 solar masses rush through life quickly and die with a supernova explosion also. In this case, the remnant may be a "black hole," a body so compact that not even light radiation can escape its gravitational pull.

Draw the portions of filaments which you see onto your map. These filaments seem to be slightly rounded. Can you determine the approximate center from which the filaments originated? Astronomers have found a radio and x-ray source behind the dust cloud which forms the Milky Way rift. These observations may indicate that a supernova occurred there.

VI. PLANETARY NEBULAE

Another chapter in the life history of stars is the death of small stars, those below 4 solar masses. These form white dwarf stars through a limited collapse of the stellar core. The small core is very dense and hot but will slowly cool until only a cold black remnant is left. The outer envelope of the star is gently ejected. Only a few tenths of the mass of the star are released, but it is visible as an ionized nebula of low density (a few thousand particles per cubic centimeter) called a planetary nebula. The majority of visible stars in the Milky Way will probably form white dwarf stars and planetary nebulae, just as our sun will in about four billion years. The planetary nebula which results from the mild expulsion is a very large shell of gas. The most famous example is the Ring Nebula in Lyra. It appears as a ring because the observer is looking through the thin parts of the shell in front and back of the stars and along a line of sight which encounters a thicker path of gas.

Planetary nebulae received their name in 1791, when William Herschel thought they resembled planets in color and shape. They were later found to have no physical resemblance to planets. They are detected by their emission spectra, which show lines similar to those from an H II region. The gas shell, like an H II region, shines through the process of fluorescence, absorbing ultraviolet radiation from the central star and emitting visible light. The red H-alpha emission is stronger than the blue emission lines. The temperatures of the central white dwarf star are the hottest known -- up to 100,000 degrees Kelvin.

On print E-1099 locate the planetary nebula. This nebula was originally discovered by George Abell while looking at this very photograph, and it is named Abell A71. It has a well defined edge. Plot it on your map, and try to determine its approximate right ascension and declination. Can you find it on print O-1099?

VII. DUST

The interstellar medium also contains tiny particles or grains of material (about the size of smoke particles perhaps averaging about 2000 A in size). These comprise only 1 percent by weight of the interstellar medium. Their composition is not definitely known, but some particles are believed to be carbon or rocky material, others to be crystals of dirty ice. There are several methods to optically determine their presence:

1. Dust scatters light, as you can observe by looking through some smoke at a light source. The light will seem dimmer. Careful measurements and calculations have shown that short wavelength light is scattered more than longer wavelength light, e.g., blue is scattered more than red. Can you name a common example, observable daily, which illustrates scattering of light?

   The discovery of the scattering of light and its wavelength dependence answered many questions for astronomers in the 1930s. Many stars had been observed which appeared red in color but had spectra which indicated that they should have high temperatures and be blue. The interstellar dust not only dimmed the light from those stars but also scattered the blue more than the red. While the existence of these reddened stars was explained by the presence of dust, the dust creates many other problems for astronomers. As seen in unit 12, the distance to a star can be calculated if the apparent magnitude and the absolute magnitude are both known. A guess must also be made as a correction factor for the dimming caused by scattering of light, or all the stars will be measured as too distant. This absorption is low in areas of the sky perpendicular to the plane of the galaxy but very high in some areas along the galactic plane.

2. Reflection nebulae occur in H I regions with nearby stars. The dust in the cloud scatters the blue light more than the red, so we see a blue nebulous region around a star. On which prints do you expect to see reflection nebulae? Find at least one example and add it to your map. Include the coordinates of the star(s) responsible for it. The most famous example of reflection nebulae are the areas surrounding the brightest stars in the Pleiades cluster. In your reflection nebula, why doesn't the star ionize the gas associated with the dust and cause an emission nebula to be seen?

3. Dark nebulae are relatively dense clouds but have no nearby stars to excite them. The dust in these clouds absorbs a

considerable amount of light and either greatly dims or totally obscures the light of stars behind them. The area known as the Milky Way Rift or the Northern Coalsack can be seen on prints 754. We can see only the stars between us the dark nebula, none of the stars behind it. Earlier in that century, some astronomers thought that this was a hole in our galaxy and that they could see beyond our galaxy by looking through it. Astronomers in the eighteenth century thought they could gauge the size of the universe by counting stars of successively fainter magnitudes. William Herschel counted many regions and found that, in some directions, the number of stars thinned out rapidly, while in other directions there were more stars. With these observations, he could explain the shape of our galaxy as a disk with the sun at the center. Modern observations, considering the effects of absorption of light by dust, show the galaxy to be a spiral-shaped structure with the sun about two-thirds from the center, lying in the plane. The sun is located within the dusty plane of the Milky Way galaxy, and the absorption by dust is so severe that Herschel was in fact only observing stars in the immediate vicinity of the sun for his count. Distant stars were totally obscured. He was correct, however, in concluding that the galaxy is much wider than it is thick. Deducing the correct size and shape of our galaxy is one of the triumphs of twentieth-century astronomy.

4. Globules are very dense compact clouds which are sometimes completely isolated. The globules are easily seen against a background of bright gas. They are usually round or oval and are quite small compared to the large nebulae -- only a few arcseconds in angular size, with very sharp edges. Present theory of star formation suggests that these globules may be protostars, the groupings of atoms which will condense by gravitational attraction to become a star. On prints O-754 and E-754 look at several globules with a magnifier. (Do you expect them to appear black or white?) How can you tell if the spot at which you are looking is really a globule or just a defect in the print? Draw the positions of the globules on your map. Note: while globules may in fact evolve on to become stars, no one has been able to demonstrate that all stars must begin in this way.

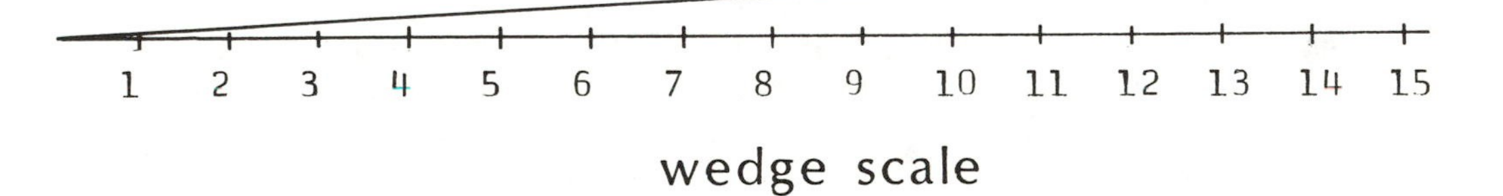

wedge scale

CYGNUS - CEPHEUS - LACERTA REGION

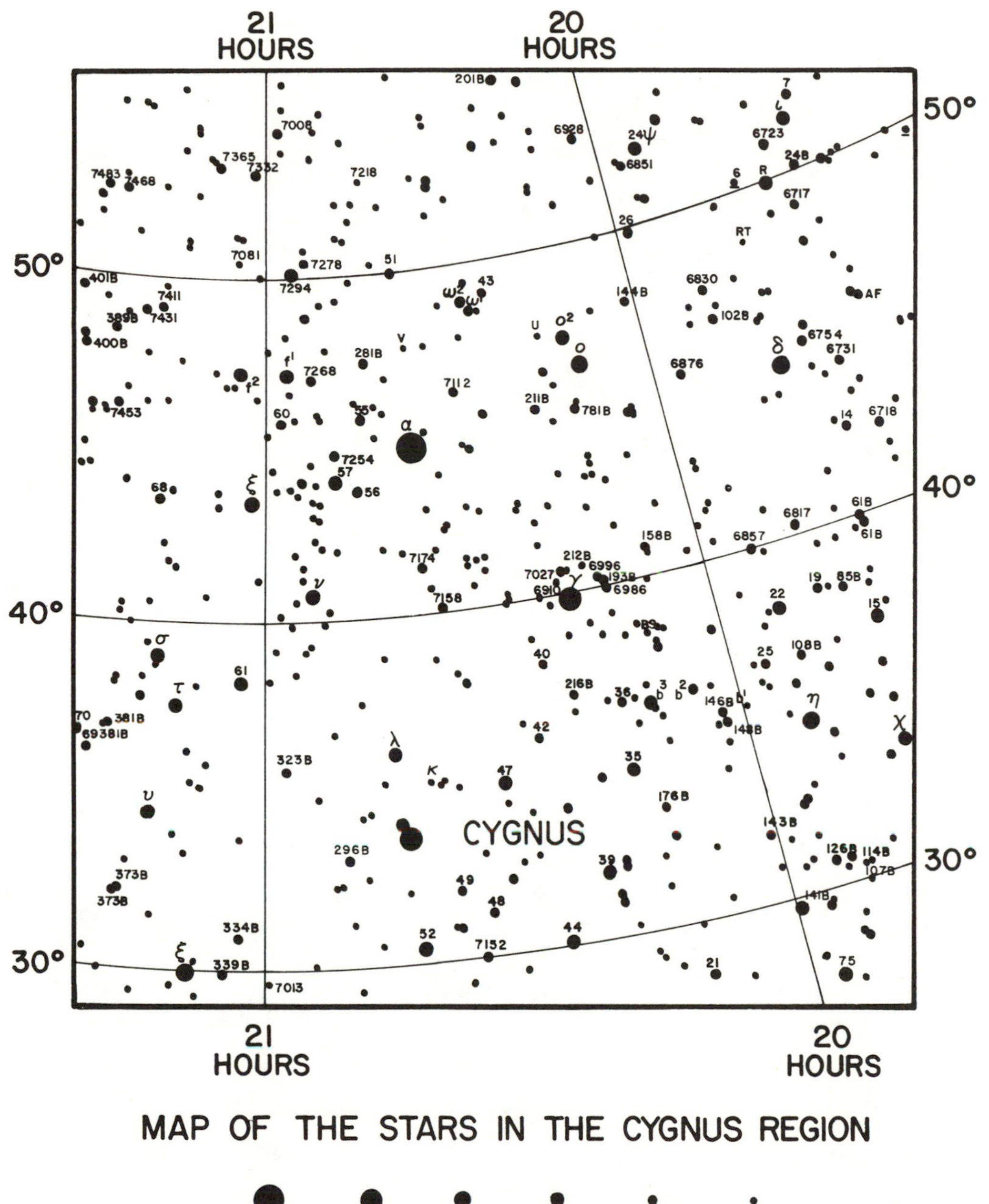

MAP OF THE STARS IN THE CYGNUS REGION

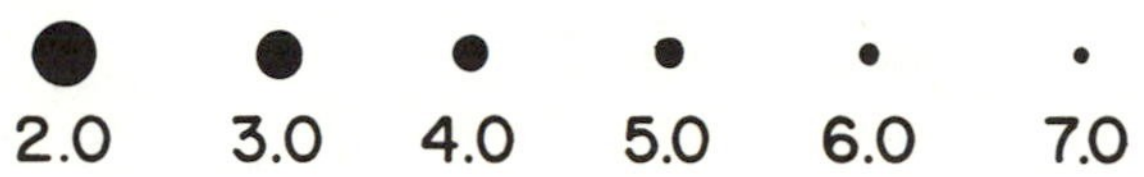

2.0 3.0 4.0 5.0 6.0 7.0

14

Studies of Galaxies

HALE OBSERVATORIES

The Andromeda Galaxy is relatively close—only about 2 million light-years away. As a result, we can resolve and study individual stars in this system of 100 billion stars. But, as more distant galaxies are studied, information is harder to gather and harder to interpret.

OBJECTIVES

1. to be able to distinguish the main classes of galaxies: spirals, ellipticals, and irregulars
2. to be able to distinguish subclasses of spiral galaxies based upon the size of the nucleus and the tightness of the spiral arm winding
3. to be able to distinguish subclasses of ellipticals based upon their deviation from sphericity
4. to count the relative numbers of spiral to elliptical galaxies in two different clusters of galaxies, one near and one distant
5. to explain why the numbers in 4 are different on the basis of observational "selection effects"
6. to explain the Doppler effect and one of its astronomical uses
7. to explain at least two methods of determining relative distances of galaxies

EQUIPMENT NEEDED

NGS-POSS prints O-83 and O-1563, a magnifier with a reticle (or a small ruler), a transparent overlay marked with thirty six small squares, each approximately 1.8 centimeters on a side.

We live in a system of 100 billion stars. It is referred to both as the Galaxy and as the Milky Way, although some prefer to use the name Milky Way to refer only to the diffuse band of illumination across the sky. Galileo, the first telescopic observer, noted in his early observations that the Milky Way consists of many faint stars which are unresolved by the naked eye. When we look toward Sagittarius we are looking lengthwise through the long dimension of our stellar system from within it. William Herschel and others documented the dimensions of our stellar system by counting stars in all directions in space, but it was not until after World War II that astronomical instrumentation had developed sufficiently to be able to demonstrate that the stars in the disk of our galaxy were arranged in extensive spiral patterns. See figure 1 for a view of our galaxy.

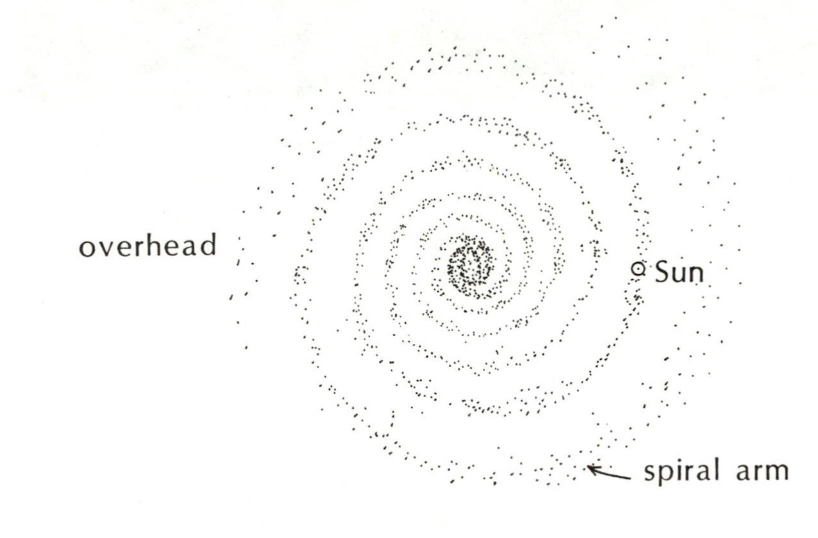

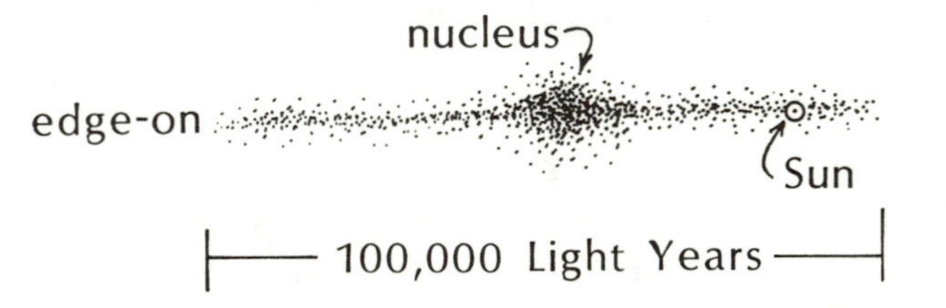

Figure 1 Schematic Views of Our Galaxy

Our Galaxy is only one of countless millions of such stellar systems. The best surveys indicate that from the earth we could presently photograph about one billion galaxies, each containing perhaps 100 billion stars on the average. This unit will guide you to discover some of the properties of galaxies external to our own. The photographic prints you will use are of research quality, identical to those used by professional astronomers. The National Geographic Society-Palomar Observatory Sky Survey consists of photographs of the entire sky north of -27 degrees. Each of the photographs is of six by six degrees of the sky, an area roughly half the size of the bowl of the Big Dipper. This area is equal to 1/1146 of the celestial sphere. The faintest stars visible are approximately magnitude 21 (which is a million times fainter than the faintest star the naked eye can perceive). The original photographs were taken on glass plates, which do not stretch, shrink, or warp as badly as ordinary photographic film. The copies you will use are NEGATIVE contact copies, which preserve more detail than do other types of copies. The dark night sky appears white and the stars and other emitting objects appear dark on the prints.

I. CLASSIFICATION OF GALAXIES

Print O-1563 contains bright and faint stars which lie within our galaxy and, also, many bright and faint galaxies. The brightest stars have diffraction "spikes" around their images, caused by light diffracting around the supports holding the secondary mirror in place on the telescope. Galaxies, being extended objects and not point sources, do not show these spikes. However, faint stars also will not be bright enough to have visible spikes and you must distinguish faint stars from faint galaxies by their sharpness, especially at the edges of the image. Examining a few faint images on the photograph should illustrate this difference fairly well. Use the magnifier. See figure 2 for directions on how to position the overlay correctly. At the end of this unit is a pattern for an overlay to be used in conjunction with the photographs. Each small square is 1.8 by 1.8 cm and covers an area of sky 0.33 by 0.33 degree. The overlay pattern may be turned into a transparency overlay by using a piece of heat-sensitive plastic with a thermofax machine, or a piece of plastic with a duplicating machine.

Begin your examination of the differences between galaxies by examining the three objects whose positions are given in table 1 and indicated in figure 2. Find other spiral systems on plate O-1563 and compare them to these; describe how they differ in appearance. Much of our knowledge about galaxies comes from their shapes as viewed from the earth. Make up a classification scheme and try to find at least four objects which fit each of your categories. List the category type and the position of these objects in your notebook. Write a short description of each category of galaxy, noting especially what characteristics you have used to distinguish one type from another.

A. Spiral Galaxies

Galaxies in a variety of shapes and forms appear on the photograph. The SPIRAL galaxies are one obvious type, identifiable by the various systems of arms spiraling away from a central bulge, or nucleus.

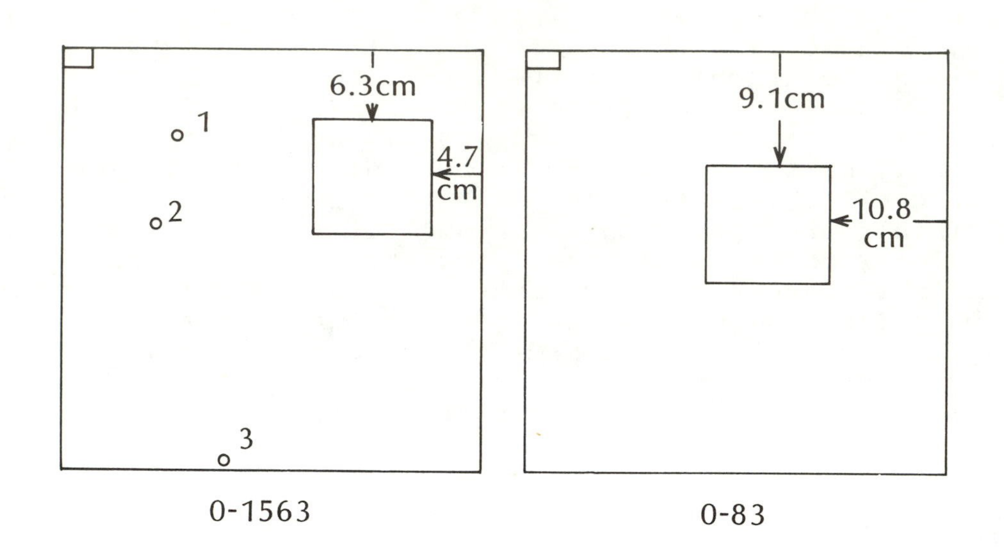

Figure 2 Overlay Positions

Table 1. Positions of three galaxies to note on 0-1563.
1) 10 cm from left, 7 cm from top
2) 8.5 cm from left, 14 cm from top
3) 13 cm from left, 1 cm from bottom

Notice the spiral galaxy numbered 3 in the figure, at the bottom of the print. This type is called Sc, indicating a spiral with a small nucleus and a very extended system of arms. At 1 is a type referred to as Sb. Here the arms are somewhat more tightly wound and the nucleus is somewhat more prominent. Finally, at 2 is an Sa galaxy -- the nucleus is even more prominent and the arms are the most tightly wound around the nucleus of any of the spiral galaxies. Seen edge-on, examples of these types may be found at the overlay locations F-5 (Sc), E-6 (Sb), and B-2 (Sa); the relative prominence of the nucleus is what allows one to distinguish these types when viewed edge-on.

Type Sb systems most closely resemble our own galaxy. Our sun is located in a spiral arm about two-thirds of the way out from the center, directly in the plane of the galaxy. Because of our location within our own system, it is difficult to study our galaxy. Some properties of our galaxy are inferred indirectly by studying other nearby Sb galaxies and assuming that our system is similar. Unit 13, "Components of the Milky Way," illustrates how astronomers investigate some properties of our own galaxy.

B. Elliptical Galaxies

Note numerous prominent objects without spiral arms; astronomers call these ELLIPTICAL GALAXIES. Notice the system just to the left of C on the overlay, which is very nearly round. The big system at B-4 is somewhat flattened and the system at A-3 is very flattened. Ellipticals are classified from E0 (round) down to E7 (quite flattened); the system at A-3 is classified as E5. Notice that even the most flattened ellipticals are not nearly as flat as edge-on spirals. However, spirals which are between edge-on and face-on may have an apparent eccentricity (deviation from circularity) comparable in appearance to that of elliptical systems.

Answer the following questions in your notebook. How might one distinguish an elliptical from a spiral galaxy seen at an angle? For which types of spiral galaxies would this work best? What will happen when galaxies are farther away and thus smaller?

C. Irregular Galaxies

Note that in any classification scheme there are objects which do not fit into a standard box, so there is always a category called miscellaneous or, in the case of galaxies, IRREGULAR. Can you find any examples of irregulars on your print? Write down their location with respect to the overlay grid, or corner of the print. Note that there is no way to give a general description of an irregular galaxy, except to say that it is anything that does not resemble either a spiral or an elliptical.

D. Sizes of Galaxies

Notice that there are large and small spirals on the print and, also, large and small ellipticals. Does this tell us anything about the distances of these galaxies, or does it tell us something about their size variations?

E. The Astronomical Classification Scheme

Now that you have examined examples of galaxies is some detail, a description of the criteria by which astronomers today classify extragalactic objects is appropriate. In figure 3 are shown examples of the major categories of galaxies - spirals, barred spirals, ellipticals, and irregulars.

Spirals are classified on the basis of: the relative sizes of the nucleus to the disk (or arm) region, and on the openness of the arm structure. For example, type "a" spirals (designated Sa) have large, dominant nuclear regions and a fairly modest disk region in which the spiral arms are very tightly wound; see NGC 4594 in figure 3. This edge-on view clearly shows the large nucleus and the dust concentration in the plane. Type "b" spirals have less prominent nuclei and more open arms (as for example in the frontispiece photograph of this unit), while type "c" spirals have tiny nuclei and widely spread arms (see the face-on view of NGC 5364 in figure 3). Measurements of the colors of spiral galaxies show that their nuclear regions are reddish, while the spiral arms are blue. The blue color in the arms is because these are the regions in the galaxy where star formation is currently going on. The color is dominated by the light from a few very massive, luminous, hot stars that are strongly blue in color. Star formation is no longer taking place in the nuclear regions of spirals. The young, blue stars that were once in the nuclear region have long since evolved into middle-age and old-age, a state characterized by lower surface temperature, and hence, redder color. Refer to unit 13 for a longer discussion of stellar evolution.

The elliptical galaxies may have shapes from round to highly flattened. They are characterized by the degree of flattening, with the round objects classified as E0 and the most flattened as E7. Notice that an object which is classified as E0 may not necessarily be spherical in form -- why is this so? Elliptical galaxies are typically reddish in color, indicating that there is little or no recent star formation in them. They are generally devoid of dust or loose unassociated gas between the stars.

Figure 3 also shows an example of an irregular galaxy, one of the Magellanic Clouds which orbits our own giant spiral system as a small satellite galaxy. It is only about 200,000 light years away, one of the nearest external systems. Irregulars have a wide variety of shapes and forms which defy any sort of regular classification.

Note that spiral galaxies may also have a "bar" region crossing their nucleus. In the barred spirals (SBa,SBb, and SBc), the spiral arms typically begin at the end of the bar.

There are also two other categories of galaxies which we will not be concerned with in any detail: S0 galaxies, a transition type between the spiral and elliptical shapes, and the ring galaxies, which show a circular ring of stars outside the nucleus. The origin and significance of these galaxies is a subject of hot debate among modern astronomers.

II. COUNTING GALAXIES

On your print, count the relative number of spiral to elliptical galaxies in all thirty six squares of the overlay. Count only those systems which you can DEFINITELY classify as either spiral or elliptical. It will probably take a little time and practice before you can be sure in some cases; for many there will be uncertainty. As a guideline, the smallest spiral the author of this unit can reliably identify can be found in the upper right corner of A-4; the smallest elliptical identified by the author is the small companion to the big spiral in E-6. You may not feel that you can reliably identify objects this faint, in which case your counts should reflect only objects that you can identify. (Or, you may be able to identify even fainter objects, in which case please do so.) These examples are given only to allow you some basis for comparison with one other observer. As you count, also keep an approximate tally of the TOTAL number of galaxies within the grid, whether you can classify them or not.

Figure 3

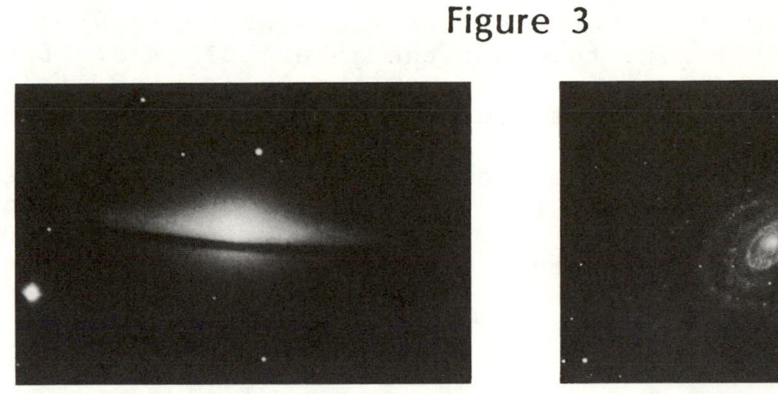

NGC4594 (Sa)

NGC5364 (Sc)

NGC4486 (E0)

NGC205 (E7)

NGC7741 (SB)

Small Magellanic Cloud (Irr)

Answer the following questions in your notebook. Clearly the biggest difficulty in the counting is for the small, faint objects. As the galaxies get smaller and fainter, what type of misidentification do you feel is most likely -- confusing spirals for ellipticals, or vice-versa?

If one type of object is easier to identify, then your counts will be biased in the direction of that type. This is called a selection effect. What do you feel are the most serious selection effects in trying to count the galaxies? What effect do you think these had on your counts? In your opinion, is the ratio of spirals to ellipticals you have determined greater or smaller than the true value?

Estimate the total number of galaxies in the Virgo Cluster, using your total counts. Note that the cluster covers the entire photograph, whereas the overlay you counted comprised only 10 percent of the area of the photo.

III. THE SIZE DISTRIBUTION OF GALAXIES IN A CLUSTER

A new piece of information is about to be added to your consideration of the photographs. Other lines of research have demonstrated that all the galaxies on the photograph are at approximately the same distance from you. What you are actually seeing is an extensive CLUSTER of galaxies in the constellation Virgo. Such a cluster would have formed from an immense gas cloud fairly early in the history of the universe, approximately 10 billion years ago. Thus, all the objects you are seeing (with the exception of the individual stars, which are foreground objects in our own galaxy) are at approximately the same distance. Therefore, the size difference between the various objects must be real, intrinsic differences and not just a function of variable distance. To gain some idea of the range of the size differences, measure, using the reticle in a magnifier (or a small ruler), the sizes of the largest and smallest spiral galaxy and the largest and smallest elliptical galaxy you can find. You can use the objects at positions A-4 and E-6 discussed above (if you feel you can actually discern their type); if you cannot, or if you can do better, measure the smallest objects that you can distinguish.

Answer the following questions in your notebook. How do the size distributions of the two types of galaxies compare? What are the sizes in millimeters on the print of the largest and smallest examples of each type of galaxy?

IV. COMPARING A SECOND CLUSTER OF GALAXIES

Now examine the photograph labeled O-83. It contains a grouping of galaxies forming a cluster in the constellation of Hercules. Notice that because this cluster is farther away, the cluster itself is smaller in angular size than the Virgo cluster, and the individual galaxies are likewise smaller (and harder to distinguish as a consequence). Where is the Hercules cluster on the photograph?

Measure the size of the largest galaxies in the Hercules cluster and compare them with the largest objects in the Virgo cluster. How much farther away is the Hercules cluster, according to this comparison? What assumptions must be made in this comparison? Place the overlay in the position indicated in figure 2. Consider C-3, C-4, D-3, and D-4 only. In this field make a count of the number of spiral galaxies, number of elliptical galaxies, and total number of galaxies of all kinds

that you can see. Calculate the ratio of spiral to elliptical and compare it to the ratio you obtained for the Virgo cluster.
How does the ratio of spiral to elliptical systems compare in the two clusters, and can you explain the difference in the value?

V. RECENT RESEARCH ABOUT THESE CLUSTERS

The very brightest cluster members are giant ellipticals. In the entire Virgo cluster, 50 percent of the galaxies from 12 to 14 magnitude are spirals. From 14 to 16 magnitude, only 25 percent are spirals. Of the fainter galaxies, George Reaves found only 11 out of 76 NOT elliptical. (Faint galaxies are difficult to study; thus, although 1000 faint galaxies in Virgo were discovered, only 76 were studied in any detail. Even fainter galaxies may be present but not seen with today's technology.)

Virgo and Hercules are both irregular clusters of galaxies. There is little or no symmetry and no marked concentration about a unique cluster center. Figure 4 shows how the number of galaxies increases as one observes to fainter limits. The magnitude scale shows divisions of one magnitude.

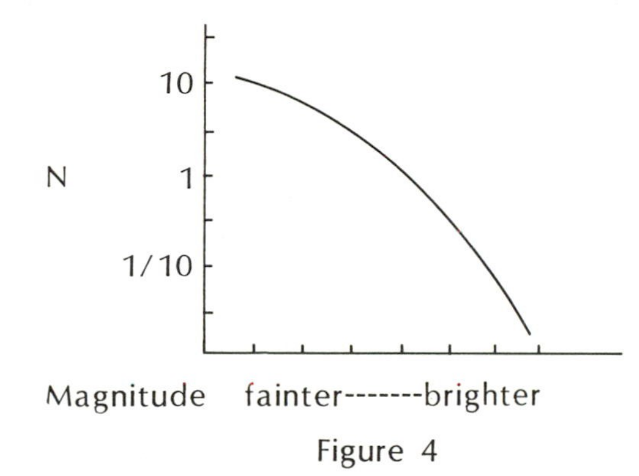

Figure 4

Using your counts for the Hercules Cluster above, and assuming that the distribution of galaxies of various sizes in Hercules follows the same distribution shown in figure 4, estimate the total number of galaxies in the cluster. Assuming that all the galaxies in the Hercules cluster lie within the four squares which you counted, make an estimate of the total number of galaxies in Hercules. Compare the numbers of galaxies for the two clusters. Note: you are not expected to come up with an accurate number for the Hercules cluster, just an estimate. But in your notebook, list explicitly the assumptions and the method you used.

VI. THE DOPPLER SHIFT

One characteristic of the light received from any object in the universe is its wavelength. Light radiation is often shown as a series of waves coming from a source. (See figure 5.)

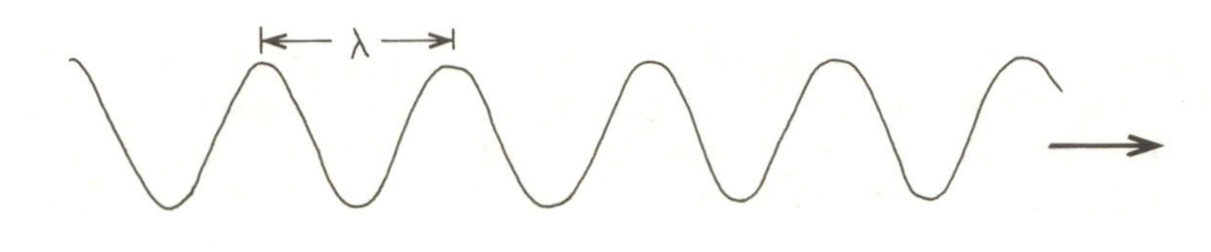

Figure 5

The distance between adjacent crests is one wavelength. Astronomical objects produce radiation at many different wavelengths (from gamma radiation at one-billionth of a centimeter to mile-long radio radiation), but here we are concerned with visible light with wavelengths about ten to the minus fifth (10^{-5}) centimeter. All electromagnetic radiation travels at the same speed, known as the speed of light. This quantity has been measured very precisely. We shall use a value of 3000,000 kilometers per second.

In 1842 Christian Doppler discovered that the motion of an object either away from or toward an observer produced a change in the observed wavelength. This phenomenon, known as the Doppler effect, is true for other types of waves (such as sound) as well. As an example, consider a pond into which water is regularly dropping while you watch the pattern of waves formed. The distance between the crests is one wavelength. (Note figure 6.) Consider how the pattern changes when the source of water drops moves to the right. The distance between the crests is altered. To an observer at the right, the wavelength will be measured as smaller. To an observer at the left, the wavelength appears larger. This change in wavelength is the Doppler shift.

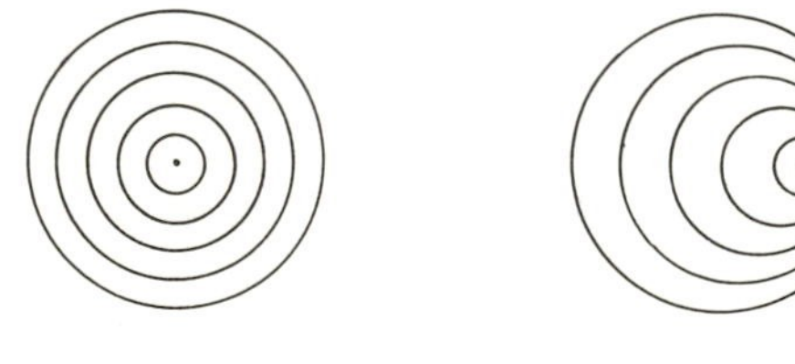
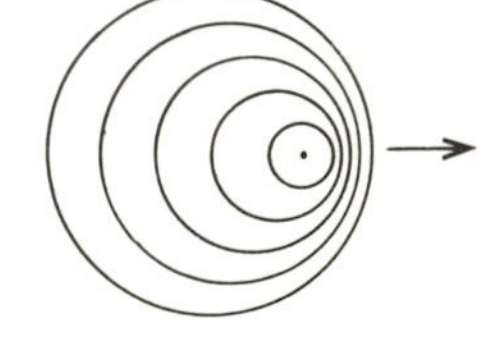

stationary source   moving source

Figure 6

For light, we express this relation with a formula giving the change in wavelength ($\Delta\lambda$) produced by a relative motion between source and observer (v) when the original wavelength (λ) is known:

$$\frac{\Delta\lambda}{\lambda} = \frac{v}{c}$$

The light emitted by galaxies is a composite of the light given off by millions of stars. One might expect that such a blend would lose all characteristic detail or identifying features, but in fact this is not the case. When the light is broken up into its component wavelengths by a spectrograph (unit 11), specific features at specific wavelengths are seen. In figure 7, you see pictures of several galaxies and their spectra. The lines above and below the galaxy spectra are spectra formed by a stationary light source at the telescope. They may be used to determine the wavelength shifts shown by the moving galaxies. In the galaxy spectra between them, the two dark features labelled H and K are calcium absorption lines seen in the spectrum of many galaxies.

Measure the diameter of each galaxy pictured. Then measure the displacement of the calcium lines with respect to the comparison lines. If galaxies are all approximately the same intrinsic size, the diameter is inversely related to a galaxy's

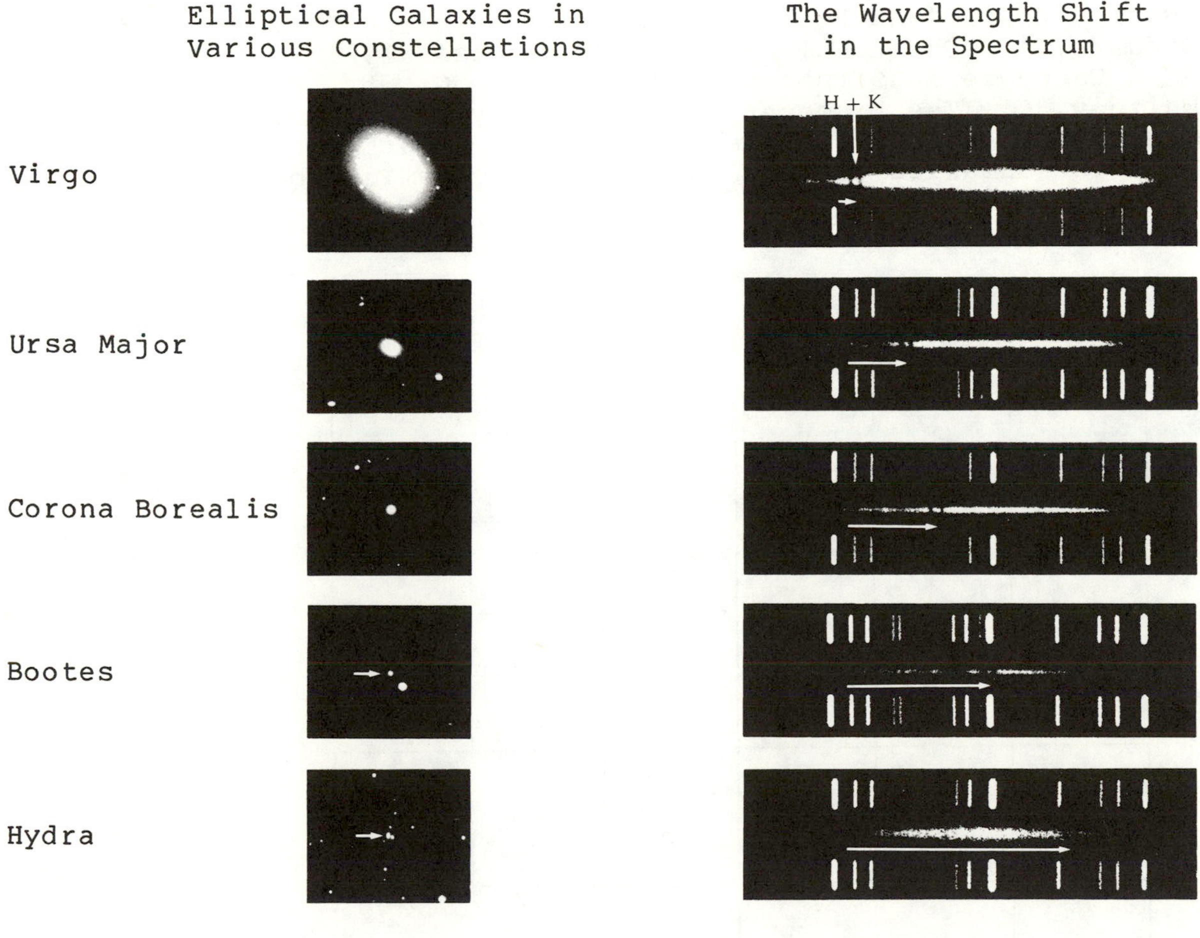

Figure 7

H and K designates a pair of absorption features of ionized calcium seen in the spectra of many galaxies. If the shift in wavelength of these features is interpreted as a Doppler Effect, the bottom galaxy is receeding from the earth at approximately one-fifth the speed of light.

distance. Plot the calcium line displacement versus the galaxy's approximate distance (the inverse diameter = one divided by the diameter). Do you see any correlation? This type of activity was originally done by Edwin Hubble and the relationship is called Hubble's Law. By plotting the displacement of the calcium lines, you have plotted the line of sight velocity. Hubble's Law is a relation between the recession velocity and distance for galaxies. Your plot should show that more distant objects move away from us at greater speeds than nearby objects. These large red shifts are seen in the spectrum of virtually all external galaxies and constitute the basic evidence for the expansion of the universe and for the so-called Big Bang cosmology. Extrapolating the observed expansion backward in time reveal that our universe began with a fantastic primeval explosion about 10 to 15 billion years ago.

For Further Reading

Hubble, Edwin "The Realm of the Nebulae", 1936 Yale Univ. Press
Sandage, A. "The Hubble Atlas of Galaxies", 1961 Carnegie Institute of Washington, D.C.
Whitney, C. "The Discovery of Our Galaxy", 1971 Knopf
Berendzen, R., Hart, R., and Seeley, D. "Man Discovers the Galaxies", 1976 Neale Watson Academic Publishers

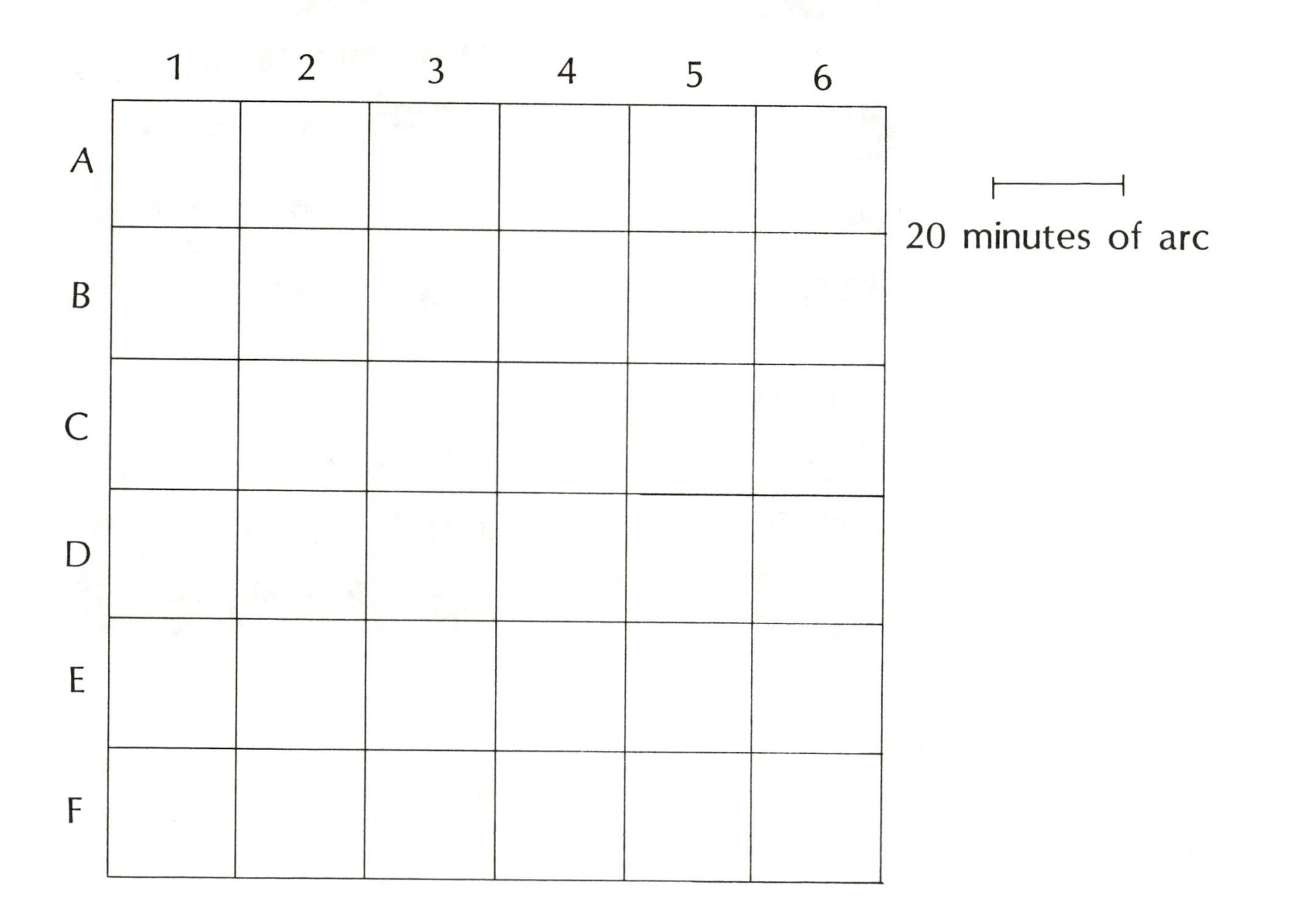

15

Introduction to Computers and the BASIC Language

A computer is very fast but also very dumb. Its programmer must tell it every single thing it is to do and must tell it in absolutely precise statements. Fortunately, the languages used to communicate with computers are simple and easy to learn.

OBJECTIVES

1. to successfully interact with a modern computer and communicate instructions to it
2. to learn a simple set of commands in BASIC language, especially: LET, PRINT, INPUT, DATA, FOR-STEP, NEXT, GENERATE, LIST, and SAVE
3. to program some simple calculations useful for astronomy -- this will involve constructing a set of commands and storing them in the computer, testing and correcting them, and then successfully executing your program and printing out correct results for the calculations

EQUIPMENT NEEDED

A home computer system (e.g.,Radio Shack TRS-80, an Apple, a Pet, etc.) or a central computer system (available at many schools) accessed remotely through a terminal of some kind.

Computers have touched the everyday life of each of us in the modern world. Their use has extended out of the scientific and business world into the classroom (at every level) and home. Many languages are available for computers to use. Among the most common are FORTRAN, a scientific language, COBOL, a business language, and PASCAL, a structured language. The language used in this unit, BASIC, is probably the easiest one to use, and the most available to the non-specialized user.

This unit will give guidelines for learning to use a modern programming language to communicate with a computer system. There are many different computer languages, each with different strengths, but for an introduction, the so-called BASIC ("Beginner's All Purpose Symbolic Instruction Code") language has many advantages. It is versatile, simple, very interactive in its design, and widely available. This unit will direct you to learn a set of commands in BASIC, write two programs in this language, and execute them on a computer. The printout of your programs and the printed results of the calculations, if correct, will be the criteria for successful completion of this unit.

It must be noted that there are a wide variety of different computer systems in existence and that the specific operating instructions for each of them are different. Therefore, it will be necessary in this unit to give only an outline. For detailed instructions on accessing and using a specific computer, you will have to obtain directions written on the system you are using. If possible, carry out this activity on a computer or terminal that produces a paper printout of your sessions with the machine. This printout will allow you to review your progress, assess your mistakes, and check your programs in detail.

I. MAKING CONTACT WITH THE COMPUTER

Read this section entirely before commencing.

If a self-contained small computer system is being used, consult the instructions for that system to learn the start-up commands that activate the computer and get it ready to receive and execute your instructions. If a remote central computer is being accessed through a terminal device, you will typically need to obtain a user-number and secret password. When you contact the system, it will first ask you to supply these numbers to identify yourself as a legitimate user. You will in many cases contact the computer by telephone, literally dialing its number, and when you receive a response tone over the telephone line, putting the telephone receiver into the cradle of the interface device (called a "modem") which transmits signals into your terminal. A busy signal on the telephone line means you try again later; all the access lines into the computer are busy.

Thus, for a central computer system, your initial startup sequence should be: turn on the terminal, turn on the modem, dial the computer phone number, and when you get a response, put the receiver into the cradle of the modem. At this point, you will generally be in contact with the Monitor of the system, that part of the computer which is programmed to act as an answering service, switchboard, and overseer of all computer activities. It will typically give some characteristic response on your terminal to indicate that it is in contact with you and awaiting further commands. Your first interaction with it will be to supply your user number and password. Hit the carriage return key on the terminal to send this information to the computer. Note that a line of type is not sent from the terminal to the computer until you hit the carriage return; therefore, you can correct mistakes before transmitting.

In central computer system, it is quite common for the main computer to interact with many different types of terminals made by numerous manufacturers, so it may also be necessary for you to supply certain information to the computer about your terminal to allow it to communicate properly back to you. For example, if you note that your terminal is printing out double letters, or if the first few letters of each line are missing, there are simple "housekeeping" commands you can transmit that will clear this up. Learn whatever commands are necessary for your terminal.

To correct mistakes, terminals will have an erase key of some sort. There is usually a key that erases one character at a time (it may be called RUBOUT, ERASE, or DELETE), and a command that erases an entire line. If you do these operations before hitting the carriage return key, the computer will never know about your mistakes. Learn correction procedures for your terminal before going any farther.

When interacting with a computer, it is sometimes possible to get "stuck" in the computer and not see an obvious way of quitting or getting off. This could occur in a program of your own construction, if you accidentally set up some kind of looping instructions that had no termination, or it could occur while running someone else's program where you have not been supplied with information on how to stop it. There are instructions that will allow you to escape the computer, and you should learn these for your system before embarking farther.

When you finish a session with the computer, you must sign off in a specific fashion recognizable to the system. Learn the sign-off commands for your system before going any farther. After a proper signoff, you simply hang up the telephone and turn off the terminal.

Having read this section and learned the specific commands for the situations discussed above, do the following introductory exercise: start up your terminal and contact the computer, identify yourself, execute the housekeeping commands for your situation, type a little, and then use the erase commands to undo what you typed.

Next, play some games on the computer if any are available to you. The computer will give you instructions on what games are available and how to play them. These simple activities will make you comfortable with the practice of interacting with a computer. After some game playing, sign off the computer and shut down the system. At your next session, you will begin to learn programming language and will be giving serious commands to the computer.

II. THE BASIC LANGUAGE

Many computers today come equipped with tutorial programs that will teach you the BASIC language; this is true of both home systems and many central systems available at schools and colleges. If you have access to such tutorial routines, they will provide a step-by-step introduction to the statements and rules of the language. Typically, they will provide you with some information and then question you about what you have learned. You type your response to the questions into the terminal and if you are correct, the computer will continue to supply you with more information. If you are not correct, the computer will continue to drill you on the old information. If you have access to such tutorial programs, learn how to call them up ad begin to work your way through the necessary material.

If such pre-programmed material is not available, there are many excellent books available which teach the BASIC language, since it is the most widely used language for instructional purposes. Obtain one of these books and begin to read it.

Continue your studies until you have covered the commands: INPUT, DATA, PRINT, LET, and END. At this point, you can write and execute a simple program. Other commands you may use in setting up your program are: NEW, LIST, RUN, SAVE, and UNSAVE. The next section discusses the program you should write.

III. WRITING YOUR FIRST PROGRAM

In unit 1 of this book, it was stated that for sufficiently small angles, the angular size of an object and its true size were linearly related through the distance of the object. That is, the true size equals the distance times the angular size in radians. This equation actually involves the approximation that the tangent of an angle is equal to the angle itself, for sufficiently small angles. The tangent of an angle θ is illustrated in figure 1.

By definition, the tangent of θ is the length of side A divided by the length of side B. Thus, $\tan\theta = A/B$. We want to test the proposition that for small angles, the value of the tangent of an angle is approximately equal to the value of the angle itself, if the angle is expressed in radians.

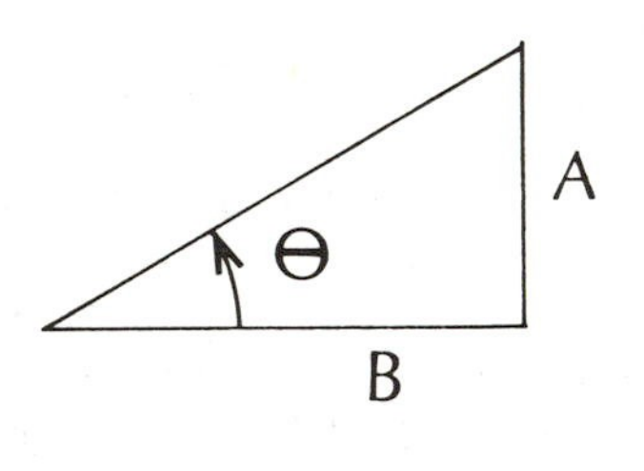

Figure 1

That is, $\theta = \tan\theta$ if θ is small and expressed in radians.

Write a program that takes values of θ equal to 30, 10, 2, and 0.5 degrees, converts each of these angles to radians, computes the tangent of the angles, and then evaluates the percentage difference between the angle and its tangent by computing the quantity

$$\frac{\theta - \tan\theta}{\tan\theta}$$

Note that to get the computer to compute the value of the tangent of a quantity, all you have to do is say something like

```
LET X = 0.7
LET N = TAN(X)
```

The computer will calculate the value of the tangent of 0.7 radians and store the result in a location such that it knows how to find it again. If you need the value of N, you simply write N into an expression and the computer goes and finds the value that was stored earlier.

IV. EXECUTING YOUR PROGRAM ON THE COMPUTER

To get the computer to carry out your program, you must enter your command statements into the computer and then tell it to carry them out. Armed with your program, activate the computer system as before and sign in properly. You must now tell the computer what language you are going to speak (in this case, BASIC), and that you are creating a new program. Learn the statements for these communications. The computer will ask you what the name of your program should be, and you may tell it whatever name you wish (usually less than 6 characters). You then type in your program, line by line, with each line followed by carriage return. If your first command to the machine is GENERATE, it will automatically create statement numbers for you, a considerable convenience. The last line of your program should be END. Transmitting the LIST command will then instruct the machine to print out your entire program and you can check for typing errors, or other problems. To change a line, type the number of that line and the new statement, followed by a carriage return. To insert a new line, choose an unused number between the line numbers where you wish the new line. Type that number and the contents of the new line, followed by a carriage return. To delete a line, type the line number and a carriage return.

To get the machine to carry out your sequence of instructions, transmit the command RUN. If you have made any programming mistakes that the computer is trained to recognize (e.g., mistakes in syntax), it will tell you at this time. Mistakes are very common, even for experienced programmers. Correct any errors using the instructions in the preceding paragraph.

Note that just because the computer makes no further comments on your program, that does not mean it is a correct program. The computer cannot recognize bad thinking. For example, if you tell it to multiply A times B when you really wanted to divide A by B, the computer will have no way of knowing what you wanted as long as you give it a grammatically correct statement. So you must still examine the results from your program carefully.

Continue running your program and checking the answers it gives until you are satisfied they are correct. Try to design a print statement into your program that outputs the answers in a readable fashion, with some kind of heading to identify what is being printed. Read over what the tutorial routines (or your BASIC manual) have to say about the format of print statements as you write yours.

V. USEFUL ADVANCED COMMANDS IN BASIC

In your first program above, it was probably necessary to repeat some commands when calculations were repeated. BASIC has two very useful commands, FOR and NEXT, which allow a repetitive calculation to be done in a continuous fashion by establishing a loop within the program. It is looping calculations of this type that make effective use of the tremendous speed of modern computers.

If you have access to BASIC tutorial programs, continue studying them until you complete the required material. If you are learning your BASIC from a book, continue reading until you cover the material involving the FOR and NEXT statements.

Equipped with these more powerful statements, rewrite your earlier program making use of them. To illustrate their usefulness, do the computations for a much larger number of angles this time: start with a very small angle (such as zero) and keep incrementing the angle by 1 all the way up to 45 degrees. Produce a program that does all these calculations in one continuous sequence and then prints out the results, with well organized headings identifying the columns of your printout.

The successful completion of this second program achieves the objectives of this unit. You might think of other simple calculations you would like to program on the computer; feel free to do so. A suggested program that relates to the concepts in this book might be to compute the luminosities for stars of various masses, using the mass-luminosity relationship discussed in unit 12. In mathematical form, this relationship states that the luminosities of stars are given by $L = M^{3.5}$ where M is the mass of the star. For the sun, M = 1 and L = 1. Inserting numbers for stellar mass in terms of the solar mass gives luminosities in terms of the solar luminosity (a number you determined observationally if you did unit 7).

16

Using a Solar Telescope to Study the Sun

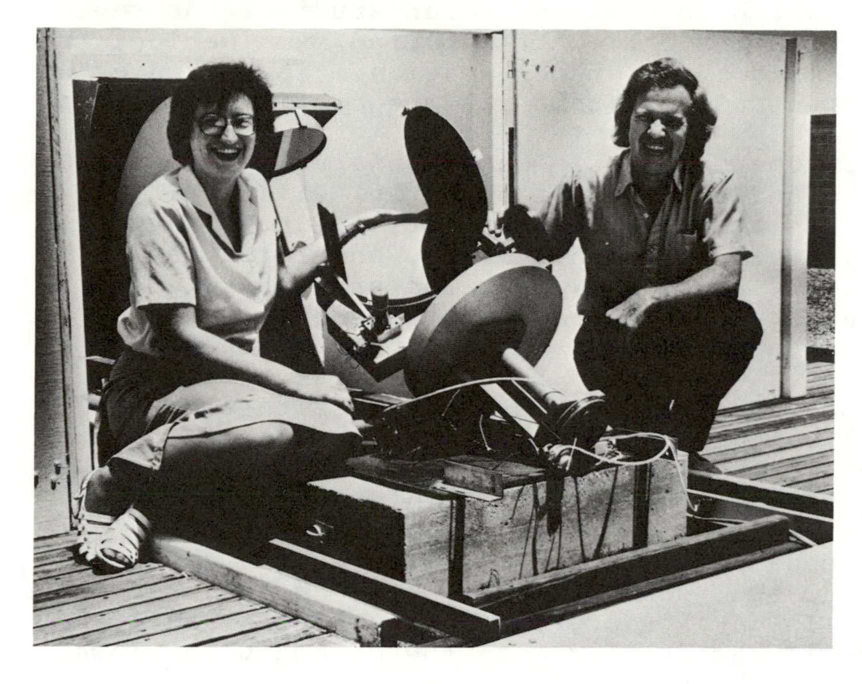

The authors pose beside the University of Texas solar telescope. The sun is so bright that even a telescope of modest size can form a very large image or a highly dispersed spectrum, allowing detailed studies possible for no other star.

OBJECTIVES

1. to set up the solar telescope and explain its optics by drawing a schematic diagram
2. to calculate for the sun the rotational period and the equator's postion through observations of sunspots
3. to produce a list of wavelengths of solar absorption lines and identification of the associated element for many of them
4. to describe prominences, their appearance through an H-alpha filter, and their changes over short time scales
5. to define these terms: solar limb, sunspots, prominences, photosphere, chromosphere, corona, limb darkening, and active sun

EQUIPMENT NEEDED

Solar telescope, accessory lenses, grating, H-alpha filter, meterstick.

In this unit you will be studying some physical properties of the sun in order to continue the solar investigations begun in unit 6 and 7. You will also examine the special optical system called the coelostat, which is used for solar studies. As the only star close enough to the earth for detailed examination, the sun is the subject of many types of observation. Your studies of optics (unit 8) and spectroscopy (unit 11) are continued in this unit.

Galileo was one of the early telescopic observers of the sun. He discovered sunspots and solar rotation in 1610. In 1814 Fraunhofer published his study of the absorption lines in the solar spectrum. He visually noted 574 lines, and formed a notation system, using A, B, C, D... for the strongest lines. Several lines have retained their Fraunhofer designation, such as the sodium D lines at 5890 and 5896 Angstroms. The use of photography expanded this work, with Rowland's list of about 20,000 absorption lines and their estimated intensities in 1897. The presence of the lines indicates which chemical elements occur in the sun's photosphere. Detailed studies of the position, shape, and intensity of the lines provides information of the physical state of the sun's atmosphere, e.g., temperature, pressure, relative abundance... Since hydrogen is the most abundant element in the sun, some studies center on the Hydrogen-alpha line. In particular, prominences appear as streams of chromospheric gas jetting into the corona when viewed in the hydrogen-alpha emission line.

WARNING: NEVER ALLOW YOUR EYE TO ENTER THE LIGHT PATH

PERMANENT DAMAGE OR BLINDNESS COULD OCCUR

I. THE TELESCOPE

The most common arrangement for detailed observations of the sun's surface is a system called a coelestat. It employs two flat mirrors to gather sunlight and direct it down a vertical shaft to a lens which focuses a solar image farther on down. Refer to figure 1 for a diagram of the optical system of a coelostat. (This figure is not to scale.) The drive mechanism moves the first flat mirror to compensate for the earth's rotation; with such a drive, the sun's image should stay within the field of view. If it does not, small corrections with the guiding controls should re-center it. In a coelestat the field of view does not rotate.

The image formed at the prime focus of the objective lens can be focused by moving the lens up or down the vertical shaft. The image can also be magnified and projected by employing additional projection lenses to direct an enlarged image to a screen or wall, where it can be viewed by a group of observers. Generally, the image formed at prime focus is relatively small, and it is so intense that it is uncomfortable to look at. The image is quite capable of igniting a fire if it falls on something combustible.

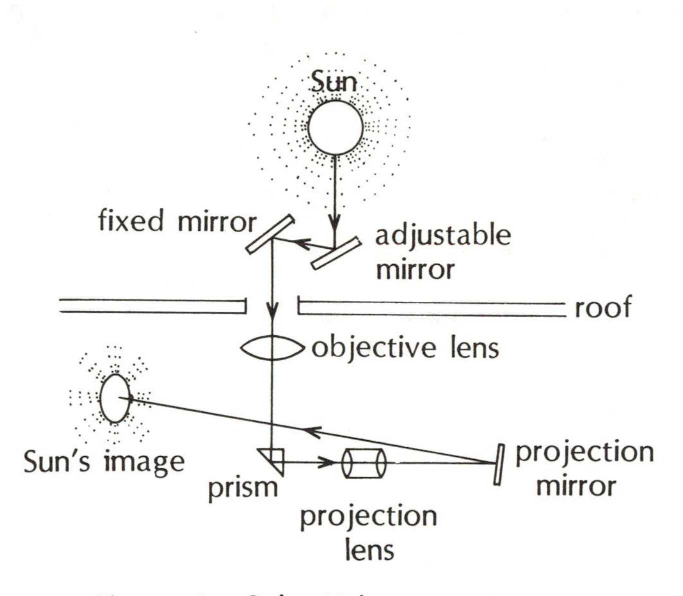

Figure 1 Solar Telescope Optics

Figure 1 also illustrates a projection lens used with a coelestat. Note that even if the first image of the sun is in focus, the projected image will generally need additional focus adjustment; this is accomplished by moving the projection lens horizontally.

Carry out the following preliminary activities with your solar telescope. Learn the procedures to uncover its mirrors and lenses, point the system safely at the sun, engage the drive, and focus the image. Measure or estimate the diameters of the mirrors and lenses in the system, estimate the focal length of the objective lens, and compute its f-ratio. Measure the diameter of the first image formed by the system. Set up the various projection lenses available and make a table of their characteristics versus the size of the solar image. Do some of the lenses invert the image and some not invert it?

If you have a solar telescope other than a coelestat available to you, learn its features and draw a diagram of its components to scale. Carry out as many of the activities in the previous paragraph as you can.

II. THE DISK OF THE SUN AND SUNSPOTS

Examine the sun's image in detail on at least two different days within one week. NOTE THAT WEATHER CAN BE A FACTOR. Draw a sketch of the sun and any details you see on its surface, such as sunspots and any variations in the brightness of its surface. Are all parts of the solar disk equally bright? Compare the center with the edge (limb). Make a drawing of any structure you see within a sunspot. Count the number of spots -- how many appear as pairs? Make the first drawing as accurate as possible (you may wish to attach a large piece of paper to the wall or screen and trace the sunspots), so that you can identify the sunspots on your second observation. On another date, measure the location of the sunspots and try to match them to your previous observations -- they will probably have changed their shapes slightly. Determine the rate at which the sun is rotating and explain your method of doing so in your notebook. If there are only a few sunspots, average the rotation rates. In either case, determine the direction of the solar equator and spin axis. At which solar latitudes do most of the spots occur? How many spots did you detect? Calculate the diameter of the largest spot you saw. Note: if spots are abundant and seen at various distances from the equator, you might attempt to demonstrate the fact that different latitudes on the sun rotate at different rates. (Given this information, can the sun be a solid body?)

III. THE SPECTRUM OF THE SUN

In figure 1, if the projection mirror is replaced by a reflection grating, one can project the solar spectrum in considerable detail. The beam of solar radiation is directed through a slit and then onto a reflection grating which disperses the light into a spectrum. To use such a system, first place a transparent slit in the slit holder which is next to the prism while the projection mirror is in place. Adjust the projection lens for a clearly focused image of the slit. Then move the objective lens to obtain a sharp image of the sun. Replace the transparent slit with an opaque slit. Replace the projection mirror with the reflection grating. Note that some projection lenses give a smaller but more intense image, so the lines may be easier to see. You may wish to experiment with several lenses at this point. Also, note the effect of using the different width slits. When you are satisfied with your lens-slit combination, you may wish to adjust the distance of the projection screen so that one Angstrom equals one millimeter, or a similarly convenient scale.

Using figure 2 for a comparison, identify several strong absorption lines, particularly hydrogen-alpha at 6563 A and hydrogen-beta at 4861 A. Having identified these lines and their wavelengths, calculate the scale in Angstroms/mm -- then measure at least 20 lines you can see clearly and calculate their wavelengths. Estimate the strengths of the lines by some criteria of your own devising, or at least draw them in some way as to indicate their relative intensities. These lines may all appear in the brief list in Allen's "Astrophysical Quantities", and in the MIT Wavelength tables. Identify the chemical elements causing the absorption features you measured. Suggest a research procedure to determine whether any of these features are caused by the earth's atmosphere.

If there are many sunspots visible, you may be able to center one in the slit. If you do this, you may be able to observe the spectrum of a spot surrounded on the top and bottom by the spectrum of the bright disk of the sun. Compare the two very carefully -- can you see any differences between them, as for example in the RELATIVE intensities between various lines? Sunspots are regions of lower temperature and as a consequence, spectral differences are expected.

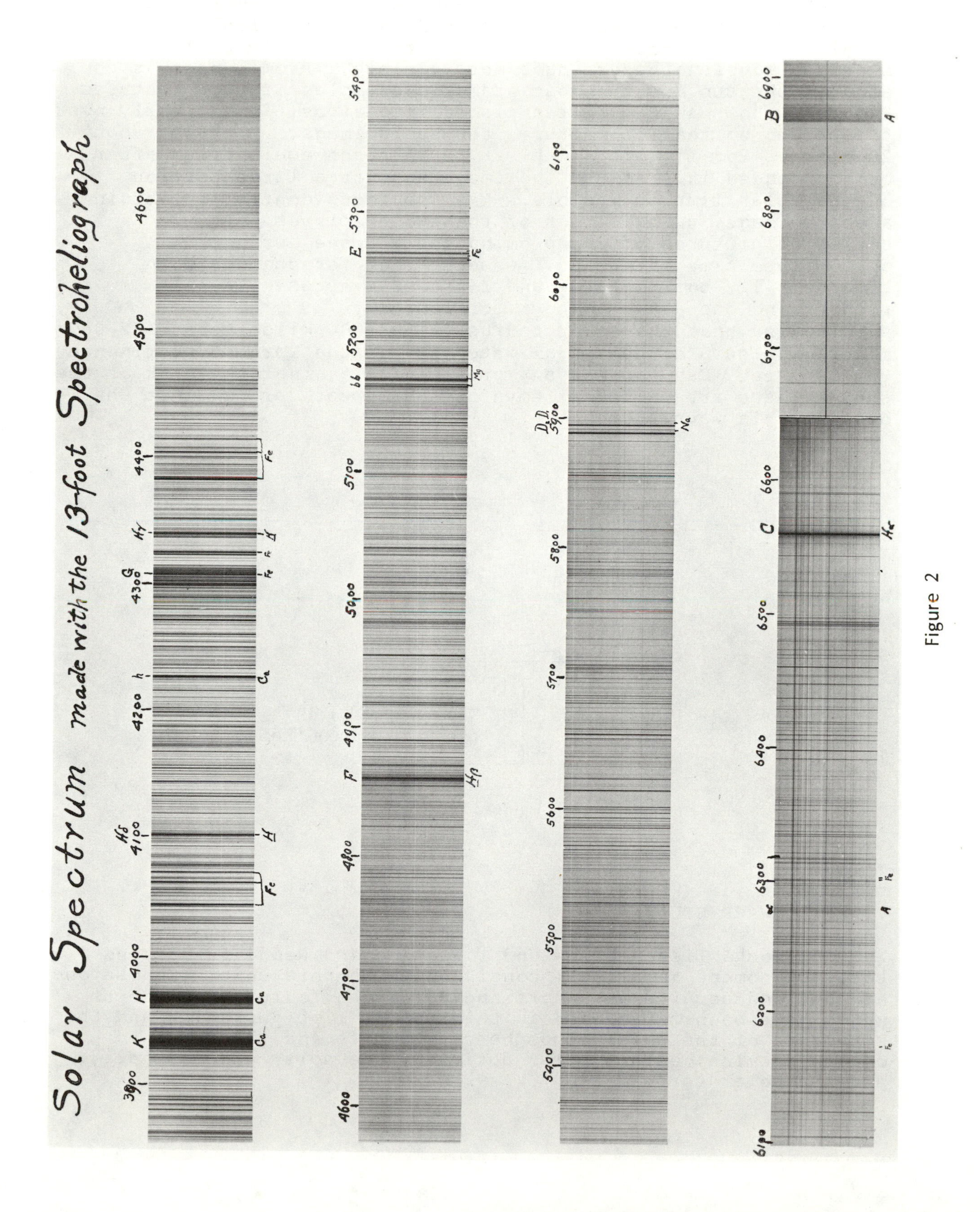

Figure 2

IV. PROMINENCES

A H-α filter passes less than 1 Angstrom of the spectrum centered on a wavelength of 6563 A, the strong hydrogen absorption line. All radiation that is not between 6562 and 6564 A is rejected. By isolating hydrogen radiation in this fashion, certain features on the sun that strongly radiate hydrogen wavelengths are made more visible. Prominences at the edge of the sun are the most dramatic examples of such features. Figure 3 shows an H-α filter placed into a position directly over the prism, so that it captures the sun's image. It takes about 20 minutes for filters of this type to reach equilibrium after being plugged in. Adjust the image so that a large portion of the solar limb is visible. Use the focus control to obtain a sharp image, using either of the two eyepieces. Draw a series of pictures of any prominences you see, using appropriate time intervals. Observe them for changes with time over 1/2 to 2 minutes and 15 to 20 minutes. Try also reobserving one after an hour (or longer).

Knowing that the solar radius is 696,000 kilometers (432,000 miles), calculate the approximate size of the largest prominence you can see. Can you see any prominences against the disk of the sun? (These are called filaments, and appear dark against the disk at this wavelength.)

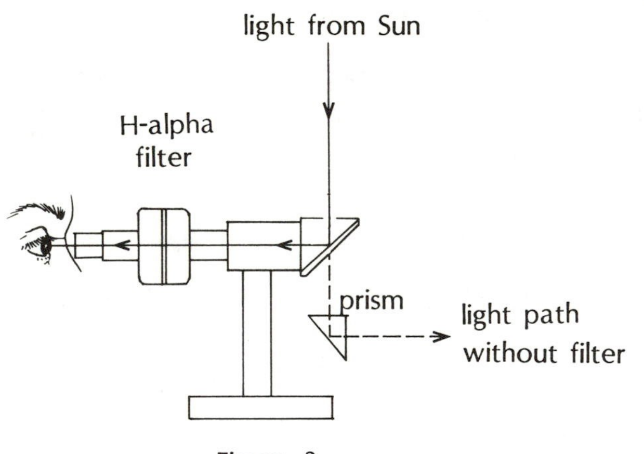

Figure 3

V. FOLLOW-UP STUDY

The books listed below under Recommended Readings discuss solar phenomena on a level consistent with this unit. Choose one of them and acquire an understanding of the following concepts which have been introduced in this unit: limb darkening and the structure of the solar atmosphere, sunspots and the sunspot cycle, prominences and solar activity, the solar chromosphere, and the solar corona.

REFERENCES

Allen, C.W. "Astrophysical Quantities" third edition, Athlone Press, 1973.

MIT Wavelength Tables. John Wiley, 1939.

RECOMMENDED READINGS

Abell, George "Realm of the Universe," Holt, Rinehart and Winston, 1976. Chapter 17.

Eddy, John A. "A New Sun: The Solar Results From Skylab," NASA SP-402, 1979. Chapters 1 and 2.

Gibson, Edward G. "The Quiet Sun" NASA SP-303, 1973. Chapters 1 and 2.

Jefferys, William and Robbins, R. R. "Discovering Astronomy," Wiley, 1981. Chapter 17, section 8.

Kuiper, G.P. (editor) "The Sun", volume 1 of the Solar System, University of Chicago Press, 1953. Chapters 1 and 3.

Menzel, Donald H. "Our Sun," Harvard University Press, 1959. Chapters 2, 6, 7, and 8.

Pasachoff, J. and Kutner, M. "University Astronomy," Saunders, 1978. Chapter 8, sections 1, 2, 3, 4, 6.

Zeilik, M. "Astronomy: The Evolving Universe," Harper and Row, 1979. Chapter 8.

17

A Spectral Comparison of the Sun and Beta Draconis

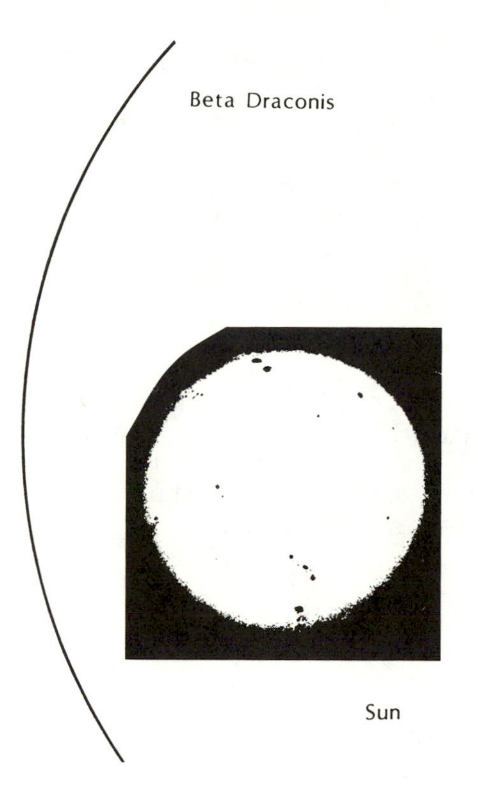

Dispersing the light into a spectrum vastly increases the amount of information we can extract from starlight. Even size differences between stars of the same surface temperature reveal themselves in a fine analysis of the absorption line profiles.

OBJECTIVES

1. to carry out an analysis of the atmospheres of two stars -- the sun, a main sequence star, and Beta Draconis, a giant star with the same surface temperature as the sun. You will determine the wavelengths of the absorption lines in the spectra and identify the chemical elements present in the atmospheres of these two stars
2. to measure the strengths of selected lines in the spectrum by calculating their equivalent widths and central depths, and to relate the line strengths to the abundances of the chemical elements for a star
3. to study the physical principles that determine the great diversity of appearance in stellar spectra, and how detailed spectral analyses can be used to differentiate conditions in stellar atmospheres
4. to understand the larger significance of the chemical abundances in stars in terms of long-term evolutionary processes in the stars of our galaxy

EQUIPMENT NEEDED

None.

In unit 11 you examined the three basic types of spectra that radiating sources emit (continuous, bright-line, and dark-line) and the underlying physical conditions which produced the different spectra. You also studied the pattern of emission from a number of different chemical elements, leading to an appreciation of the fact that each element in the periodic table, because of its unique atomic structure, shows a different spectral "fingerprint" in the pattern of wavelengths that it is capable of emitting and absorbing.

In this unit, you will carry out a series of activities to understand in greater depth the dark-line spectra given off by stars.

I. THE ASTRONOMICAL SPECTROGRAPH

In unit 11 you constructed a simple spectroscope to visually examine the spectrum produced by dispersing light with a grating. A professional spectrograph that attaches to a telescope is designed to record the spectrum permanently for future study. Figure 1 indicates schematically how such a spectrograph operates.

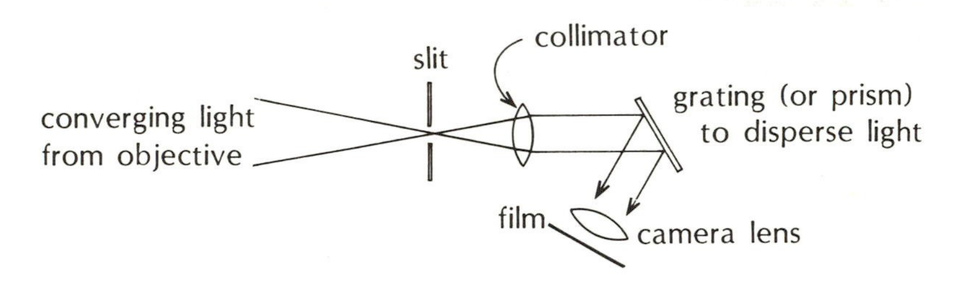

Figure 1

In the focal plane of the objective (main) mirror or lens of the telescope, a slit is placed which serves the function of shaping the light gathered by the telescope conveniently for imaging by the rest of the system. In unit 11, you observed that the sharpest line images (and hence the greatest resolution of features in the spectrum adjacent in wavelength) were obtained by a narrow slit. Thus, from the fuzzy image of a star that is formed by a telescope, the slit rejects most of the image and passes a thin rectangle of light on through the system to achieve the greatest possible spectral resolution. This wastes some light but greatly increases the quality of the final data. See figure 2 for a view of the star image as reflected off the jaws of the slit. This "rejected" light is collected in an eyepiece by the astronomer, who uses it to make fine corrections in the tracking of the telescope during the exposure that captures the spectrum.

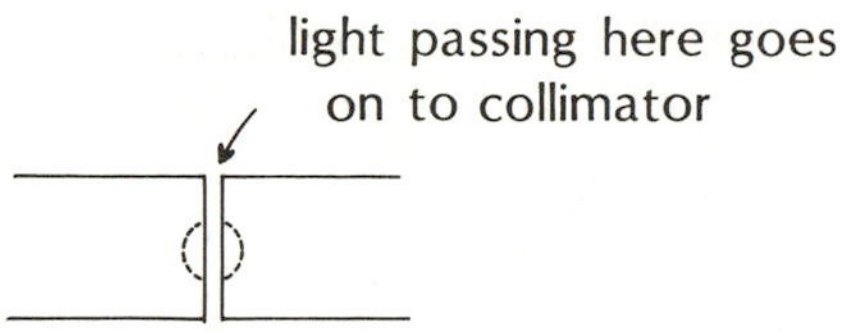

light passing here goes
on to collimator

Figure 2

Beyond the slit, the light begins to diverge until it is gathered and made parallel by a lens called the COLLIMATOR (Latin: to make parallel). Unless the light entering the dispersing unit (either a prism or a grating) consists of parallel rays, resolution and clarity in the spectrum are again compromised. After the light is dispersed, it is focused by a camera lens onto a photographic glass plate and permanently recorded.

On the photographic negative, the darkest portions show where the most light has fallen. It is possible to obtain an intensity reading from a photographic plate by passing a constant beam of light through it and reading the current from a photoelectric detector on the other side, as in figure 3.

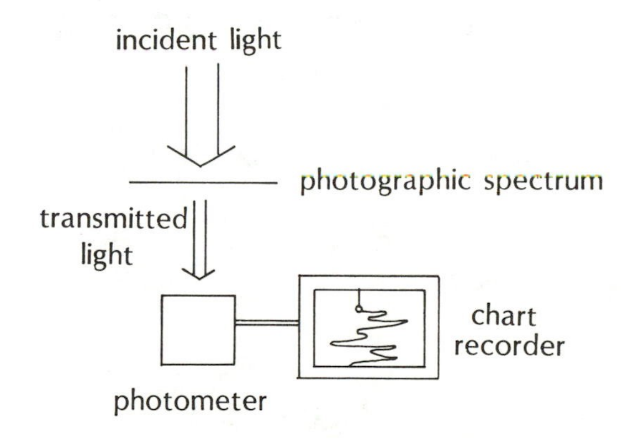

Figure 3

When this current is recorded on a strip-chart recorder, one obtains a quantitative representation of the energy curve of the star as a function of wavelength. An absorption feature in the spectrum of a star will appear as in figure 4 -- a reduced amount of radiation leaving the star at a certain wavelength due to the absorption by a certain chemical element in the atmosphere of the star. The strength of an absorption line is measured relative to the nearby continuous emission.

continuous
spectrum
absorption line
λ

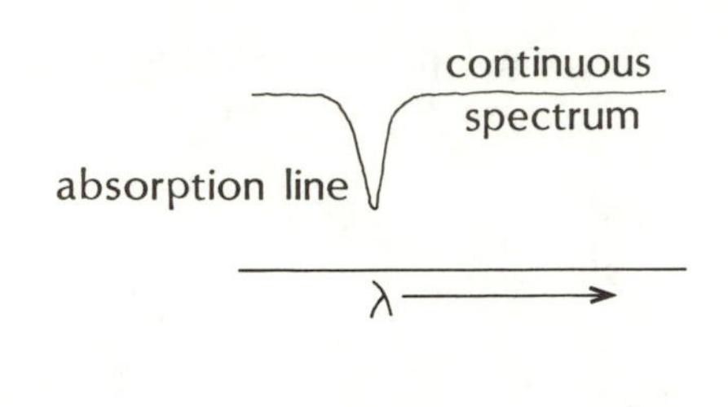

Figure 4

The following pages contain a tracing of the spectral output of the sun between the wavelengths of 6400 and 6450 Angstroms. This figure is taken from the Utrecht Atlas of the Sun, a standard reference work used by astronomers. The wavelength scale is indicated on these tracings. In practice, this scale would be obtained by photographing the spectrum of a known element (for example, an iron arc) onto the same photographic plate as the stellar spectrum, directly above and below it. Figure 5 shows a typical stellar spectrum with its comparison spectrum included.

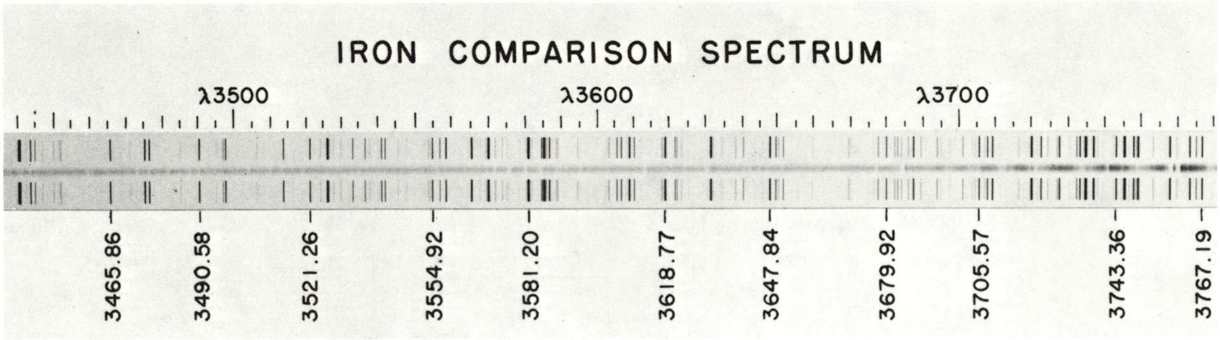

IRON COMPARISON SPECTRUM
λ3500
λ3600
λ3700
3465.86
3490.58
3521.26
3554.92
3581.20
3618.77
3647.84
3679.92
3705.57
3743.36
3767.19

Figure 5

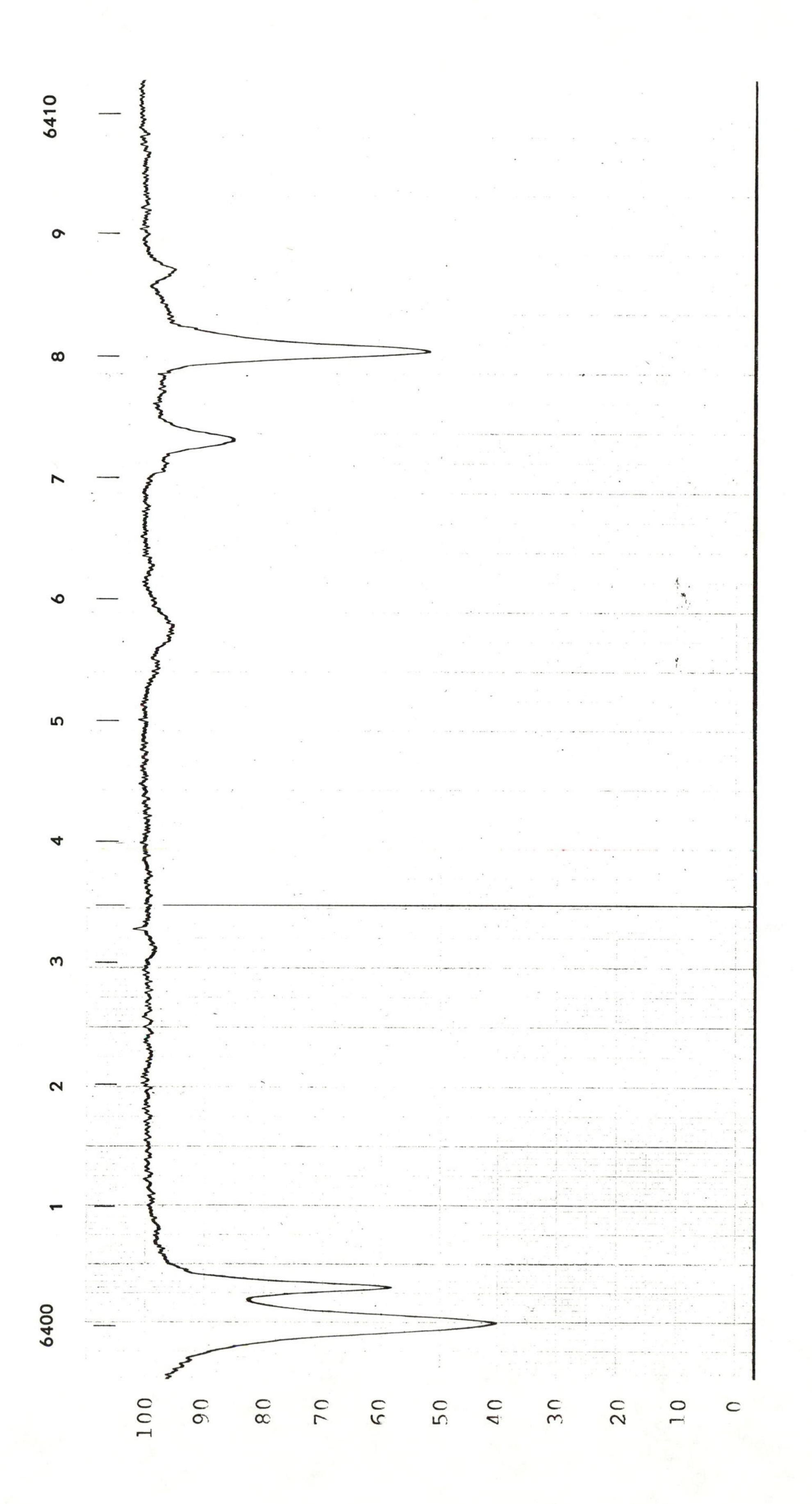
6400
1
2
3
4
5
6
7
8
9
6410
100
90
80
70
60
50
40
30
20
10
0

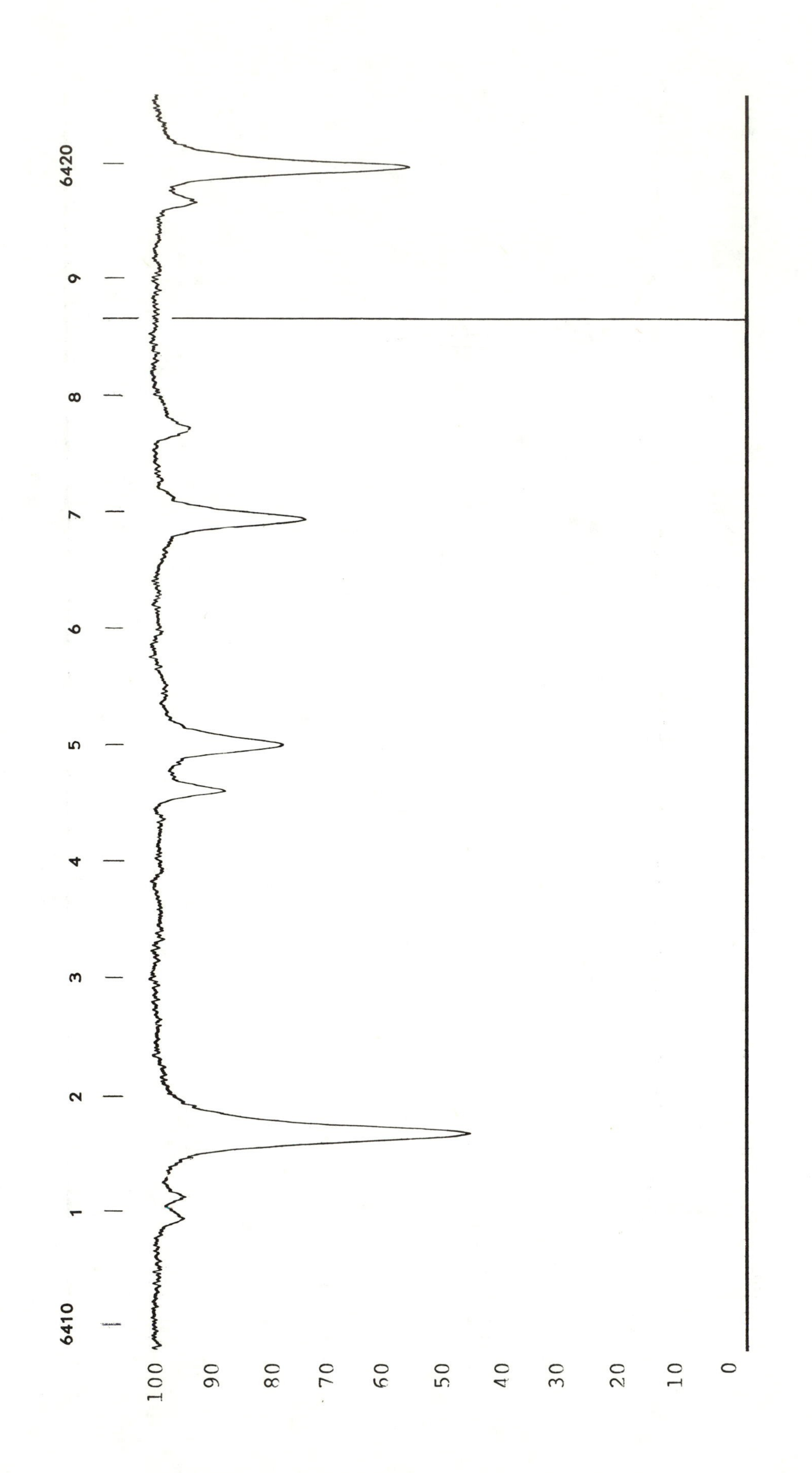
6410
1
2
3
4
5
6
7
8
9
6420
100
90
80
70
60
50
40
30
20
10
0

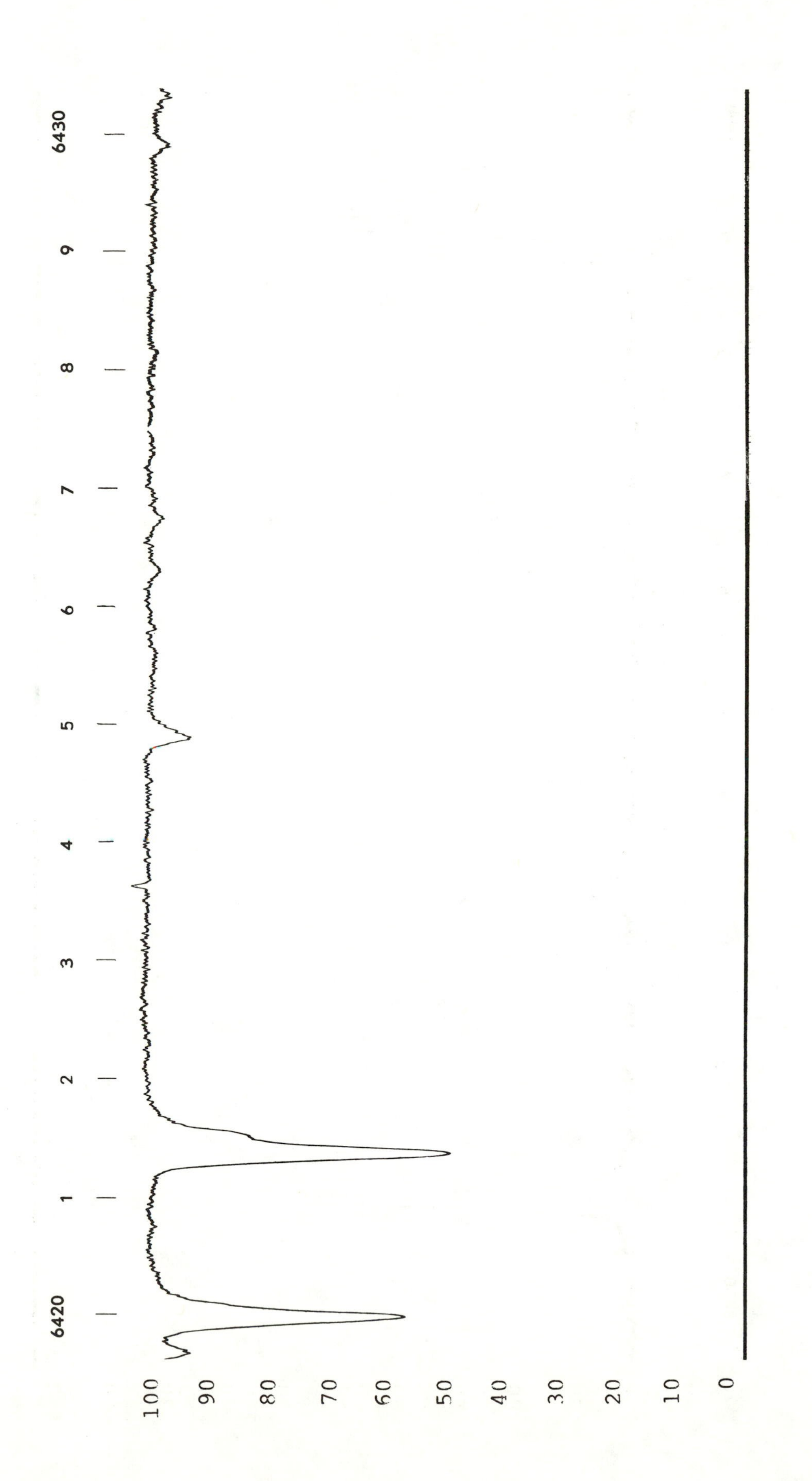
6420
1
2
3
4
5
6
7
8
9
6430
100
90
80
70
60
50
40
30
20
10
0

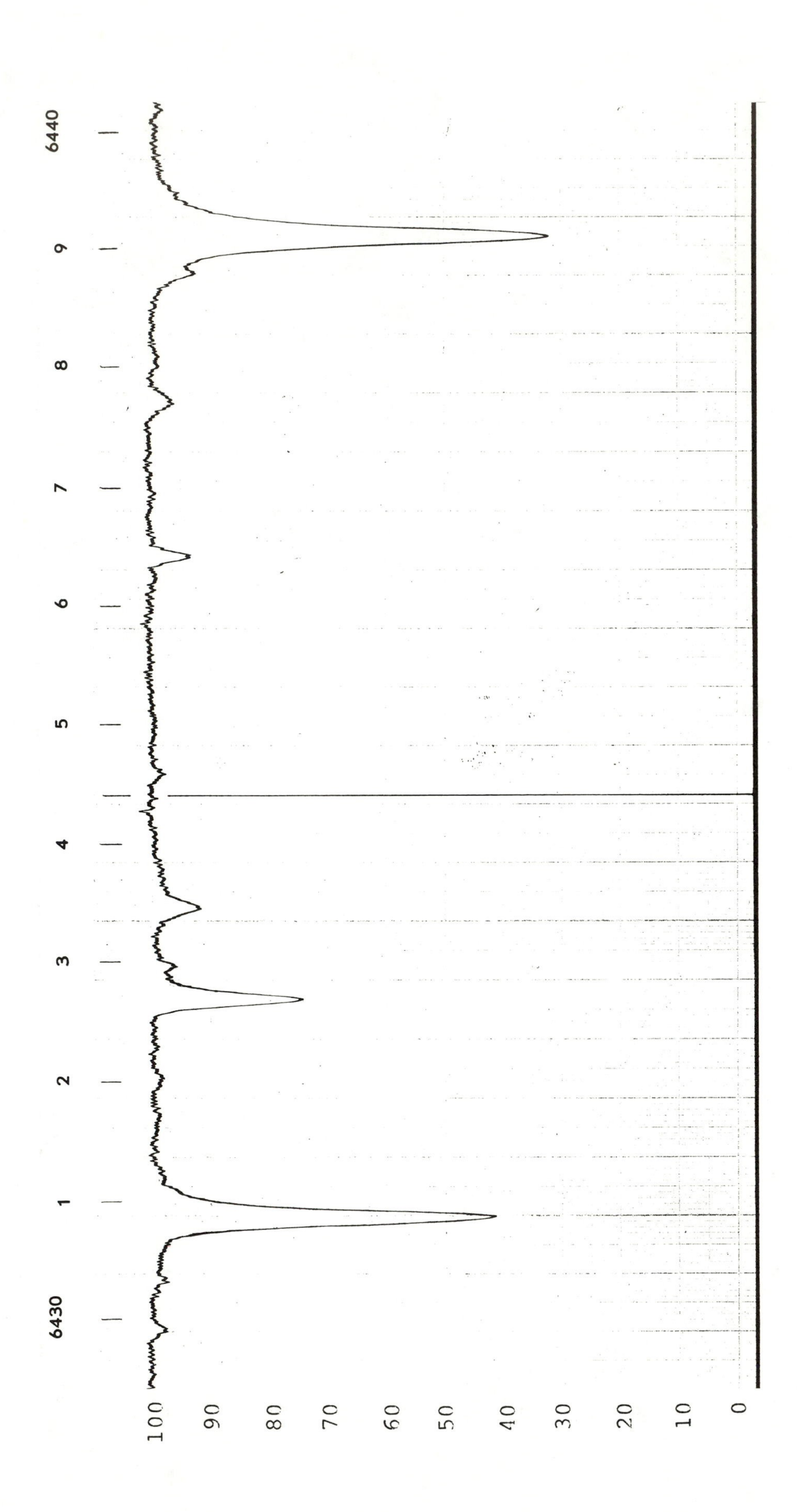
6430
1
2
3
4
5
6
7
8
9
6440
100
90
80
70
60
50
40
30
20
10
0

Table 1 (The MIT Wavelength Tables)

Wave-length	Ele-ment	Intensities Arc	Spk.,[Dis.]	R
6444.697	Co I	25	–	–
6444.610	Ta	40	–	–
6444.51	I I	–	[100]	Db
6444.25	Se I	–	[100]	Bl
6444.17	Dy	2	–	Ed
6443.96	As	–	3	Ro
6443.937	Pr	2	–	–
6443.89	A	–	[4]	Rt
6443.887	Ta	10 h	–	–
6443.71	Dy	3	–	Ks
6443.492	Mn	10	–	–
6443.47	Cu II	–	5	Sh
6443.26	I	–	30 h	Bl
6443.24	Rn I	–	[4]	Rs
6443.05	La II	8	25 h	Me
6442.55	U	2	–	–
6441.95	A	–	[6]	Rt
6441.91	Rn I	–	[6]	Rs
6441.85	Dy	2	–	Ks
6441.70	N I	–	[70]	Mt
6441.698	Cu II	–	40	Sh
6441.43	Se I	–	[20]	Rd
6441.31	Er	10	–	Ed
6441.14	Lu	40	–	Me
6441.045	Ce	3	–	–
6441.03	Tb	4	–	Ed
6440.974	Mn	60	–	–
6440.95	N	–	[25]	Du
6440.81	Yb	–	5	Me
6440.74	Kr II	–	[5 hl]	Me
6440.54	Sm	8	–	Ab
6440.22	I	–	[100]	Bl
6439.970	Ce	6	–	–
6439.97	Eu	40	–	Kn
6439.86	Nd	3	–	Kn
6439.83	Rn I	–	[2]	Rs
6439.83	Co	2 h	–	Me
6439.720	W	6	1	–
6439.72	Sm II	10 d	–	Ks
6439.318	Sm	3	–	–
6439.171	Nd	20	–	–
6439.171	Co	80	–	–
6439.073	Ca I	150	50	I
6439.03	Zr I	8	–	–
6438.96	In	–	5	Sq
6438.9	Ra I	–	[30]	Rs
6438.4696	Cd I	2000	1000	IS
6438.08	Br	–	[2]	Ks
6438.03	W	3	1	Me
6437.69	Eu I	700	–	–
6437.63	A II	–	[4]	Rt
6437.630	Sm	10	–	–
6437.540	In II	–	[12]	Ps
6437.365	Ta	2	–	–
6437.158	Yt I	4	5	–
6437.06	Te	–	[1000]	Bl
6437.01	N I	–	[30]	Du
6436.914	In II	–	[5]	Ps
6436.57	Dy	2	–	Ks
6436.412	In II	–	[5]	Ps
6436.405	Ce	12	–	–
6436.2	Au I	5	–	Ml
6435.94	Pt	3	–	–
6435.66	Dy	2	–	Ks
6435.5	P II	–	[5]	Dj
6435.318	Sm	25	–	–
6435.156	V I	2	–	–
6435.000	Yt I	150	50	–
6434.96	Dy	2	–	Ed
6434.80	Cl	–	[25]	Ks
6434.550	Ta	5	–	–
6434.396	Ce	12	–	–
6434.329	Zr I	6	–	–
6433.953	Sm	15	–	–
6433.6	bh Pb	3	–	L
6433.309	Sm	3	–	–
6433.236	Nd	12	–	–
6433.22	Cb	30	4	Me
6433.175	V I	2	[3]	–
6432.96	Dy	2	–	Ks
6432.78	Cu II	–	3	Sh
6432.732	Yb	30	40	–
6432.654	Fe II	–	2	Kn
6432.653	Nd	12	–	–
6432.50	Er	4	–	Ed
6432.07	Ca	–	6	Ad
6431.966	Cs I	–	15	Ms
6431.96	Sm II	50	–	Kn
6431.92	Kr	–	[2 h]	Me
6431.863	Pr	8	–	–
6431.8	Hf	2	–	Me
6431.711	Nd	2	–	–
6431.634	V I	4	–	–
6431.57	A I	–	[15]	Ms
6431.258	Rh	2	–	–
6431.177	Nd	2	–	–
6431.09	Co I	5 h	–	–
6431.03	Sm II	50 d	–	–
6430.97	I	–	15 h	Bl
6430.95	Tm	15	60	Me
6430.93	U	4	–	–
6430.851	Fe I	100	80	S
6430.79	Ta	150	–	–
6430.471	V I	8	–	Me
6430.46	Cb	80	10	Me
6430.45	Tl II	–	5	El
6430.337	Co I	30 h	–	Sl
6430.155	Xe I	–	[20]	IMe
6430.068	Ce	10	–	–
6429.907	Co I	50	–	–
6429.840	Nd	2	–	–
6429.645	Pr	8	1	–
6429.49	Hf	1	2	Me
6429.04	Mo	4	2	–
6428.956	Sm	15	–	–
6428.68	U	3	–	–
6428.67	Se	–	[15]	Bl
6428.645	Nd	25	–	–
6428.596	Ta	40	–	–
6428.54	I	–	[30]	Bl
6428.315	Sm	50 d	–	–
6428.28	Eu	200	–	Kn
6428.125	Cr	12	–	–
6427.79	Dy	2	–	Ks
6427.7	bh Pb	3	–	L
6427.690	K II	–	[20]	Dm
6427.57	Cu I	3	–	Me
6427.51	U	2	–	–
6427.40	Dy	2	–	Ks
6426.731	Ta	5	–	–
6426.73	Rn I	–	[8]	Rs
6426.62	Sm II	100 d	–	–
6426.614	La I	4	–	–
6426.170	Zr I	4	–	–
6425.909	Sm	10	–	–
6425.790	Nd	15	–	–
6425.64	S	–	[15 h]	Bl
6425.64	Cl	–	[5]	Ks
6425.574	Er	4	–	–
6425.442	Ta	3 h	–	–
6425.36	W	5	1	Me
6425.296	Ce	3	1	–
6425.115	Co I	5	–	m
6424.905	Ni I	2 h	–	–
6424.888	U	6	–	–
6424.841	Nd	4	–	–
6424.817	Th	4	2	–
6424.52	Ca	2 h	–	Ad
6424.51	Gd	10	–	–
6424.502	Ce	4 d	–	–
6424.43	Tb	6	–	Ed
6424.368	Mo	100	20	–
6424.256	Sm	3	–	–
6424.1	bh Yt	2	–	Me
6424.0	bh Sc	2	–	Me
6423.90	Cu II	–	30	Sh
6423.10	Er	6	–	Ed
6422.96	Te	–	[70]	Bl
6422.94	A	–	[5]	Rt
6422.93	N	–	[10]	Du
6422.90	Se II	–	[125]	Bt
6422.9	Pb II	–	[2]	Ea
6422.415	Gd	50	–	–
6422.06	Cb	6	–	Me
6421.93	Dy	5	–	Ks
6421.88	Gd	2	–	Ks
6421.743	Co I	20	–	m
6421.708	Ne I	–	[100]	Ps
6421.54	Yb	3	–	Me
6421.52	Pr	2 h	–	Kn
6421.507	Ni I	3 h	–	–
6421.48	Rn I	–	[10]	Rs
6421.368	Cr	35	–	–
6421.355	Fe I	60	40 h	S
6421.029	Kr I	–	[100]	S
6420.47	N I	–	[30]	Du
6420.3	bh Sc	2	–	Me
6420.18	Kr II	–	[300]	Me
6419.977	Fe I	18 h	15 h	–
6419.763	Mo	7	2	–
6419.633	Pr	2	–	–
6419.541	Cs II	–	[10]	Lp
6419.4	Ga II	–	[25]	Sy
6419.3	bh Zr	2	–	L
6419.27	Se	–	[15]	Bt
6419.25	Cl II	–	[8]	Ks
6419.096	Ti I	15	–	–
6419.0	bh F	50	–	L
6418.992	Mo	5	2	–
6418.98	Xe I	–	[30 h]	Me
6418.95	Hg	–	[25]	Lf
6418.928	Sm	5	–	–
6418.90	S	–	[7]	Bl
6418.88	Er	4	–	Ed
6418.58	Xe II	–	[30]	Hu
6418.60	Rh	2	–	Me
6418.477	Ta	2	–	–
6418.43	A	–	[4]	Rt
6418.41	Xe I	–	[30]	Me
6418.41	Dy	2	–	Ks
6418.340	Ir	5	–	–
6417.99	Ta	2	–	Ks
6417.97	Yb	125	3	Me
6417.824	Co I	200 r	–	–
6417.7	bh F	100	–	L
6417.69	Se II	–	[10]	Bt
6417.66	I	–	[15]	Bl
6417.568	Ru	15	–	–
6417.513	Sm II	100 d	–	–
6417.220	La I	6	–	–
6417.17	Sm	8	–	Kn
6417.05	N	–	[10]	Du
6416.95	Se	–	[6]	Bl
6416.942	Fe II	–	2	Kn
6416.61	Kr II	–	[60 hs]	Me
6416.5	bh F	100	–	L
6416.33	Tm	20	–	Me
6416.315	A I	–	[100]	IMe
6416.30	F	–	[4]	Gl
6416.101	Th	3	2	–
6416.004	W	5	1	–
6415.93	Sb	6 hl	–	Wt
6415.79	I	–	[20]	Ev
6415.65	Kr I	–	[20]	Me
6415.531	Pr	3 W	–	–
6415.51	Er	4	–	Ed
6415.50	S I	–	[10]	Ms
6415.3	bh F	100	–	L
6415.24	Si I	4 h	–	Ks
6415.18	Cu I	3	–	Az
6414.724	Rh I	50	–	–
6414.62	Cu II	–	20	Sh
6414.603	Ni I	5	–	–
6414.1	bh F	100	–	L
6414.029	Nd	3	–	–
6414.01	Ga	–	15	Kl
6413.950	Mn	25	–	Sl
6413.71	S	–	[500]	Bl
6413.699	Pr	10	1	–
6413.66	F I	–	[150]	En
6413.612	Th	5	–	–
6413.59	Er	6	–	Ed
6413.353	Sc I	10	25	–
6412.995	Rh I	8	–	–
6412.9	bh F	80	–	L
6412.53	Kr II	–	[4 h]	Me
6412.389	Mo	15	4	–
6412.38	Xe I	–	[10]	Me
6412.3	bh Zr	12	–	L
6412.15	Te	–	[70]	Bl
6411.893	Th	10	–	–
6411.8	bh F	50	–	L
6411.664	Fe I	100	80 h	–
6411.593	U	6	–	–
6411.467	Re	20	–	–

Table 1

Wave-length	Ele-ment	Intensities Arc	Spk.,[Dis.]	R
6411.403	Sm	10	–	–
6411.344	Eu	500	–	–
6411.34	Ba	9	3	Lr
6411.29	I	–	[30]	Ev
6411.18	Cu II	–	10	Sh
6410.995	La I	100	–	–
6410.99	Se I	–	[15]	Rd
6410.7	bh F	30	–	L
6410.660	Pr	6 w	–	–
6410.598	Nd	2	–	–
6410.344	Sm II	4	–	–
6410.327	Er	4	–	–
6410.32	Br I	–	[30]	Ks
6410.216	Rh I	4	–	–
6410.17	Kr I	–	[5]	Me
6410.103	Eu	500 W	–	–
6409.84	Kr II	–	[10 hs]	Me
6409.753	Ne I	–	[150]	Ps
6409.7	bh F	20	–	L
6409.54	I	–	[15]	Bl
6409.52	Hf	3	5	Me
6409.4	Te I	–	[5]	Rd
6409.109	Mo	25	4	–
6408.598	Th	4	–	–
6408.555	Gd	2	–	–
6408.473	Sr	50	20	–
6408.458	Co I	3 h	–	–
6408.423	Er	4	–	–
6408.4	bh Sc	2	–	Me
6408.13	S I	–	[5]	Ms
6408.042	Sm II	10 d	–	–
6408.029	Fe I	50	30 h	–
6407.7	bh F	10	–	L
6407.606	Nd	5	–	–
6406.997	Zr I	18	–	–
6406.966	Mo	3	2	–
6406.7	bh F	20	–	L
6406.462	Mo	8	2	–
6406.24	Sm II	30 d	–	Kn
6406.16	Re	20 W	–	Me
6406.110	Eu	100	–	–
6406.08	Sb	–	[6]	Lg
6405.97	Tb	6	–	Ed
6405.95	As II	–	10	Ro
6405.9	Te I	–	[18]	Rd
6405.87	I I	–	[5]	Bl
6405.8	bh F	10	–	L
6405.6	bh Yt	5	–	Me
6405.54	Er	4	–	Ed
6405.406	Sb	4	–	Wt
6405.15	F	–	[4]	Gl
6404.9	bh F	5	–	L
6404.69	Kr	–	[3 wh]	Me
6404.62	Yb	2 h	–	Me
6404.618	Sm II	2 h	–	–
6404.53	Se I	–	[15]	Ms
6404.485	U	3	–	–
6404.395	Pr	4 w	–	–
6404.30	Zr I	4	–	–
6404.204	W	25	2	–
6404.117	Sm	5	–	–
6404.0	bh F	5	–	L
6403.98	Sm	2	–	Kn
6403.885	Ir	3	–	–
6403.70	Cu II	–	5	Sh
6403.58	S I	–	[2]	Ms
6403.2	bh F	5	–	L
6403.196	Nd	3	–	–
6403.151	Sc	2 h	2	–
6403.150	Os	15	–	–
6403.15	Tb	4	–	Ed
6403.10	A	–	[2]	Rt
6403.01	Eu	3	–	Kn
6402.758	Nd	3	–	–
6402.4	bh F	5	–	L
6402.33	Rh I	3	–	–
6402.31	Dy	3	–	Ks
6402.31	Gd	3	–	Ks
6402.246	Ne I	–	[2000]	I
6402.23	Sm	2	–	Kn
6402.07	W	5	1	Me
6402.005	Yt I	12	7	–
6401.7	bh F	2	–	L
6401.45	Sm	10 d	–	Kn
6401.45	Tm	40	5	Me

Wave-length	Ele-ment	Intensities Arc	Spk.,[Dis.]	R
6401.295	Mo	3	–	–
6401.17	Se I	–	[15]	Ms
6401.076	Ne I	–	[100]	Ps
6401.070	Mo	20	6	–
6400.932	Eu	700 W	–	–
6400.9	bh F	2	–	L
6400.590	V	–	2	Me
6400.59	Cu	3	–	Az
6400.40	Yb	200 h	4 h	Me
6400.355	U	3	–	–
6400.318	Fe I	2	–	–
6400.018	Fe I	200	150 h	–
6399.99	Se I	–	[15]	Ms
6399.907	Ce	4	–	–
6399.86	Sm	2	–	Kn
6399.79	Er	4	–	Ed
6399.736	W	5	1	–
6399.6	bh F	2	–	L
6399.415	Sm	3	–	–
6399.41	Cl II	–	[10]	Ks
6399.23	A II	–	[8]	Rt
6399.053	La II	15	200	–
6399.0	bh F	2	–	L
6398.858	Pt	6	–	–
6398.857	Os	3	–	–
6398.752	V	–	2	Me
6398.64	Cl	–	[40]	Ks
6398.295	Sm	12	–	–
6398.259	Ir	4	–	–
6398.13	Er	6	–	Ed
6398.05	S	–	[300]	Bl
6397.996	Pr	12	–	–
6397.99	Xe II	–	[50]	Hu
6397.69	Se I	–	[15]	Rd
6397.346	V	–	2	Me
6397.30	S	–	[300]	Bl
6397.185	U	12	–	–
6396.876	Nd	3	–	–
6396.63	A	–	[2]	Rt
6396.61	Ga	–	20	Kl
6396.61	Dy	4	–	Ks
6396.54	S I	–	[15]	Fh
6396.524	Co I	10 h	–	–
6396.46	Te	–	[50]	Bl
6396.373	Sc	2 h	–	–
6396.244	Ce	6	–	–
6396.21	Sb	4 h	[2 h]	Lg
6396.0	bh Zr	4	–	L
6395.446	U	100	–	–
6395.427	Sm	2	–	–
6395.26	I	–	[30]	Bl
6395.195	Co I	125	–	–
6395.07	S I	–	[15]	Fh
6395.0	bh Cr	3	–	L
6394.967	Sm	2	–	–
6394.94	Hg II	–	[25]	Ps
6394.80	Nd	4	–	–
6394.7	bh F	10	–	L
6394.28	Kr II	–	[4 hs]	Me
6394.234	La I	150	–	–
6394.1	bh Zr	5	–	L
6393.605	Fe I	100	80 h	S
6393.275	V I	4	2	–
6393.191	Pr	25	–	–
6393.023	Ce	5	–	–
6392.781	U	20	–	–
6392.445	Sm	2	–	–
6392.209	Ta	15	–	–
6392.175	Sb	8	2 h	Wt
6392.103	Pr	10 W	–	–
6391.96	Se	–	[15]	Bt
6391.323	U	2	–	–
6391.215	Mn	3	–	Sl
6391.14	Kr II	–	[30]	Me
6391.118	Mo	12	4	–
6390.99	Hf	1	2	Me
6390.838	Sm II	100	–	–
6390.661	Dy	2	–	–
6390.484	La II	70	100	–
6390.321	Ce	8	–	–
6390.30	Hf	1	2	Me
6390.228	Ru	9	–	–
6390.19	Te	–	[15]	Bl
6389.997	Nd	15	–	–
6389.870	Sm II	100	–	–

Wave-length	Ele-ment	Intensities Arc	Spk.,[Dis.]	R
6389.804	U	18	–	–
6389.589	Pr	8 W	–	–
6389.595	Eu	9	–	–
6389.447	Ta	100	–	–
6389.111	Mo	15	4	–
6388.973	Mn	2	–	–
6388.94	I I	–	[30]	Bl
6388.91	Hf	1	2	Me
6388.38	Dy	2	–	Ks
6388.324	Sb	2 h	–	Wt
6388.239	Sr I	35	10	–
6388.20	Hf	1	2	Me
6388.194	Er	12	–	–
6388.068	Sm	3	–	–
6387.991	Ta	3	–	–
6387.972	V	–	2	Me
6387.8	bh Zr	4	–	L
6387.72	Yb	3	–	Me
6387.59	Ca	2 h	–	Ad
6387.56	Dy	2	–	Ks
6387.1	bh Yt	10	–	Me
6387.07	Yt	8	5	–
6386.94	Cs	25 l	–	Me
6386.864	Ce	12 w	–	–
6386.81	Dy	4	–	Ks
6386.768	Sm	10	–	–
6386.69	Co I	5 h	–	–
6386.56	Rh I	4 h	–	Me
6386.501	Sr I	35	10	–
6386.48	S	–	[5]	Bt
6386.23	Hf	15	20	Me
6386.102	Ce	5	–	–
6385.196	Nd	100	–	–
6384.915	U	2	–	–
6384.89	S	–	[300]	Bl
6384.825	Sm	2	–	–
6384.739	W	3	1	–
6384.719	A I	–	[100]	Ms
6384.697	Ni I	5 h	–	–
6384.669	Mn	25	–	–
6384.633	Nd	6	–	–
6384.6	bh Zr	4	–	L
6384.487	Co	2 h	–	–
6384.303	Sm	2	–	–
6384.13	Cl II	–	[5]	Ks
6384.04	Er	4	–	Ed
6383.861	Eu	350	–	–
6383.731	Mo	3	2	–
6383.591	U	8	–	–
6383.34	Hg	–	[15]	Lf
6382.9914	Ne I	–	[1000]	S
6382.944	Re	15 W	–	–
6382.93	Yb	1	2	Me
6382.741	Eu	200	–	–
6382.487	W	3	1	–
6382.188	Gd	60	–	–
6382.169	Mn	20	–	–
6382.069	Nd	20	–	–
6381.416	Ti I	10	–	–
6381.262	V I	2	–	–
6380.974	Gd	100	–	–
6380.747	Fe	25 h	8 h	–
6380.746	Sr I	30	8	–
6380.709	Sm	3	–	–
6380.45	Rn I	–	[12]	Rs
6380.19	Hf	3	6	Me
6380.115	V	–	20	Me
6380.045	Sm	3	–	–
6379.75	Ce	2	–	Ks
6379.636	U	15	–	–
6379.63	N II	–	[70]	Fl
6379.364	V I	8	2	–
6379.069	Ta	8	–	–
6378.956	Mn	20	–	–
6378.91	Ba II	3	[5]	Rs
6378.824	Sc I	8	15	–
6378.80	I	–	[30]	Ev
6378.623	Pr	8 W	–	–
6378.32	Tl II	–	[10]	El
6378.3	bh Zr	15	–	L
6378.263	Ni I	20 h	–	–
6378.075	Sm	2	–	–
6377.84	Cu II	–	20	Sh
6377.72	Dy	3	–	Ks
6377.617	Pr	5	–	–

II. IDENTIFYING THE FEATURES IN THE SOLAR SPECTRUM

Determine the wavelengths of all the easily identifiable absorption features in the solar spectrum given in this unit. Following the solar spectrum is table 1, a standard reference list of the absorption features of all of the chemical elements which you can use in your attempt to identify the features you will measure. Such a reference is constructed from careful investigations of the emission and absorption spectra of each of the elements in the periodic table, studied one at a time in a laboratory.

As you do your identifications, you may run into situations where several possible features are near your measured wavelength and you have to decide between them. There are two criteria you can use in making such decisions. One is to consult the intensity column in the tables and try and judge whether the line is of the proper strength. This can be tricky, because the intensity column was tabulated from laboratory sources and not from stars; it can only be used as a rough guide.

You can also use the following commonsense principle to guide your identifications. If a certain feature is indeed present in the spectrum, then other features of the same or greater intensity due to that SAME ELEMENT should also be present. Checks of this kind, in combination with the information in the following two paragraphs, should usually allow your identification to be definitive.

Making line identifications will be aided by keeping in mind the data that have been gathered by astronomers concerning the relative abundances of elements in the universe. Studies of stars and interstellar gases throughout our galaxy and in other galaxies have demonstrated the presence of a rather uniform "cosmic abundance" throughout the universe. That is, most objects that have been studied prove to be close to 90 percent hydrogen, around 10 percent helium, and only approximately 1 percent EVERYTHING ELSE. In other words, 9 out of every 10 atoms in the universe are found to be hydrogen atoms; the rest are mostly helium and only about 1 out of every 100 atoms is anything else (lithium, beryllium, boron, carbon, nitrogen, oxygen, iron, etc.). This situation is interpreted as follows: in the "beginning," the universe was all hydrogen (with perhaps some helium -- depending upon whose model you subscribe to). As time passes and stars form and evolve, however, heavier elements are created inside stars by nuclear reactions going on at the core of the star. For example, the sun is powered by a chain of reactions called the proton-proton cycle, which takes four hydrogen nuclei and creates from them one helium nucleus. This is the process of nuclear FUSION. These reactions are the source of the energy of the sun.

Such fusion reactions in general combine light elements (i.e., low mass elements) into heavier ones; for example, some evolving stars arrive in a situation where they can take three helium nuclei and fuse them into a carbon nucleus, thus creating carbon atoms. The evolution of stars is continually creating heavy elements out of lighter ones and their abundance is growing slowly with time in the evolution of the galaxy. In general, the heavier the element the rarer it will be, since the lighter elements out of which a heavy element could be synthesized must themselves be created by the fusion of even lighter elements yet. Astronomers reason fairly generally that the abundance of an element declines with increasing atomic number (or weight); gold and praseodymium are far rarer than carbon, nitrogen, and oxygen. Note, however, that this is not a hard and fast rule, because there are many complicated twists and turns in the total picture of how the chemical elements are created (for example, lithium, beryllium, and boron are rare even though their atomic numbers are low). For the purposes of this project, a useful guideline is: the greater the atomic weight of the element, the less abundant (on the average) it is in the cosmos.

A note on the nomenclature: a neutral atom has sufficient electrons to balance the positive charge of the nucleus, giving overall neutrality; such elements are spectroscopically designated by a roman numeral I. If one electron is removed from carbon, the ion thus created is called C+, since it has a net charge of +1; it is also designated as C II. Similarly, carbon with two electrons removed is designated C III.

When your identifications are finished, check them with the first and third columns of Table 2. Then proceed to section III.

III. MEASURING SOLAR LINE STRENGTHS

Using the solar intensity plots, measure the CENTRAL LINE DEPTH as a percentage of the neighboring continuum strength and measure the EQUIVALENT WIDTH of the following lines: Fe I, 6408.0 A; Fe I, 6411.7 A; Fe II 6416.9 A; Fe I, 6420.0 A; Fe I, 6430.9 A; and Fe II, 6432.7 A. Figure 6 illustrates what is meant by equivalent width and central line depth.

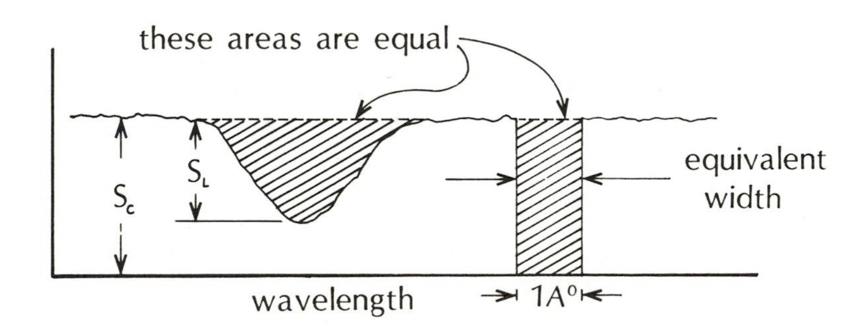

Figure 6

$$\text{Central Line Depth} = \frac{S_L}{S_C} \times 100 \qquad \text{(in percent)}$$

The equivalent width (E.W.) of a line is the width of the RECTANGULAR ABSORPTION LINE of central depth = 100 percent that has the same area as the observed line. The equivalent width is given in Angstroms or milli-Angstroms (1 mA = 0.001 A). The easiest way to determine the E.W. of the above lines is to trace them on transparent graph paper and count the little squares, obtaining, say, M for an answer. Then count the squares in a rectangle of width equal to 1 A in the continuous emission region adjacent to the line, obtaining, say, N. The equivalent width of the line is then M/N. Now plot the equivalent widths of the Fe I lines versus the central line depth values. Are the quantities correlated? Does a straight line fit the points? What use can be made of this correlation?

Now compare your equivalent widths for these lines with the published values. The correct values for the equivalent widths of the lines are listed in Table 2, taken from "Revised Solar Spectrum" by Charlotte Moore. Note that it gives equivalent widths in milliAngstroms. Unless your values are within 10 percent of the accepted values, you should remeasure.

Table 2. Solar Equivalent Widths

6400.009	181	FeI	6414.594	15	NiI	6433.452	15	
6400.323	46	FeI	6414.987	45	SiI	6433.737	5	
6405.45	6	CN	6415.424	5	CN	6436.413	7	FeI
6405.763	13		6416.928	47	FeII	6437.698	6	EuII
6407.113	6	SiI	6417.685	9	CaI	6438.773	7	FeI
6407.291	26	SiI FeII	6419.650	9	FeI	6439.083	156	CaI
6408.026	80	FeI	6419.956	80	FeI	6446.400	5	FeII
6408.375	5	CN	6421.360	87	FeI	6449.127	34	
6408.682	8	H_2O	6421.526	16	NiI	6449.820	98	CaI
6410.926	6		6424.862	11	H_2O NiI	6450.179	24	CoI
6411.113	6	FeI	6430.856	106	FeI	6450.325	20	
6411.658	129	FeI	6432.683	38	FeI			

IV. DIFFERENCES IN STELLAR SPECTRA

A. What Determines the Wavelengths of the Spectral Lines?

The wavelengths of the various features in the solar spectrum are determined by the atomic structure of each chemical element. Consider, for example, the various absorption lines of Fe I, neutral iron. Since an absorption feature in the spectrum is due to numerous iron atoms in the stellar atmosphere having absorbed energy at that wavelength, it is apparent that the individual iron atoms have moved from lower to higher energy levels in the process. The amount of energy absorbed by each such transition, and hence the wavelength, depends upon the difference in energy between the two energy levels of the iron atom. Iron has 26 positively charged protons in its nucleus (and 30 neutrons, for a total atomic weight of 56), and it has 26 orbital electrons for overall electrical neutrality. It is the complex interaction of the 26 positive charges with the 26 negative charges that creates the structure of the energy levels of the iron atom and hence determines the energies (and wavelengths) at which it can absorb.

B. Ionization in Stellar Spectra

You will have noticed that iron exists in the solar atmosphere in both a neutral and a singly ionized form (Fe II). The temperature in the outer parts of the sun is sufficiently high that some of the iron atoms have had one electron torn away, either by absorbing too much energy or by suffering an energetic collision with some other atom. The relative proportion of Fe I to Fe II is determined by the temperature. The higher the temperature, the more ionization occurs and the smaller the fraction of iron atoms that will be found existing as Fe I. In stars somewhat hotter than the sun, we find that Fe III and Fe IV also appear in the spectrum.

In the early part of the twentieth century, astronomers examining the appearance of stellar spectra were at first quite bewildered by the great diversity they exhibited. Even neutral iron has thousands of absorption lines, and when all the other elements and their various stages of ionization were considered, stellar spectra seemed incredibly complex. However, when atomic physics progressed to the point where the effects of temperature on spectra were understood (in the 1920s), astronomers soon found that this bewildering complexity reduced itself to an impressive simplicity. It turned out that almost all the stars had basically the same chemical composition -- by number of atoms, the proportions were about 85 to 90 percent hydrogen, approximately 10 percent helium, and just a few percent at most of all the other elements in the periodic table! The tremendous diversity shown by stellar spectra came to be understood as a straightforward manifestation of the differences in surface temperatures of the stars.

C. Temperature Effects in the Spectrum

Consider the very coolest of the normal stars, which have surface temperatures on the order of 3,000 degrees Kelvin. (The Kelvin temperature scale is simply Celsius plus 273 degrees. In this scale, there are no negative temperatures. A star with a temperature of 3,000 degrees on the Kelvin scale has its temperature denoted as 3,000 K.) If we examine the spectra of these stars, we find that they are laced with absorption features due to molecules like C , CH, and CN. This is because at such low temperatures, molecules can indeed form and survive in the stellar atmosphere, and molecules are such effective absorbers of radiation that they cover the spectrum with dark lines. In stars of higher temperature, high energy radiation and energetic collisions tear the molecules apart into component atoms. Thus, for those stars, we will see the absorption features due to those atoms rather than the now nonexistent molecules. As we go to stars of higher and higher temperature, we see an increasing ionization apparent in the spectrum. Stars of surface temperatures above 7,500 degrees or so exhibit mostly absorption lines due to ionized elements, and the hottest stars, with surface temperatures on the order of 30,000 degrees show absorption features due to ionized helium. (It requires very high temperatures to remove an electron from its tight bonding to the helium nucleus.) The diversity of spectral types is illustrated in figure 7.

D. Spectral Differences at the Same Temperature

While the gross differences between stellar spectra inform us about the surface temperatures of the stars, it is also possible to examine fine differences between stars of the same surface temperature and learn even more details about the stellar atmospheres. Reproduced on the following pages is a tracing of the spectrum of the star Beta Draconis, a giant star having the same surface temperature as the sun. These observations, graciously provided by Dr. Earle Luck of Louisiana State University, were obtained at the coude spectrograph of the University of Texas McDonald Observatory 82-inch Struve telescope.

Identify the lines in the Beta Draconis spectrum using the solar spectrum as a guide. Note that the wavelength scale in this tracing is not the same scale as in the solar tracing, so you will have to locate your place in the spectrum by recognizing PATTERNS in the line spacings. Label the lines you identify with element names and wavelengths, and measure the equivalent widths of the same Fe I and Fe II lines you studied in the solar spectrum.

Finally, calculate the ratio of the strength (equivalent widths) of the Fe II lines to the Fe I lines both for the sun and for Beta Draconis. Use several different pairs of Fe II and Fe I lines for this, and make a table of your results. How does the Fe II/Fe I ratio compare for the two stars?

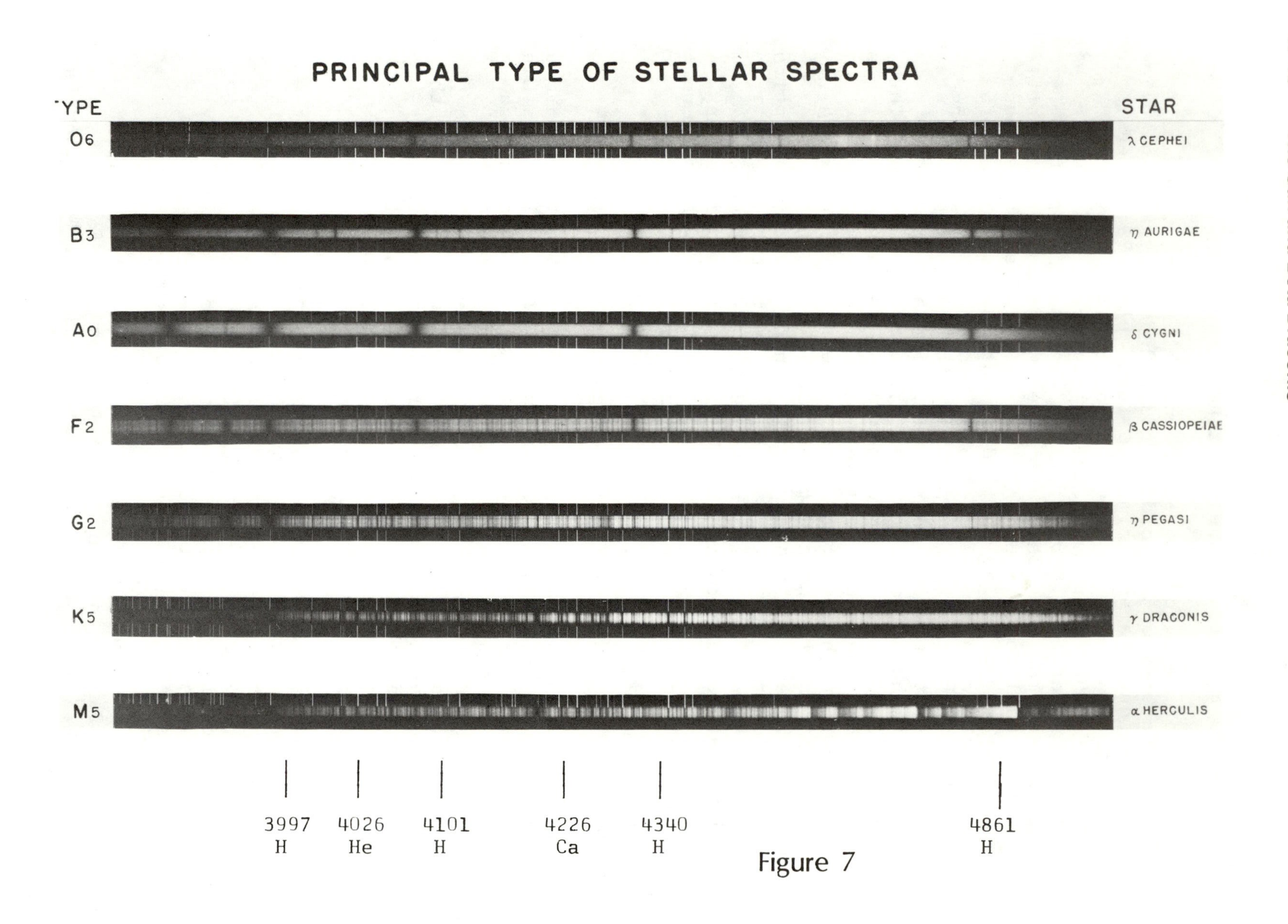
PRINCIPAL TYPE OF STELLAR SPECTRA
TYPE
STAR
O6
λ CEPHEI
B3
η AURIGAE
A0
δ CYGNI
F2
β CASSIOPEIAE
G2
η PEGASI
K5
γ DRACONIS
M5
α HERCULIS
3997 H
4026 He
4101 H
4226 Ca
4340 H
4861 H

Figure 7

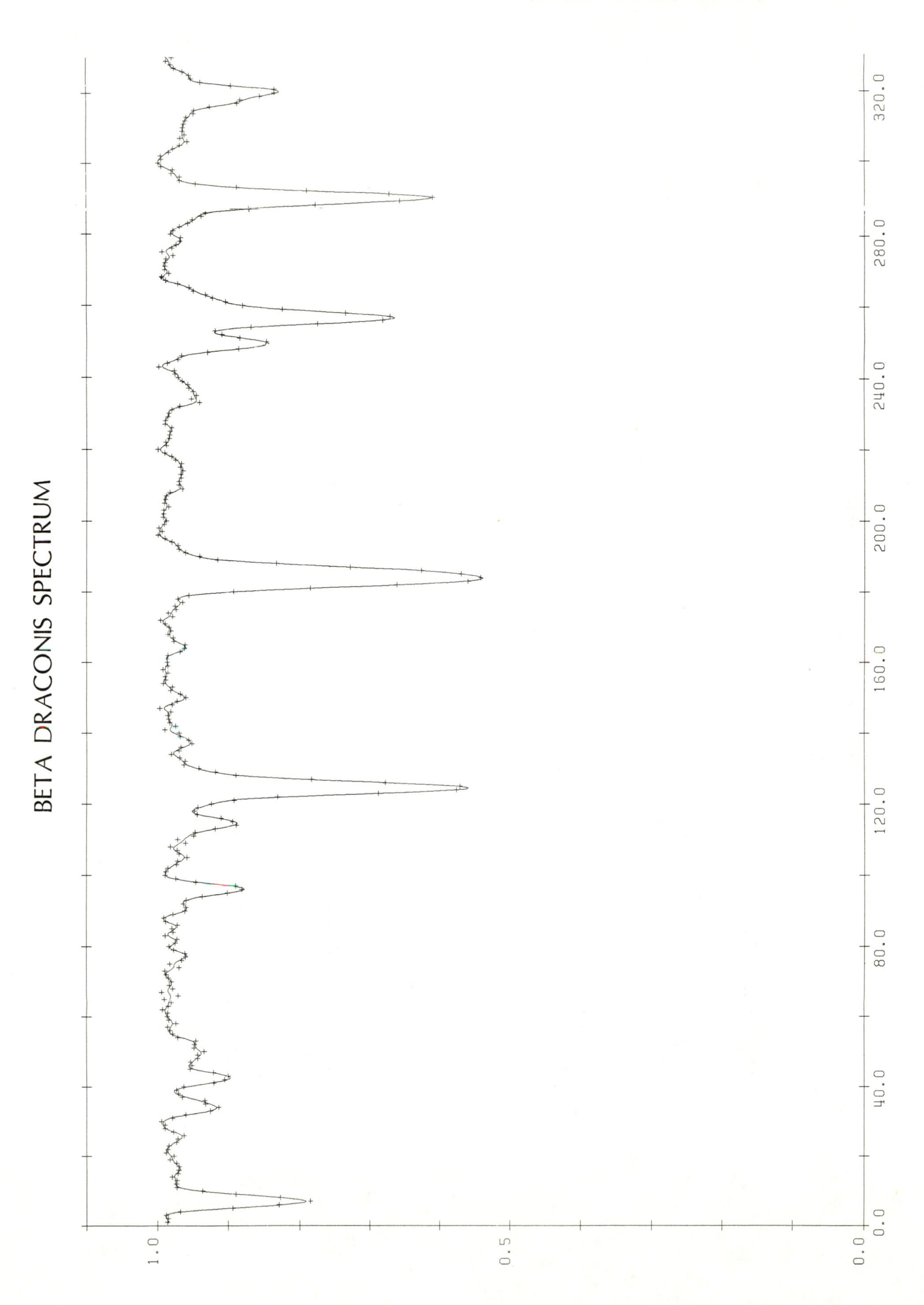
BETA DRACONIS SPECTRUM
1.0
0.5
0.0
0.0
40.0
80.0
120.0
160.0
200.0
240.0
280.0
320.0

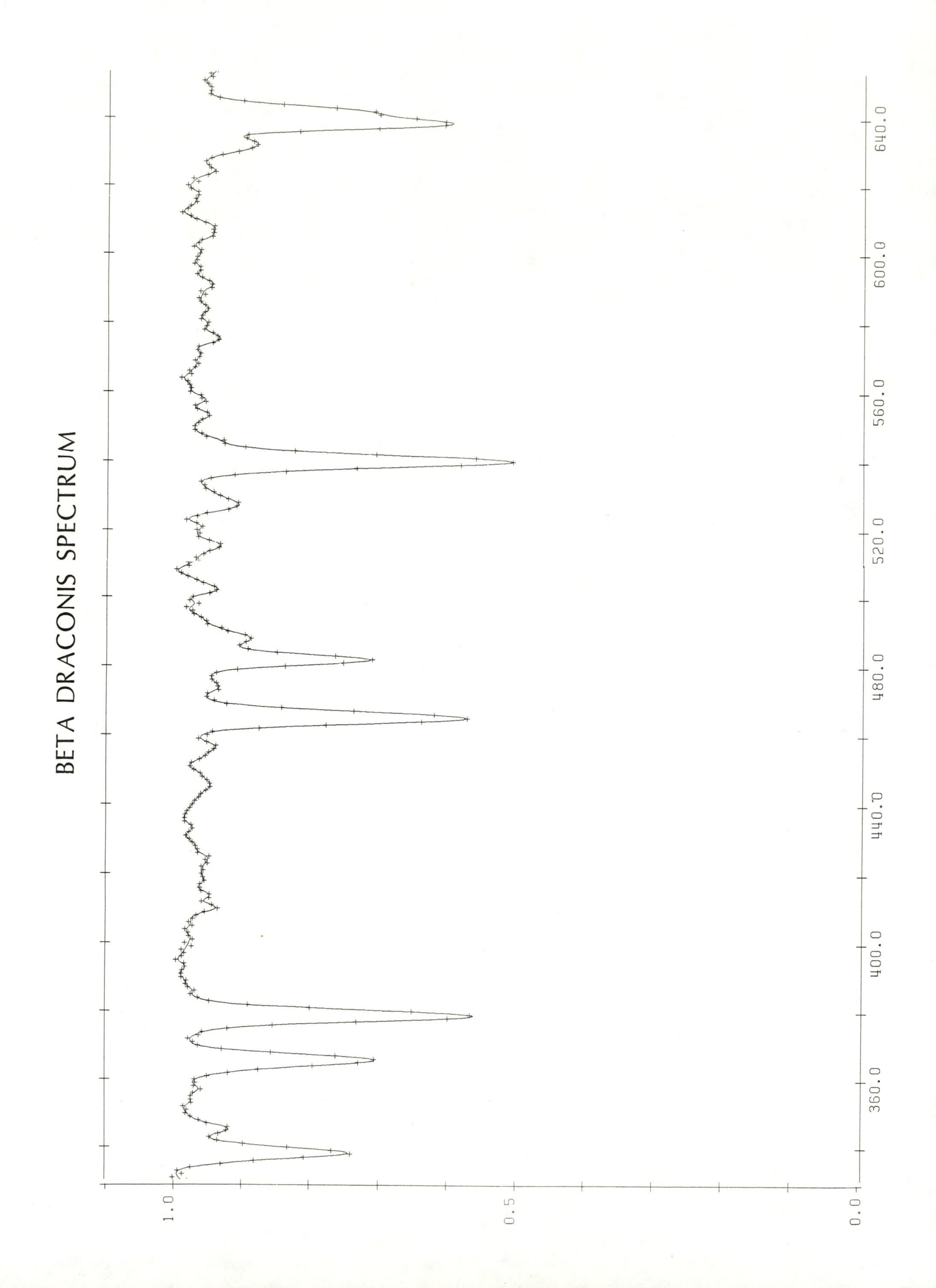
BETA DRACONIS SPECTRUM
360.0
400.0
440.0
480.0
520.0
560.0
600.0
640.0
1.0
0.5
0.0

V. EXPLANATION OF THE DIFFERENCES BETWEEN THE STARS

Was the ratio of EW(Fe II)/EW(Fe I) larger in Beta Draconis than in the sun, or smaller?

Consider your results in the light of the following explanation. Consider an atom of carbon in the outer atmosphere of a star. From the interior regions radiation floods out of the star. The carbon atom may absorb some of this radiation and acquire so much energy thereby that one of its orbital electrons is removed, leaving behind a carbon ion. This process is called PHOTO-IONIZATION, because the atom is ionized (an electron removed) by absorbing a photon of light. The carbon ion C , however, will have a tendency to recombine with free electrons it finds moving around in the stellar atmosphere because of the attraction of opposite charges. Clearly this tendency will depend upon the density, because the more packed the atmosphere, the more likely RECOMBINATION is.

The fraction of an element which is ionized thus depends upon a balance between factors. The temperature of the star determines how much radiation floods through and the tendency for electrons to be removed. The temperature and density determine the tendency to recombine. Thus, the ratio of C I/C (total), or C I/C II, or C III/C (total), or whatever, depends upon this interaction. The sun and Beta Draconis have the same temperature, so the ionizing tendency should be the same. But the giant star has a much more rarefied, extended, and tenuous (low density) atmosphere. As a consequence, recombination will be less effective and the ionization of the atom will be greater in the giant star. If we consider some ratio like C I/C II, this number will be larger in the sun and smaller in the giant. Is this what your measurements showed? If not, you should reexamine them for errors and, if necessary, remeasure until the proper results are obtained.

There is also another way in which the difference in atmospheric density between the sun and the giant star can be detected. In a rarefied giant star's atmosphere, the lower density of atoms means that the atoms suffer fewer collisions. They are able to absorb light at their characteristic wavelengths without much perturbation from neighboring atoms. But in the denser atmosphere of the sun, more frequent collisions cause disturbances in the energy levels of the atoms and instead of absorbing at one sharply defined wavelength, they absorb over a range of wavelengths. In spectral terms, this means that absorption lines in the giant star should be NARROWER than those in the sun. Unfortunately, the spectrum of Beta Draconis shown above was not obtained with sufficient resolution to show this effect. The line widths in that star are determined by the resolution of the spectrograph system, and they are larger than the actual widths in the stellar spectrum. (Note that this problem does not affect the values of EQUIVALENT widths, however.)

18 Celestial Photography

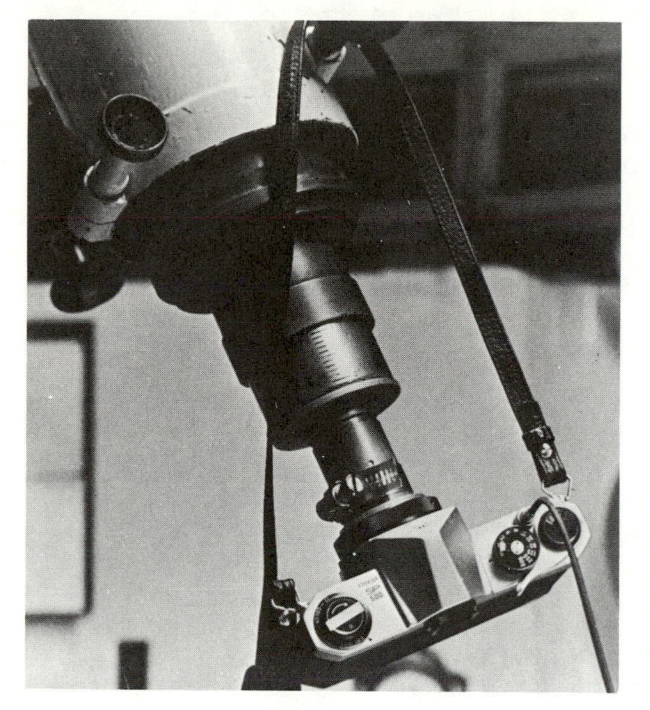

In addition to giving a permanent record, the camera's great advantage over the eye is its ability to take a time exposure. Since the telescope is the lens, an astronomical camera is sometimes quite simple—often no more than a light-tight box which holds a piece of film.

OBJECTIVES

1. to use a telescope with a camera attached
2. to produce satisfactory photographic prints of the moon, one planet, and one Messier object

EQUIPMENT NEEDED

A telescope with an equatorial drive, a camera, film (such as Tri-X or a special astronomical film like 103a-O), and an equipped darkroom.

For about 250 years after the invention of the telescope, astronomers used their naked eye and notebooks to record information. In the beginning of the nineteenth century, the photographic process had its birth. Sir John Herschel in 1839 was the first to make a photograph on a photosensitive glass plate. The first astronomical photograph was a daguerreotype of the moon taken in 1840 by John Draper. Although throughout the remainder of the century only the brightest astronomical objects could be photographed because of low sensitivity of the photographic emulsion and poor tracking of the telescope, the stability of glass emulsions marked their superiority over other mediums for astronomical uses. The convenience and accuracy possible with a permanent record of the sky justified the long exposure times -- sometimes over 60 hours for a single plate -- which were common at the close of the century.

The use of photography changed the character of telescopes also, as reflectors improved in design and increased in use over the refractor. In particular, reflectors are not troubled by chromatic aberration, the tendency of a lens to focus different colors of light at different distances behind the lens. Even today, a century after the invention of the "dry" plate, photography is an important tool for the professional astronomer.

In unit 19 you will use a plate holder and film or glass plates for astrophotography, but in this unit you will attach a conventional camera to a telescope and use 35-mm roll film. Conventional emulsions such as Plus-X or Tri-X may be used in this activity (see unit 9 for a description of these films), but astronomers have also devoted much effort to the development of special emulsions more suitable for the long exposures of astrophotography. Conventional film suffers from reciprocity failure. As explained in more detail in unit 19, reciprocity failure refers to the emulsion's failure to darken in direct proportion to the amount of incident light. If you can obtain an astronomical film, such as the 103a series, you should use it in this activity.

I. PHOTOGRAPHING ASTRONOMICAL OBJECTS

A. Selecting Objects to Photograph

Choose a set of objects to photograph by examining a standard star atlas such as Norton's, seeking relatively bright objects that will be high in the sky at your observing session. (Low altitude angles result in more atmospheric absorption and distortion.) For the placement of planets and the moon consult the Astronomical Almanac or a recent astronomical periodical. Note that near full moon, the sky is brightened considerably and it will be difficult to photograph fainter objects.

Norton's Atlas has a list of Messier objects (non-stellar celestial sources) given by their right ascension and declination, along with a brief description of their appearance. These objects are characterized by lower surface brightnesses than stars, and photographing them will be the most challenging part of this unit. The brighter Messier objects include: M42 - the Orion nebula, M31 - the Andromeda galaxy, M3 and M13 - globular clusters. Many of the open clusters listed are rather faint with widely scattered stars. For these objects you may wish to make a finding chart for use at the telescope, marking the object's position with respect to nearby bright stars.

B. Guiding

The mechanics of attaching the camera to the particular telescope you will use will depend upon the specific instruments available to you. Some telescopes will be equipped with a separate guidescope to facilitate astrophotography; this is usually a smaller telescope riding piggyback on the larger one. While you are photographing your object, you can follow visually how the telescope is tracking by looking continuously through the guidescope. Better guidescopes are equipped with illuminated reticles to facilitate setting on the object. If you are photographing a relatively faint object, you may wish to offset the guidescope until it is centered on a nearby brighter star and use that star for your tracking corrections. Should your telescope stray during your exposure, you can apply guide corrections through the slow motion drive controls until the object is recentered. If your telescope has no provision for a separate guidescope, or if the field of the guidescope is so large that tracking errors can not easily be detected, you will have to depend upon the drive mechanism tracking properly during the exposure. Note that some very successful pictures can also be taken by piggybacking a camera with a telephoto lens attached (200-mm to 400-mm focal lengths work well) to a larger telescope which serves as a guidescope.

C. Mounting the Camera

If you work at prime focus (see figure 1a), you will attach your camera body at the first focus of the objective lens or mirror. In this arrangement, the camera is simply a film holder which replaces the eyepiece. The real image formed by the objective is formed in the plane of the film. You view your object through the camera finder and adjust the position of the camera for sharpest focus.

If the proper adapters are present, it is possible to achieve a large image scale in the focal plane of the film by using an arrangement called eyepiece projection (figure 1b). An eyepiece is left in the optical path and the camera photographs the magnified image produced by the eyepiece. The light gathered by the telescope is now spread over a larger image size and exposure times will be much longer. The f-number for an eyepiece projection setup is given by

$$f = \frac{d\ f_o}{A\ f_e}$$

and may typically be ten times the f-number at prime focus. Recall that exposure time is proportional to $1/f^2$, what exposure time increase does this imply? Telescope tracking and guiding will be more critical in this situation, and the images on the camera viewfinder will be much fainter. What is the f-number of the arrangement you will use?

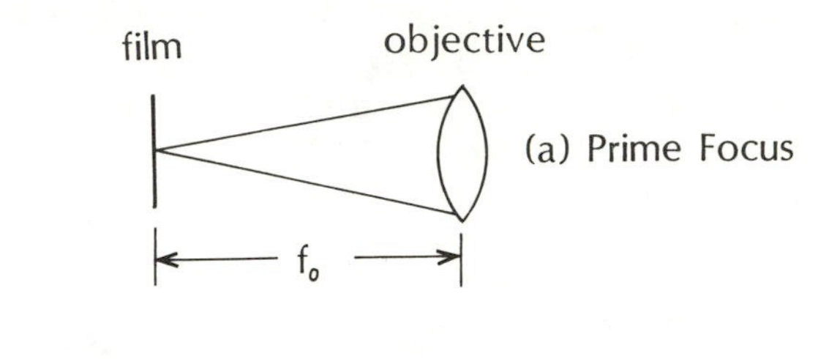

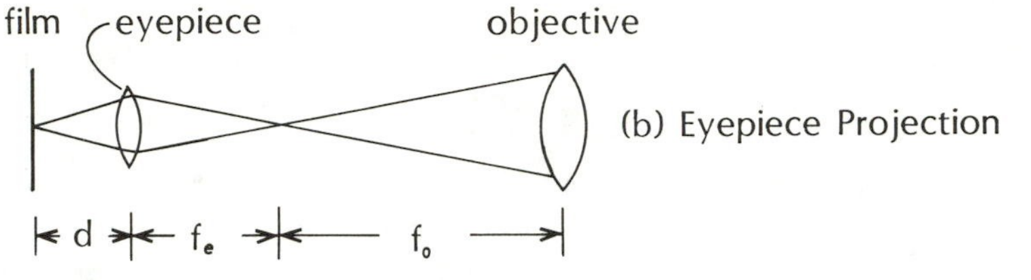

Figure 1

D. Exposure Times

Your exposure time will also depend upon the type of emulsion, brightness of the object, and local sky brightness. The best procedure to determine exposure times is to take a series of photographs at different exposures, over a great enough range that you are sure to have bracketed the proper exposure of your object. Thus, at a minimum, you will have two photographic sessions: the first to do a lot of bracketing in order to determine the proper exposures after developing your negatives (there will be no need to print the negatives from this session), and the second to take a number of exposures of each of your objects with times close to your calculated optimal time. Once you have determined proper exposure times, you can calculate the effect of variations. For example, using a finer grain film which is twice as slow would lengthen the exposure by a factor of two.

E. Additional Considerations

Note also the following camera considerations. You should use a locking cable release for time exposure (or else be prepared to stand there with your finger on the camera all through the duration of the exposure). A camera that releases the viewing mirror before activating the shutter will improve the quality of short exposures by minimizing camera shake. If you can use a camera with inter-changeable viewscreens, do so. A conventional ground glass screen works well for bright objects such as the moon, planets, and a few bright stars, but it will diffuse the light from fainter objects and make them difficult to see. A clear glass viewing screen will allow you to see several magnitudes fainter; this will be essential if you use eyepiece projection. If you do interchange viewscreens, you should focus on the ground glass screen and then substitute the clear glass screen with no change in the focal position of the camera since you cannot focus properly with the clear screen.

Estimate the scale of your telescope by measuring the size of the moon's image. (Remember the moon is 30 arcmin in diameter.) Use this image scale to determine the angular size of all your objects.

Remember to keep a complete and detailed log concerning every exposure. You cannot rely on memory when the photographs begin to accumulate.

For more information on celestial photography, refer to units 9 and 19.

II. PRINTING INSTRUCTIONS

1. Cut the paper.

Under only a red safe-light, cut the photographic paper into the desired sizes. (Typical sizes are 4X5, 5X7, 8X10, and 11X14 inches.) Replace the paper in a light-tight box.

2. Fill the developing trays.

Fill one tray with paper developer (usually DEKTOL), and another with STOP bath (or water), and a third with FIXER. The fixer is the same kind as used in film developing.

3. Place the negatives in the carrier.

Put the negatives in the carrier with the emulsion (dull) side toward the paper. Put the carrier in the enlarger. Be careful not to slide the strip of film through the carrier at any time, as this may scratch or damage the film.

4. Enlarge the image.

With the enlarger turned on, raise or lower the head until the image is the desired size. The image can be projected on a white easel.

5. Focus the image.

Looking through the focusing magnifier, adjust the focus control of the enlarger until the image is sharp.

6. Make a test strip.

Hold a mask of cardboard over 2/3 of the test strip of photographic paper, and expose the first wedge for 5 seconds. The shiny side of the paper should be facing up. Reset the timer, move the cardboard mask in order to expose the next 1/3 of the paper, also for 5 seconds. Repeat this for the paper completely uncovered. One third of your paper received 5 seconds of exposure, one third received 10 seconds, and one third received 15 seconds.

7. Develop the paper.
Place the exposed paper into the developer tray so that it is completely covered. Agitate the paper in the developer. When the sheet is immersed, gently grasp it on an edge and agitate it back and forth. Full development can be halted when the desired print has been reached. (You can judge this stage just by watching the picture develop, since your safe-light may remain on. Some papers may take as long as several minutes.)

8. Use a stop bath.
When the development has been completed place the paper in a stop bath for 1 to 2 minutes.

9. Fix the print.
Put the print into the fixer and agitate it for 3 to 4 minutes. At this time, IF ALL LIGHT SENSITIVE MATERIALS HAVE BEEN PUT AWAY, the light may be turned on and the print checked.

10. Choose the best exposure.
Choose the best exposure time. Expose and develop the actual print as you did the test print.

11. Rinse the finished prints.
Put the prints into a running water rinse for 20 minutes in order to remove any chemicals which might yellow or stain the finished prints.

12. Dry the prints.
Dry the prints in a blotter roll or with a mechanical dryer. RC papers may be hung to dry.

13. Clean up.
After any session in the darkroom, clean up after yourself very throughly. Put everything away and wash off everything, including the counters, trays, and sinks.

References

Astronomical Photography at the Telescope
Thomas Rackham
Faber and Faber:London 1972.

Tools of the Astronomer
G. R. Miczaika and William M. Sinton
Harvard University Press:Cambridge 1961.

19 Advanced Astronomical Photography

Two types of cameras employed by astronomers are shown here: a large-format Cassegrain camera that can hold emulsion-covered glass plates and a very fast wide-angle Schmidt camera that rides piggy-back on a telescope and carries out survey studies of large areas of sky.

OBJECTIVES

1. to determine the limiting magnitude of your emulsion-filter-telescope combination for an arbitrary exposure time
2. to describe the general characteristics of a photographic emulsion
3. to plan a session of astrophotography for a selected object with a given telescope
4. to produce photographic plates of at least six astronomical objects such as galaxies, nebulae, clusters, comets, minor planets, and planets
5. to complete one advanced observing project, such as those listed in section III of this unit

EQUIPMENT NEEDED

A large telescope (6-inch or larger objective), a camera designed for that telescope, several plate-holders, sheet film or glass plates, OR a piggy-back Schmidt camera with 35-mm film, and an equipped darkroom.

Prior to the midnineteenth century, astronomers relied on visual observations of celestial objects in order to study them. The introduction of photographic techniques into astronomy resulted in the sudden increase of information on a large variety of topics. The first astronomical daguerreotype was of the moon, taken in March 1840 by John Draper in New York. In July 1850, George Bond and John Whipple took the first photograph of a star using the Harvard University Observatory 15-inch refractor. Bond recognized that photographic plates represent a lasting record which contains stored information about many thousands of stars. Since the photograph can accumulate light over a period of time, it can make faint objects visible. Many new areas were opened for study. Although new electronic techniques are presently being developed, photography still holds its place as a major tool of astronomers.

In this unit you will use the telescope as a camera instead of fastening a standard camera to an eyepiece (as in unit 18). A plate-holder can be attached directly to the telescope at the focal plane. A camera situated at a Cassegrain focus consists mainly of a device to contain the plate-holder. The film or photographic plate is protected by a dark slide. A separate shutter on the telescope is desirable.

It is also possible to attach a separate telescope-camera combination piggy-back on another telescope used for guiding during the exposure. Schmidt cameras are frequently used for this due to their great speed. In a piggy-back Schmidt camera, the 35-mm film is curved within its holder to fit the curved focal plane. The frontespiece of this unit shows such an arrangement.

I. PRINCIPLES OF ASTRONOMICAL PHOTOGRAPHY

A. Exposure Time

In order to record the image of a celestial object, both a telescope and an image detector are needed. The telescope collects the light and brings it to a focus, producing an image of the object which is brighter and more detailed than the unaided eye can see. The image detector, such as the photographic emulsion or electronic device, then records the telescopic image. A permanent record is made. This allows the object to be studied in more detail later and to be compared with other objects or other records of itself. The "strength" or EXPOSURE of the recorded image is the product of both the apparent brightness of the image and the length of time for which the detector is allowed to build up (INTEGRATE) the image's light. In equation form, $E = I\ t$, where E is the exposure, I is the intensity of the image, and t is the time. The intensity of the image is in turn dependent upon both the intensity of the object and the amount of light collected by the telescope. For extended objects, the amount of light focused per unit area on the film is proportional to the square of the focal ratio. (Remember that the focal ratio equals focal length divided by aperture of the objective.) Telescopes with the same aperture size but different focal lengths will require different exposure times. A telescope whose focal ratio is f/10 requires four times as long to produce an image as a telescope of f/5. The image will be four times larger. For point sources, the exposure is proportional to the square of the aperture. If the detector is a photographic plate, we assume in these calculations that no reciprocity failure is present in the type of emulsion used. (Reciprocity failure is explained in section E.)

B. Spectral Sensitivity

Astronomers commonly use photographic materials produced by Eastman Kodak for low light levels. They are designated by a roman numeral, a lower case letter, and an upper case letter. The roman numeral designates emulsion type. This indicates the granularity, contrast, and resolution to be expected. For example, type II plates are coarser and have lower contrast than type III, but they are faster. The letter "a" on astronomical plates means these plates have special reciprocity effects (see below, section E). The upper case letter indicates spectral sensitivity. Type "O" is sensitive at wavelengths shorter than 5000 Angstroms, type "I" goes to 5500 Angstroms, and "D," "E," "F," and "N" go to successively longer wavelengths with "N" reaching to 9000 Angstroms.

Figure 1 shows the spectral sensitivity of several types of film. The developer recommended for astronomical emulsions is D-19 or D-76.

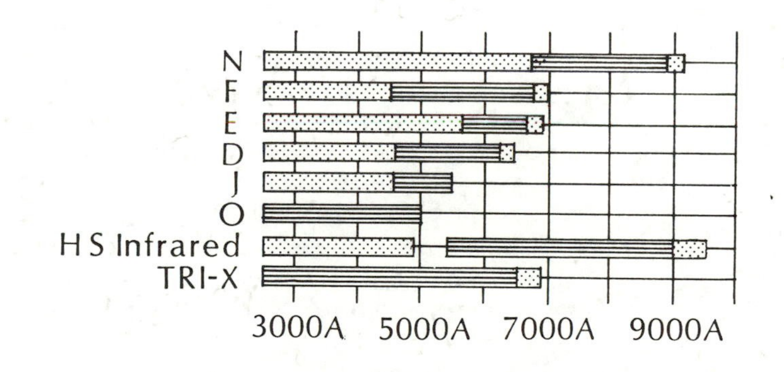

spectral region of useful sensitivity

especially valuable for this region

Figure 1 SPECTRAL SENSITIVITY

C. Plate Scale and Resolution

The plate scale of your telescope, measured in arcsec per millimeter, is 206.265 divided by the focal length in meters. The photographic plate has a resolving power measured in lines per millimeter. The plate used should have a resolving power greater than the plate scale divided by the seeing disk. For earth-based telescopes, overall resolution is usually limited by the atmosphere, not by the emulsion, since the typical seeing disk is usually one arcsec at best and frequently much larger. An astronomical plate may have a very high resolution, like IIIa-J with 200 lines/mm. For comparison, Tri-X plates have a low resolution of 50 lines/mm.

For the telescopes listed below, calculate the focal ratio, plate scale, and the minimum film resolution that should be employed with each. Which telescope is faster? By what factor? How wide would the image of the moon (1/2 degree) be on the film for each telescope? If the grain size is 0.01 mm in diameter, how many grains will the image cover on each film? (This is a measure of the "graininess" of the photograph since fewer grains produce a coarser image.)

Telescope: 40-cm reflector with a 447-cm focal length
14-cm Schmidt camera with a 22.9-cm focal length

The field of view will be the plate scale divided by the plate or film diameter. Will the moon's image fit on the 35-mm film (whose size is 24 mm by 36 mm) used for the Schmidt camera? Will it fit on 35-mm film used on the 40-cm telescope? Would a plate-holder for 4X5-inch film be more appropriate for the 40-cm telescope?

D. Characteristic Curve

The sensitometric properties of plates are often represented graphically by plotting photographic densities versus the exposures used to produce them. In equation form, we define density as equal to log (intensity of incident light/intensity of transmitted light). A plot of density versus log of exposure time, as shown in figure 2, is called a CHARACTERISTIC CURVE. The density is used because the eye judges brightness variations on an approximately logarithmic scale.

The characteristic curve demonstrates the manner in which plates or film repond to exposure and development. There are several distinct stages.

The FOG LEVEL, parallel to the horizontal axis in figure 2, is a stage in which density does not change with increasing exposure. The TOE is the non-linear portion of the curve (from A to B in the figure). Point A is the threshold exposure -- the minimum required to produce density above the fog. The STRAIGHT LINE is the linear midsection of the curve. The slope of this line is called gamma. Gamma is the usual indication for selecting proper exposure and development times. The value of gamma varies with emulsion. Films with high values of gamma produce stronger contrast. For astronomical purposes, if you are measuring magnitude as a function of density, you should be measuring within this straight line portion of the characteristic curve. The SHOULDER is the upper portion of the curve where the curve approaches a horizontal line. Exposure differences no longer produce density differences. Details will be difficult to separate in this region.

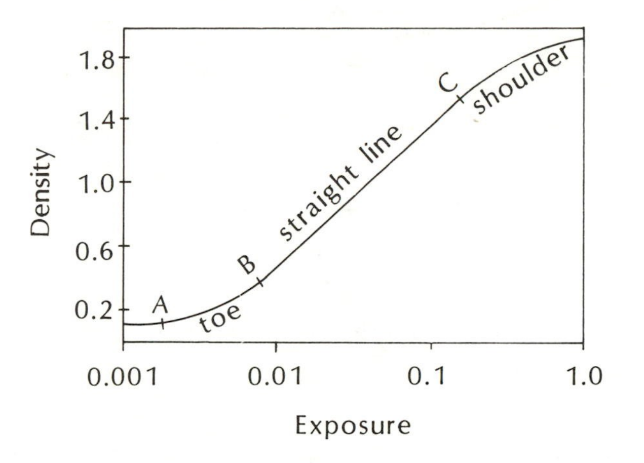

Figure 2 Characteristic Curve

E. Reciprocity Failure

Reciprocity failure refers to the fact that photographic grains have memory. With long exposures, the grains become less efficient. The characteristic curve changes for very long or very short exposures. (In astronomy, the problem is usually due to the length of the exposure, so the rest of this material will not consider extremely short exposures.) Figure 3 illustrates the reciprocity effect. The parallel, 45-degree lines on the plot are lines of constant time. A film with no reciprocity failure would appear as a line parallel to the horizontal axis. For such an emulsion, the amount of darkening that occurs would be proportional to the intensity of light. The upward turn at the left indicates that for longer exposure times the sensitivity of the material is decreased, so that more total exposure is needed.

Although reciprocity curves differ in shape for different products, each has a low point or low region at which the emulsion is most responsive to light. Astronomical films designed for low light levels, such as 103a-O, are designed to have a reciprocity curve flatter to the left of the low point.

As a comparison, consider the standard films such as Tri-X or Plus-X. These are designed for use at 1/1000 to 1/10 second exposure times. The exposure time adjustment needed to compensate for reciprocity effect is 50 seconds for an indicated exposure of 10 seconds and 1,200 seconds if the indicated exposure is 100 seconds.

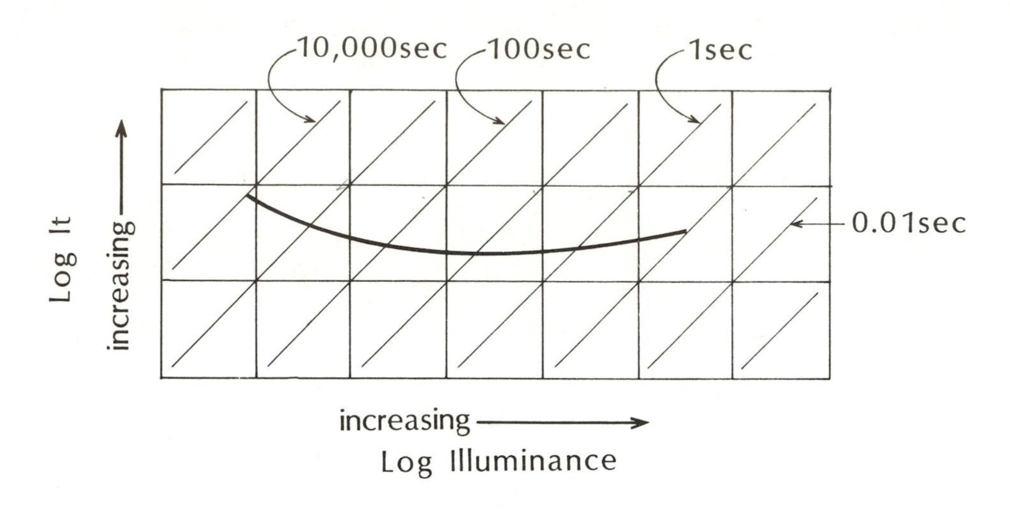

Figure 3 Reciprocity Curve

Several techniques can alter the reciprocity characteristics of an emulsion. The temperature and the environment in which it is exposed are important. Using a coldbox during exposure has gained favor in recent years. Hypersensitization of plates before exposure may be accomplished by controlled baking in a nitrogen or hydrogen atmosphere of the plate (24 hours at 50 degrees Celsius reduces exposure time by a factor of two or three). A technique particularly useful when working with red-sensitive or infrared plates is bathing the plate in an aqueous solution, such as ammonia. Occasionally plates are preexposed to a uniform level, so that latent images will emerge more rapidly.

II. OBSERVING PROCEDURE

A. System Considerations

The above information can be used to determine the optimum system to use in photographing a desired object. If you plan to photograph an extended object, a plate scale different from that used for a compact object may be desired. The brightness of the object may determine which telescope you use (if you have a choice).

Taking several exposures of a field will allow you to judge the exposure times needed for various magnitude objects. In other words, in a trial run, bracket your exposures. Look for the effects of sky brightness in your location, especially if it is not a dark site. There will be a limiting observable magnitude set by the brightness of the sky. Examine the spectral characteristics shown in figure 1. Consider the emissions from any local sources of unwanted light (the common types of streetlight include sodium and mercury). Can you choose a photographic material which minimizes the effects of streetlights? Using an appropriate filter-emulsion combination may lessen the problem.

B. Guiding

Obtaining an astronomical photograph through a telescope is simple in principle, but obtaining a good one is difficult because in practice, no telescope tracks perfectly as it follows a star. The goal is to obtain clean, unsmeared images on the emulsion. If, during the exposure, the image is allowed to move, a new image will begin to build up adjacent to the original one. The result is elongated or multiple images. (In the eye, if the image moves, the original is erased by the brain.) The photographic emulsion does not erase image movement.

One may prevent image movement by keeping the image of a guide star centered on the crosshairs of a reticle eyepiece. This may be done in two ways. Off-axis guiding allows you to use the main objective to form the image of your guide star. A small prism views a star which is outside the field of view to be photographed, and sends its image to the crosshair eyepiece. An off-axis guider is usually part of the camera assembly. In unit 18, where a standard camera was attached to the back of a telescope, the finder telescope was used to guide the exposure. A Schmidt camera can be piggy-backed to a large telescope and the large telescope used as the finder telescope. A reticle eyepiece should be used.

The technique of guiding is quite simple but requires practice to be done well. The desired astronomical object is located and centered. Either the off-axis guider is moved around the field of view and the telescope moved slightly, or the guide telescope is moved around until a star of reasonable brightness is found and centered on the crosshairs. If your crosshairs can be illuminated, you should adjust the illumination until the star can be seen centered on the dimly illuminated crosshairs.

Next, the loaded plate-holder should be inserted into the camera. After noting the time, pull the dark slide or open the shutter. If the image of the guide star begins to drift off the crosshairs, guide it back into position with the telescope controls. The task will be easier if the observer is comfortable and the eye unstrained.

After the emulsion has been exposed for the required length of time, the shutter or dark slide is closed, plate-holder removed, and time noted.

C. Finding Charts

Finding charts are extremely useful, particularly if you have a telescope which does not point accurately. The field size and magnitude range of the finding chart should depend on the characteristics of your telescope. In general, a field of 1 to 2 degrees is adequate. Useful finding charts may be made either photographically or by hand-copying the star field from an appropriate sky chart. Suitable charts include the National Geographic Society - Palomar Observatory Sky Survey, the Bonner

Durchmusterung charts, the Smithsonian Astrophysical Observatory charts, Norton's Star Atlas, the various Becvar atlases, or similar compendiums.

III. OBSERVING PROJECTS

A. General Advice

Everything you do at the telescope should be recorded in an observing log. This will aid you in relating your exposures to the star fields you are studying, as well as providing a record of any anomalies or problems which occurred during your observing session. Typical entries might include the date, the object, the telescope settings and instrument used, an exposure number, the start and stop times of exposure, the emulsion and any filters, the temperature and seeing conditions, and any remarks. A separate processing record should be kept. In one corner of the plate a brief pencil notation may be made of the exposure number.

Your first night at the telescope should be spent taking several exposures of a test field at various exposure times. This will allow you to practice using your telescope and provide you with an estimate of desired exposure times. Your test field should be one in which a large magnitude range is accurately known, such as an open cluster. Typical exposures could be 20 seconds, 1 minute, 3 minutes, 10 minutes, and 30 minutes. Changing to a different type of emulsion requires repeating your test exposures.

After developing the plates or film, use them to determine how faint an object you can observe within any given time and when the local sky brightness renders further exposure useless. Identify the star field and determine the magnitude of the faintest star on your plate. Plot this magnitude versus the exposure time.

Decide which objects to observe. Make a finding chart for each of them. After obtaining your plates or film, write a report on your findings. The sections below give a few ideas of observing projects beyond the initial goal of obtaining six exposures of a variety of astronomical objects.

B. Selected Galaxies

Your objective is to obtain two galaxy photographs, one of a spiral and one of an elliptical galaxy, of sufficient quality to be able to classify the galaxy in a standard category.

Galaxies are among the most interesting objects in the sky. Little detail can be seen with the naked eye or even with a moderate-size telescope. Photographs allow much more detail to be seen. Your first impression when looking at the galaxy photograph should be its overall shape and large-scale structure. Is it oval with no structural details (elliptical)? Does it possess spiral arms? Does it seem to fit neither category (irregular)? Don't let a casual glance answer these questions. If your photograph is underexposed, you may see only the brightest part of a spiral galaxy -- the nuclear region -- and mistake it for an elliptical. To properly classify a spiral galaxy you must gauge the relative proportions between the nucleus and the disk region and be able to judge the "openness" of the spiral arms. Can you detect any dust in your photograph?

Since regions of star formation show primarily the light from bright young blue stars, the spiral arms and blue knots will show up best on blue-sensitive emulsions. Likewise, since elliptical galaxies and nuclei of spiral galaxies are composed of older redder stars, they will be more prominent on red-sensitive emulsions. If you do not have two different emulsions available (such as 103a-O and 103a-E), you could use Tri-X with a blue filter and infrared film with a red filter.

The following list of galaxies should be considered only as a sample. They are among the brighter galaxies visible in the northern hemisphere and were chosen to provide a variety of types. Both Messier number and NGC number are given: M31 = NGC 224, M33 = NGC 598, M51 = NGC 5194, M81 = NGC 3031, NGC 6822 (fainter), M87 = NGC 4486, M32 = NGC 221, and M64 = NGC 4826.

C. Solar System Objects

Objects in the outer solar system have small angular diameters. Much of their motion is reflected earth motion. Their orbits are well known, and their positions are easy to obtain from the "Astronomical Almanac" or the yearly "Observer's Handbook" of the Royal Astronomical Society of Canada. For both of the projects suggested below, Uranus/Neptune or a minor planet, making a finding chart will be helpful. Plot the postions of the Smithsonian Astrophysical Observatory Catalog stars in the desired field, using the plate scale of your telescope. This plot will be useful when you reduce your data also.

Both the planets Uranus and Neptune have angular diameters of a few seconds of arc, giving them small image diameters. (Calculate the expected image diameter on the telescope you are using.) Unlike the nearer planets, which show perceptible disks, this image is small enough that it will probably be indistinguishable from nearby stars on a photographic plate.

In order to prove that these objects are planets, not stars, they must be shown to move with respect to the background stars. Since their periods are so long (84.01 years for Uranus and 164.8 years for Neptune), they should not produce a noticeable streak on a long exposure plate. However, they will shift positions against the background stars over a time scale as short as two weeks. You may wish to make a finding chart. Uranus, at 5.5 magnitude, or Neptune, at 7.85 magnitude, may be visible as a disk under high magnification if the seeing is good.

After obtaining plates on at least three different dates, as widely separated as possible, identify the planet. Choose at least three reference stars, and record its position relative to them.

Knowing the plate scale of the telescope, compute the amount of motion. Compare your computation with the motion obtained by using the data given in the "Astronomical Almanac."

Predicted positions for the four brightest minor planets (Ceres, Pallas, Juno, and Vesta) are listed in the "Astronomical Almanac" at two day intervals for equinox 1950. Take at least four plates on one night of one of these objects, and one plate on a later date. Measure the positions of the minor planet with respect to the reference stars in the first night's field, and determine its angular velocity. Compare this velocity with that obtained from published positions. Repeat this process for the minor planet motion between the two nights. (You will need to identify the reference stars in each field and obtain their positions.)

D. Open Clusters

Two-color photographs of an open cluster can be used to obtain a rough color-magnitude diagram for that cluster. That is, make a plot of the apparent magnitudes of all the cluster stars (use either blue or red magnitude) versus a color index for each star. Recall from unit 13 that the color index is a function of a star's temperature.

Use the techniques of unit 13 to estimate magnitude and color of the stellar images on your two photographs. Plot your magnitude function on your vertical axis and your color function on the horizontal axis. Examine this exercise for sources of error. Is there any way in which you could calibrate your photographs in order to obtain absolute rather than apparent magnitudes? If you could do this, could you obtain the distance to the cluster?

Note the following cautions to reflect on. You will want to choose a fairly bright (i.e., nearby) cluster so that you can also capture some of the fainter stars in it. Otherwise, your photograph will show only the most luminous (rare and unusual) stars in the cluster. It will not reveal the properties of the more common and more normal stars of lower mass (i.e., on the lower main sequence) whose properties are better known. Further, unless your cluster is a fairly populous one, your photograph will be contaminated by a high proportion of foreground (or background) stars, and you will not be successful in studying the unique properties of a cluster.

References

Stock, J. and Williams, A D "Photographic Photometry"
Astronomical Techniques
University of Chicago Press: Chicago, 1962.

A A S Photo-Bulletin

Kodak Plates and Films for Scientific Photography
Eastman Kodak Publication, 1973.

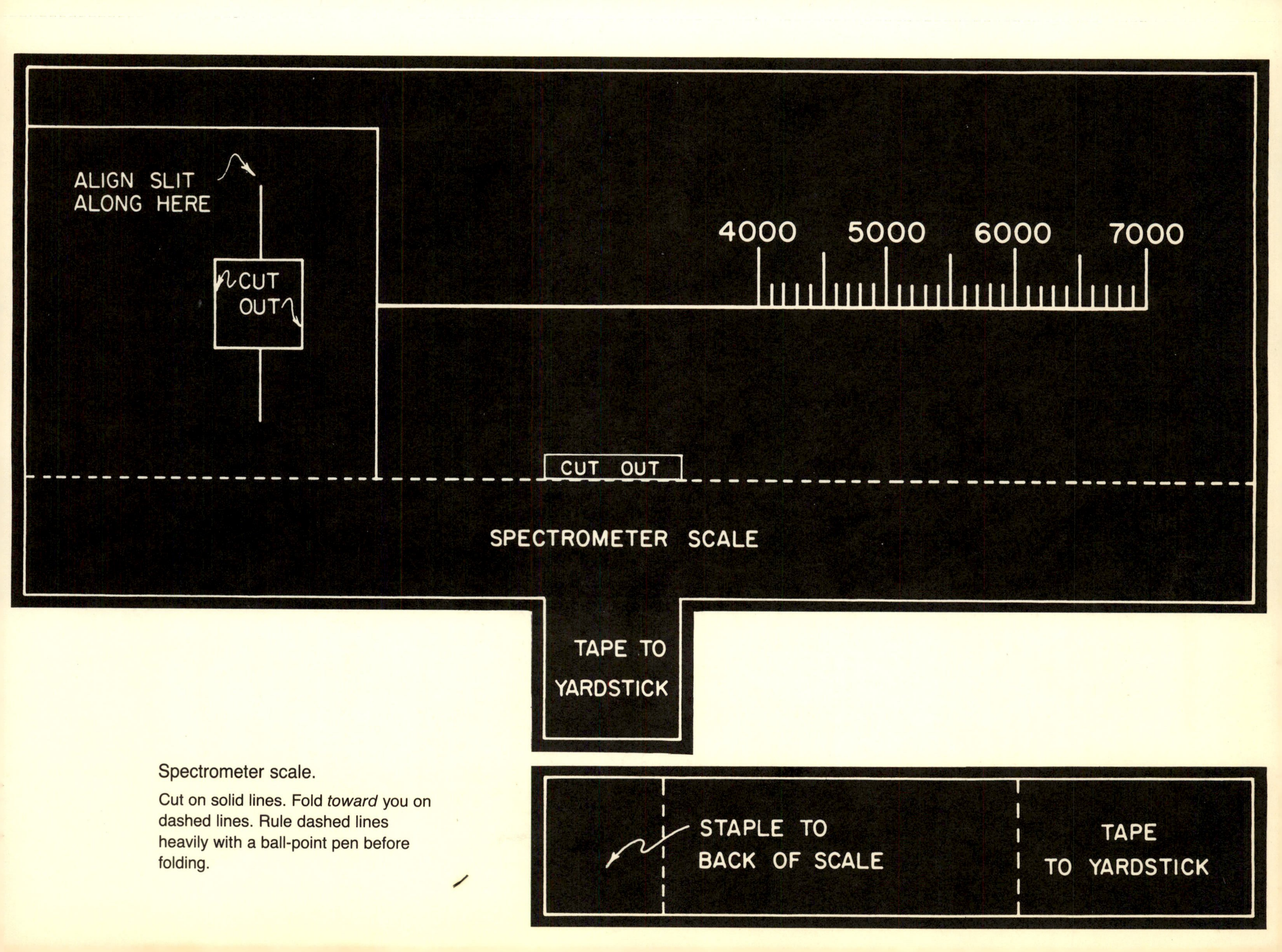

Spectrometer scale.

Cut on solid lines. Fold *toward* you on dashed lines. Rule dashed lines heavily with a ball-point pen before folding.

20

Astronomical Spectroscopy at the Telescope

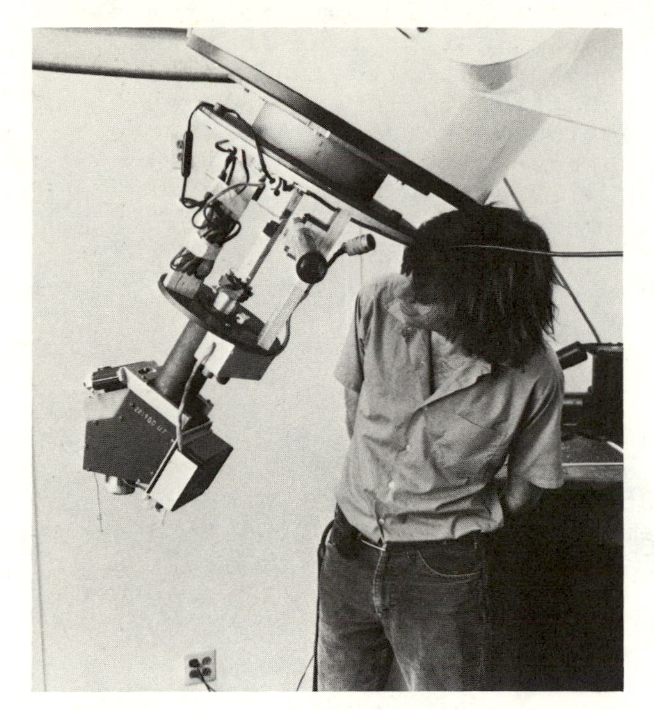

The spectrograph is an instrument which photographically records the dispersed light from celestial sources. It combines the advantages of spectroscopy and photography, and, since its invention, it has been the astronomer's most productive tool for extracting information from incoming radiation.

OBJECTIVES

1. to explain how light is dispersed into a spectrum by both prisms and gratings
2. to explain the function of the following parts of a grating spectrograph system: slit, decker, collimator, grating, camera, and detector
3. to compute the following properties of a grating spectrograph system: the angular dispersion, the demagnification factor, the image scale, the plate factor, and the spectral resolution
4. to operate a spectrograph, determining exposure times for stars of various magnitude
5. to photograph the spectra of at least 10 stars and to classify their spectra according to the criteria of the MKK system
6. to determine further fundamental properties of stars from their spectra, or, to obtain more spectra for a group of "unknown" stars

EQUIPMENT NEEDED

A spectrograph of moderate dispersion (100 to 200 A/mm) attached to a telescope of aperture 30 cm or larger, and an equipped darkroom.

The spectrograph is probably the most versatile and useful of astronomical instruments. Depending upon the sophistication and power of the system available, it is possible to study the following aspects of celestial bodies: their temperatures and atmospheric densities, their line-of-sight motions in space, their chemical compositions, and other detailed analyses of their gaseous structure and physical conditions. In unit 11 you studied the phenomenon of diffraction and observed the three basic types of spectra using a narrow slit and a hand-held spectrometer. In this unit, you will employ a research spectrograph system which is attached to a telescope (for increased light-gathering power) and which photographs the spectrum for a permanent record.

I. THEORY OF THE SPECTROGRAPH

The essential function of a spectrograph is to separate light into its component wavelengths and then record this spatial display of the energy emitted at each wavelength. A schematic diagram of a spectrograph system which uses a grating to disperse the light is given in figure 1; the individual parts are discussed in detail below.

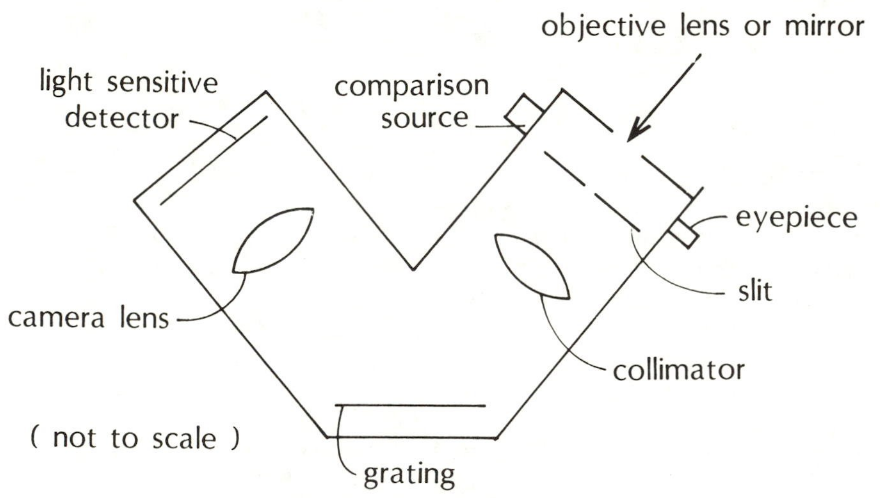

Figure 1

A. Dispersion of Light by Prisms and Gratings

Figure 2 illustrates the effect of a prism on a ray of white light passing through it. The bending of each wavelength depends upon the index of refraction of the prism's glass, and separation occurs because the index of refraction is different for different wavelengths. Notice that the shorter wavelengths are bent more than longer wavelengths.

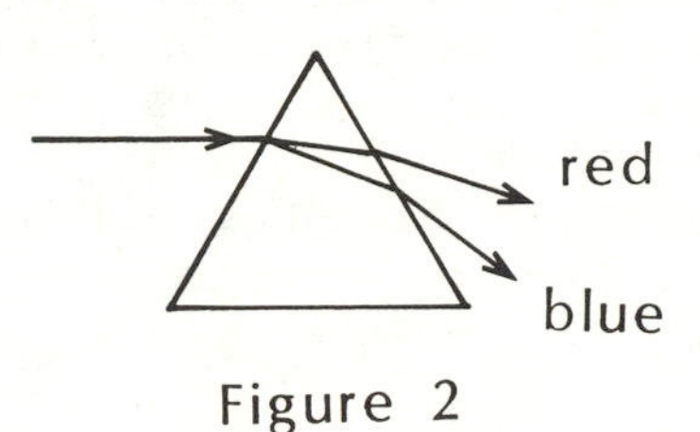

Figure 2

Figure 3 illustrates the wavelength dependence of the index of refraction for crown glass, a common prism material. The index of refraction of glass is temperature dependent, changing by a factor of 1/1000 for each 100-degree Celsius rise in temperature. As figure 3 illustrates, it is also nonlinear.

Because the variation in index of refraction with wavelength is fairly small, more than one dispersing element is usually required to get sufficient wavelength separation. An additional disadvantage of prisms is that glass absorbs short wavelength radiation. These problems, combined with the nonlinearity of the dispersion, are troublesome enough that prisms are rarely used in high-dispersion spectrograph systems.

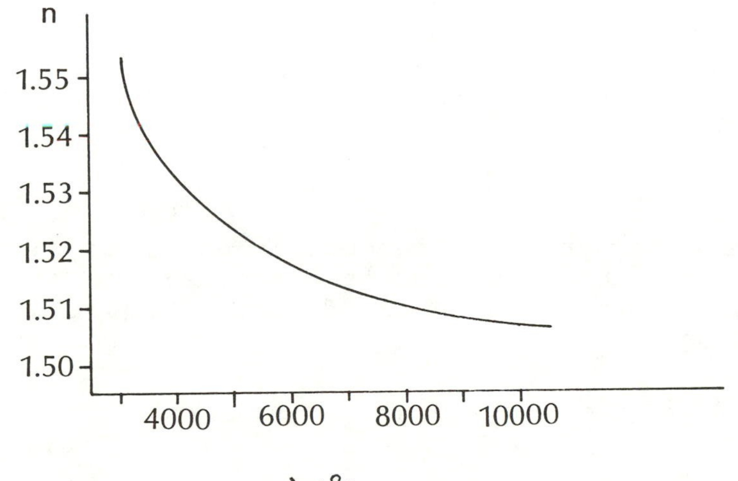

Figure 3

A grating uses the geometry of the diffraction phenomenon for wavelength separation. As seen below, this will give a linear dispersion with wavelength. Diffraction gratings are generally used in a reflection mode, so there are no absorptive losses. Figure 4 illustrates a close-up of two facets ruled on the reflective surface of a grating, with important angles marked. The dashed lines are normals to the grating and to a facet of the grating.

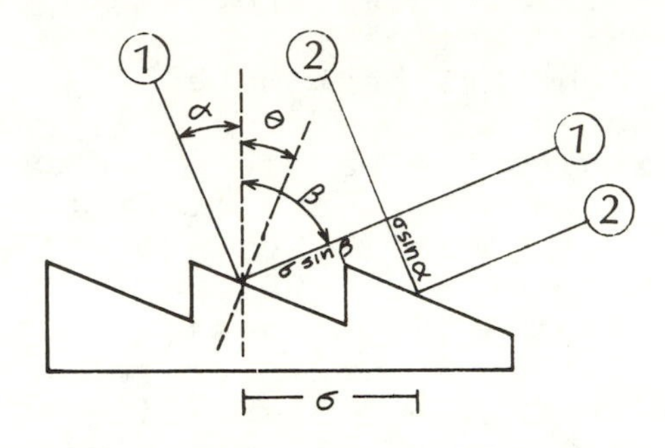

Figure 4

Alpha (α) is the angle between the grating normal and the incident beam of light. Beta (β) is the angle between the grating normal and the diffracted beam. Sigma (σ) is the distance between grooves.

From inspection of figure 4, it is apparent that ray 1 will have a path length delay of $\sigma \sin\beta$ with respect to ray 2 and that ray 2 will have a path length delay of $\sigma \sin\alpha$ relative to ray 1. This leads to a total path length delay, p, of ray 1 relative to ray 2 of

$$p = \sigma (\sin\beta - \sin\alpha)$$

When this path length difference is equal to an integral number of wavelengths, constructive interference results. This condition can be expressed in the following equation

$$m\lambda = \sigma (\sin\beta - \sin\alpha)$$

where m is zero or a positive integer. One can solve this equation for beta as a function of lambda and differentiate to get the DISPERSION RELATION:

$$\sin\beta = \frac{m\lambda}{\sigma} + \sin\alpha$$

$$\cos\beta \, d\beta = \frac{m}{\sigma} \, d\lambda$$

$$\frac{d\beta}{d\lambda} = \frac{m}{\sigma \cos\beta}$$

This equation relates the angular dispersion of the spectrograph to the groove spacing sigma and order number (m). Because m can be equal to 0, 1, 2, etc., it can be seen that a grating produces not one spectrum but a whole set of spectra, called zero order, first order, second order, etc., in different directions in space. As a consequence of this, the same wavelength will be imaged at many different places on the spectrograph detector, and different orders of spectra may overlap each other. For example, 6000 A radiation in first order will appear at the same spot as 3000 A second-order radiation. A properly designed spectrograph system will include filters that absorb certain wavelength ranges and assure that in the final photographed spectrum, each position on the film corresponds to only one wavelength of light.

Diffraction gratings are commonly BLAZED (that is, the grooves tilted on the grating), so that the reflection is more efficient near some particular value of beta (which corresponds to some set of wavelengths -- one wavelength for each order of the spectrum). The first-order wavelength corresponding to this value of beta is called the blaze wavelength.

The relation between the blaze wavelength and the angle at which the grooves are tilted is given by

$$\lambda_B = 2\sigma \sin\theta_B \cos(\delta/2)$$

where sigma (σ) is the groove spacing, λ_B is the blaze wavelength, and $\delta = (\alpha + \beta)$ is the angle between the incident and reflected rays (see figure 4).

B. The Slit and Decker

In unit 11, you observed the sharpening of resolution and detail in the spectrum when the entrance slit of the spectrometer was narrowed down. The slit (and decker) serves the function of shaping the entering light into a form convenient for dispersing and imaging by the system. They do this by controlling the directions from which light is allowed to fall on the grating. The width of the slit is in the direction of the grating dispersion and controls the spectral resolution (and also the amount of light entering the system).

The decker is an adjustable piece of metal which controls the length of the slit. If the decker is set to allow a longer slit length, then light falls on various portions of the detector (in the direction perpendicular to the grating dispersion), giving more total exposure on the detector and averaging over any inhomogeneities in the response of the detector to light. Figure 5 shows a slit-decker arrangement acting in combination to shape the light coming from the seeing disk of a star in the focal plane of the telescope.

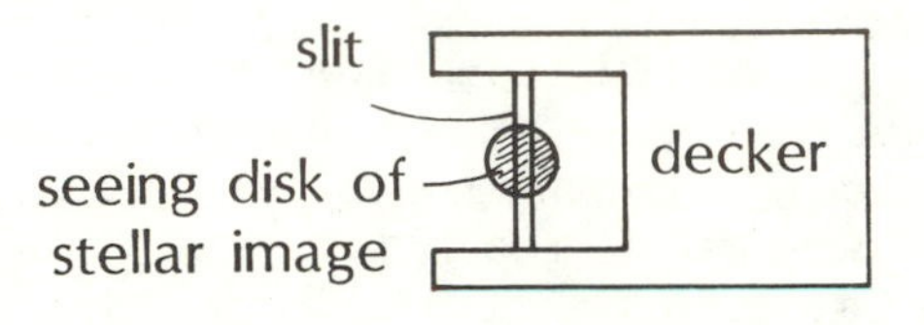

Figure 5

C. The Collimator

The collimator converts the converging beam from the telescope into a parallel beam of light falling on the grating. Unless the light reaching the grating is in parallel rays (i.e., collimated), the grating cannot function properly to disperse different wavelengths in uniquely different directions.

D. The Camera

The camera takes monochromatic collimated beams of light from the grating and images them onto the detector. The detector is frequently emulsion (usually a glass plate in larger systems), but may also be a photodiode array in some modern systems.

The spectrographic design discussed to this point is the most straightforward mode of system construction and the one most commonly used. It should be noted, however, that clever optical designs can be used to combine some of these functions. For example, a concave grating can be used, which combines the function of grating and camera. It is also possible to use a spectrographic system without a slit at all and allow the stellar seeing disk to be imaged through the system after being dispersed by a prism. This mode of observation is called Objective Prism Spectroscopy. Instead of a photograph of a region of the sky, here one obtains a photographic plate covered with spectra of stars (which may sometimes overlap for crowded star fields).

E. Detectors

For many years, the standard detector at the focus of the spectrograph camera was a photographic emulsion. As units 18 and 19 have demonstrated, the photographic process is versatile and capable, but the response of an emulsion to light falling on it is highly nonlinear and must be carefully calibrated. For precise work, each individual plate must be calibrated, since different development times and batches of films will lead to different responses. Since photographic information leaves the telescope in an analogue form, it must be somehow digitized (i.e., converted to numbers) for data reduction purposes. For very low levels of illumination on the emulsion, it is very difficult to carry out all of these processes with high accuracy. Nevertheless, for many purposes, careful work can obtain very usable information with the photographic process, and the construction and use of a photographic detector require a minimum of technical expertise.

In recent times, electronic detectors have been increasingly in use as the sensing element at the end of the light path. Such detectors have highly linear response curves to the incident light flux (i.e., an increase in light intensity of 10,000 leads to an increase in detector signal of 10,000), and this linear response obtains over a much larger range in signal than a photographic emulsion can typically handle. Further, the data leave the telescope in digitized, machine-readable form, ready for further data processing. However, electronic detectors are expensive and require a certain amount of electronic, cryogenic, and computer programming expertise in order to use them proficiently.

II. SOME USEFUL FORMULAE CHARACTERIZING SPECTROGRAPHIC SYSTEMS

A. The Demagnification Factor, M

M relates the projected slit size at the detector to the physical slit size and is related to these two quantities by

$$w_{detector} = w / M$$

where w is the physical (actual) slit width and w_d the projected width. The demagnification factor is given by

$$M = \frac{f_{coll} \cos \beta}{f_{cam} \cos \alpha}$$

where f_{cam} is the camera focal-length and f_{coll} is the collimator focal-length, and α and β have the same definitions as in the discussion in part I-A above.

B. The Image Scale, S

S is given by

$$S \text{ (arcsec/mm)} = 206265 / f\text{(mm)}$$

where f is the focal length of the telescope in millimeters. This relation is useful in figuring out what slit size to use.

C. The Plate Factor, P

The dispersion of a spectrograph is usually quoted in terms of the plate factor, P, in Angstroms per millimeter.

$$P = \frac{1}{f_{cam} \frac{d\beta}{d\lambda}}$$

where $d\beta / d\lambda$ is the angular dispersion of the grating in radians per Angstrom, and f_{cam} is the focal length of the camera in millimeters.

Note that a longer focal length camera will spread the spectrum out more. This is the reason that spectrographs which literally fill a room are built -- to get small plate factors using long focal length cameras. Note also that $d\beta/d\lambda$, or an upper limit to it, is fixed by the maximum density of grooves that can be ruled on a grating (approximately 2,000 per millimeter).

D. Spectral Resolution, $\Delta\lambda$

In the final analysis, the detail that can be captured in a spectrum depends upon the resolution of the final detector itself, and, for optimum resolution without excess loss of light, the entrance slit should be set to a size which gives a projected slit size on the detector equal to two resolution elements of the detector. If Δx is the size of a resolution element on the detector (this would be, for example, the grain size on a photographic plate), then this means that optimum resolution is achieved if

$$2\,\Delta x = w_{proj} = w / M$$

The resolution achieved can then be simply written in terms of the plate factor as

$$\Delta\lambda = P\,\Delta x$$

A larger slit than optimum will result in a degrading of the resolution. A smaller slit than optimum will result in an unnecessary loss of light falling on the detector.

III. PARAMETERS FOR A TELESCOPE-SPECTROGRAPH COMBINATION

In order to perform sample calculations, we need to consider a reasonable set of parameters for a telescope-spectrograph combination. The parameters are given for the University of Texas instrument, but you may substitute the numbers for your own instruments. Note: $\delta = \alpha + \beta = 45°$ always in the Texas design.

Telescope: Cassegrain reflector with f = 4,714 mm, D = 406 mm
Collimator: glass achromat with f = 273 mm
Gratings:
1. Bausch and Lomb 30 x 32 mm ruled surface, 200 lines/mm, blaze angle = 4°10', blaze wavelength = 7265 A
2. higher dispersion grating, 600 lines/mm, blaze angle = 8°30', blaze wavelength = 5000 A

Comparison lamp: mercury vapor with strong features at 3656, 4047, 4358, 5460, 5760, and 5769 Angstroms.

Cameras: focal lengths of 50 mm and 135 mm
Detector: Kodak Tri-X Pan or Hi-Speed Infrared roll film.

These films were selected for their high speed, reasonable grain, and widespread availability. However, they do have problems with reciprocity failure (see Units 18 and 19); astronomical 103-emulsions may be used, if available, to overcome this problem. Tri-X and Infrared emulsions have grain sizes of about 25 microns. (Grain size is what should be identified with the phrase "detector resolution element" in the discussion of spectral resolution in part II-C above.)

IV. PREPARATORY EXERCISES FOR THE PROSPECTIVE OBSERVER

To guarantee an understanding of the quantities discussed above, and to prepare the reader for the actual use of a spectrograph, it is appropriate to calculate some of the system parameters at this point. Answer the following questions before proceeding.

1. What is the size of a 5 arcsecond seeing disk in the focal plane of the telescope?
2. What is the demagnification factor of the spectrograph?
3. What is the plate factor for each order?
4. What is the wavelength resolution for each order?
5. Will the resolution be degraded by using a slit big enough to accept a 5 arcsecond seeing disk? (Remember that the ultimate resolution of the spectrograph is set by the detector resolution.)
6. Recall the overlapping-orders problem discussed earlier. What filters will be necessary to prevent two or more different wavelengths from appearing at the same position on the plate if one wishes to observe as much of the region from 4000 to 5000 Angstroms as possible? Keep in mind that the film chosen also has a certain spectral response curve (see unit 19, figure 1) and that the glass collimator and camera lenses cut off wavelengths shorter than 3700 Angstroms.

Detailed information on filter pass ranges is available from either the Wratten or Schott filter catalogs. Standard Kodak publications detail the wavelength ranges of film sensitivities.

V. A SAMPLE SPECTROGRAPH

The University of Texas student spectrograph now in use on the 41-cm telescope was built in 1972 by Richard Stover and R. G. Tull for testing new electronic detectors at the McDonald Observatory. Recent modifications have been made by attaching a Pentax f/2, 50-mm camera as the detector and replacing the original razor blade slit with a new slit and decker assembly.

The frontispiece photo of this unit shows the Texas spectrograph attached to the telescope, and figure 6 shows an optical diagram with the essential components noted. The adjustment of each of the parts of the system will be discussed in the same order as in part I.

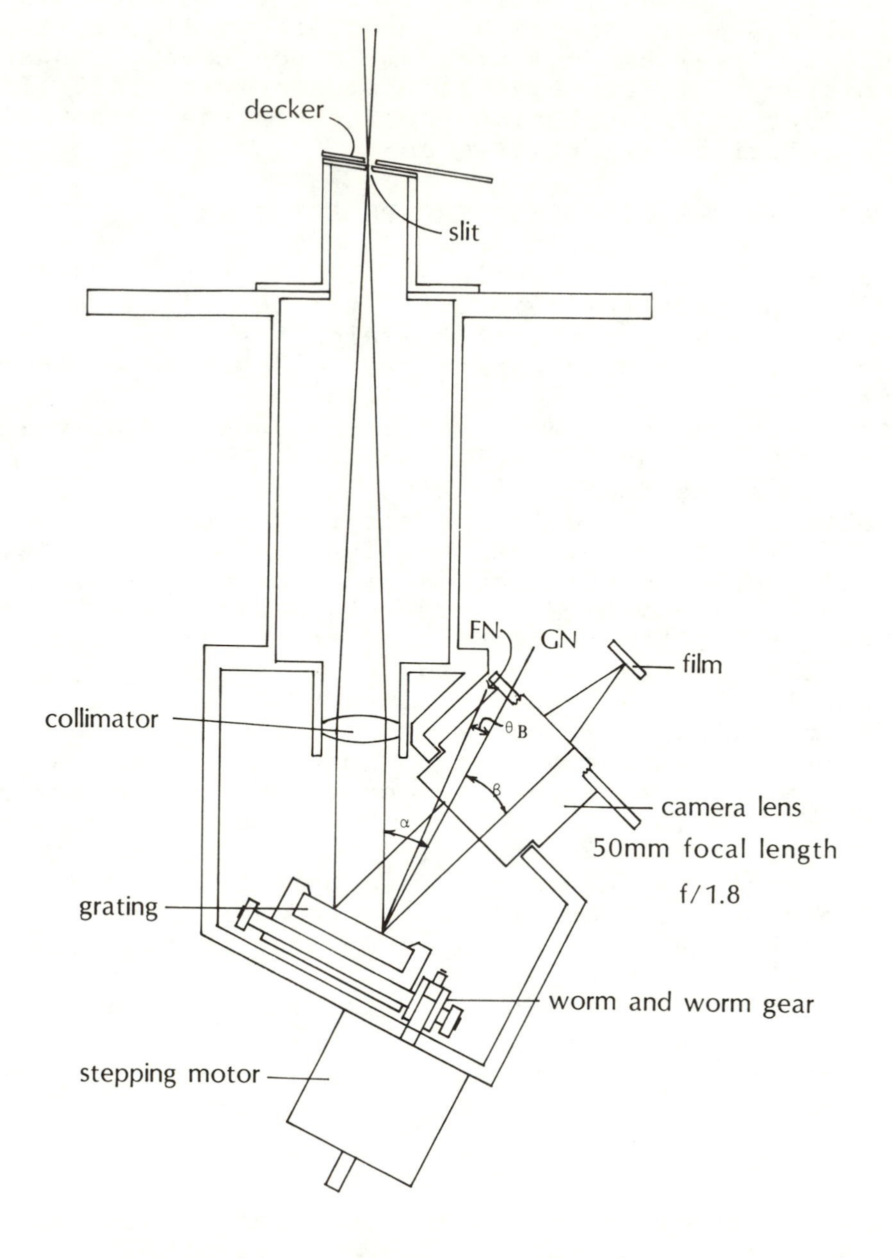
decker
slit
FN
GN
film
collimator
θ_B
β
α
camera lens
50mm focal length
f/1.8
grating
worm and worm gear
stepping motor

Figure 6

A. Grating

Do not attempt to change from one grating to another. Only an authorized technician should handle the gratings. The grating tilt is adjusted during installation of the grating, using appropriately positioned pieces of brass shim stock (which are provided with the spectrograph). Because the entire spectrum will fit on the film at one time, this adjustment should be unnecessary. The grating position is adjustable only within very small limits. Consult your instructor or technician before any grating adjustments are undertaken.

B. Slit and Decker

The slit is adjustable with two small Allen wrenches. However, there is no way of keeping the slit jaws parallel other than a spark plug gap setter. Because of this difficulty, users are requested NOT to adjust the slit. If adjustment seems essential, consult an authorized technician.
The decker is mounted in a slot behind the slit. There are two deckers available, illustrated in figure 7. When the stellar position is used, the light from the star falls on the plate. When the comparison position is used, the light from a standard laboratory light source is admitted to the photographic plate and recorded above and below the stellar spectrum. Since the standard source has emission lines of known wavelength, this superimposed spectrum facilitates the measurement of wavelengths on the unknown spectrum. Also, by making half of the comparison spectrum exposure before the stellar exposure is made, and half afterward, any internal movement of the instrument (i.e., flexure) can be discovered.

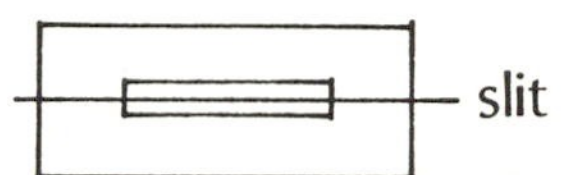

stellar position

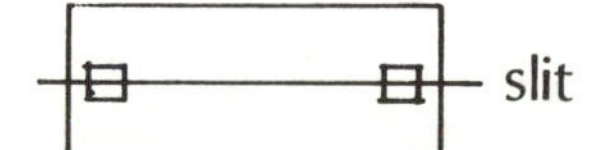

comparison spectrum position

Figure 7

C. Collimator

The collimator focus can be adjusted by moving the collimator in or out. This is done by turning the threaded collimator tube (after loosening the retaining ring). After the slit is focused, its orientation on the sky can be changed by rotating the aluminum slit assembly and tightening its retaining rings.

These adjustments are somewhat difficult and should be done very infrequently. Please do not touch the slit.

D. Camera

The camera focus should be set at infinity, and the lens should be open all the way (i.e., f/2 not f/16). The camera position adjustments should be satisfactory. However, if they are not, the camera can be moved by loosening the screw threaded into its base. Vertical adjustment is accomplished with pieces of shim stock, while the horizontal adjustment can be made by merely moving the camera in the slot. The camera position is optimized during focusing tests and should not be altered without great care. The quality of the finished spectrum is critically dependent upon this camera adjustment. Ideally, adjustments are not made at the telescope; using a point light source in a darkened room should produce satisfactory results.

VI. HOW TO TAKE A SPECTRUM

STOP AND THINK BEFORE YOU CARRY OUT EACH STEP!

1. Acquire the object in the telescope system and make fine adjustments in the telescope pointing until the starlight is falling on the slit. There is an eyepiece for viewing the slit, and the starlight which does not pass on through the slit to the grating will be reflected from the back of the slit jaws into this viewing eyepiece. During exposure you will watch that portion of the starlight reflected from the slit jaws continually. If the telescope does not track so as to keep the starlight falling exactly through the slit, you will have to make fine adjustments to the telescope pointing to reposition the image. These tracking corrections are carried out as the exposure continues; if the starlight is allowed to drift off the slit for a considerable period of time, adjustments in the exposure times will be required.
2. Trail the telescope east-west once or twice to be sure that the slit is aligned in an east-west direction, and leave the star at one end of the slit.
3. Insert the comparison lamp in front of the slit and put the decker in its comparison position.
4. Open the camera shutter and turn the comparison lamp on and off quickly, to achieve an exposure of about 1/8 second.
5. Put the decker in stellar position, remove the comparison lamp, and expose the star for the desired length of time, trailing the star up and down the slit to widen the exposure.
6. Close the shutter and advance the film in the camera for

the next exposure. At the end of a session, rewind the film and remove it from the camera.

Some guidelines are given below to assist you in your early experimentations with exposure times. They are for the higher dispersion grating in the first order. But keep in mind that exposures depend upon a great many factors, including the brightness of the star, the film speed, the color of the star, the color response of the film, the transparency of the sky, etc. Use the entries below only to give you a rough idea of where to start in your trial exposures.

Star	V mag	Color	Film	Exposure
Vega	0.04	blue-white	Plus-X	40 sec
γ Ori	1.7	blue	Plus-X	60 sec
α Per	1.9	yellow	Plus-X	90 sec
ε Cyg	2.5	red	103a-E	8 min

With this system, the faintest star that can be seen on the slit is about fifth magnitude.

VII. A SPECTRAL CLASSIFICATION ACTIVITY

Classification of stars into spectral types can be accomplished even with very low dispersion spectrographs. Make a list of about ten bright standard stars of various spectral types to observe, making finding charts if necessary. (See unit 21 for a discussion of finding charts.) A good source of standard stars is the Bright Star Catalogue (D. Hoffleit, Yale University Press); choose stars which will be high in the sky during your observing periods. Choose stars from the catalog which have spectral types classified on the MKK (Morgan-Keenan-Kellman) system, not HD (Henry Draper) types, which are less accurate.

Obtain properly exposed plates of your stars, recording for each in your observing notebook the time and hour angle, the exposure time, the decker setting, type of film, sky conditions, and any comments on other factors that seem relevant to the quality of your results. You will probably need to take several exposures to obtain a good one. Keep your exposed film in clearly labeled containers in some orderly fashion, so that you can properly identify each plate at development time. After you gain some experience with exposures, constructing plots of star magnitude (in both V and B) versus exposure time for good exposures will give you some guidelines for future exposures. For comparison it would be interesting to photograph the spectrum of the Orion Nebula (M 42 = NGC 1976) if it is visible, to gain experience with a bright line spectral source.

Classify your stars. It is often easier to classify them from prints than from the original film. The references listed below give the criteria necessary to classify spectra. The introduction to the Kitt Peak Atlas has many helpful hints in classifying unknown spectra.

As a second activity, you may attempt the following. Using the calibrations given in Allen's "Astrophysical Quantities" (Athlone Press, London, 1973), obtain the absolute magnitude, distance, effective temperature, surface gravity, and mass of each of the ten stars that you observed.

References

An Atlas of Low-Dispersion Grating Stellar Spectra
Also known as "The Kitt Peak Atlas"
H. A. Abt, A. B. Meinel, W. W. Morgan, and J. W. Tapscott
Kitt Peak National Observatory, 1978.

An Atlas of Objective Prism Spectra
N. Houk, N. Irving, and D. Rosenbush
University of Michigan, 1974.

Atlas of Stellar Spectra
W. W. Morgan, P. C. Keenan, and E. Kellman
University of Chicago Press, 1943.

Revised MK Spectral Atlas for Stars Earlier than the Sun
W. W. Morgan, H. A. Abt, and J. W. Tapscott
Yerkes Observatory and Kitt Peak National Observatory, 1978.

Photometric Studies with a Telescope

21

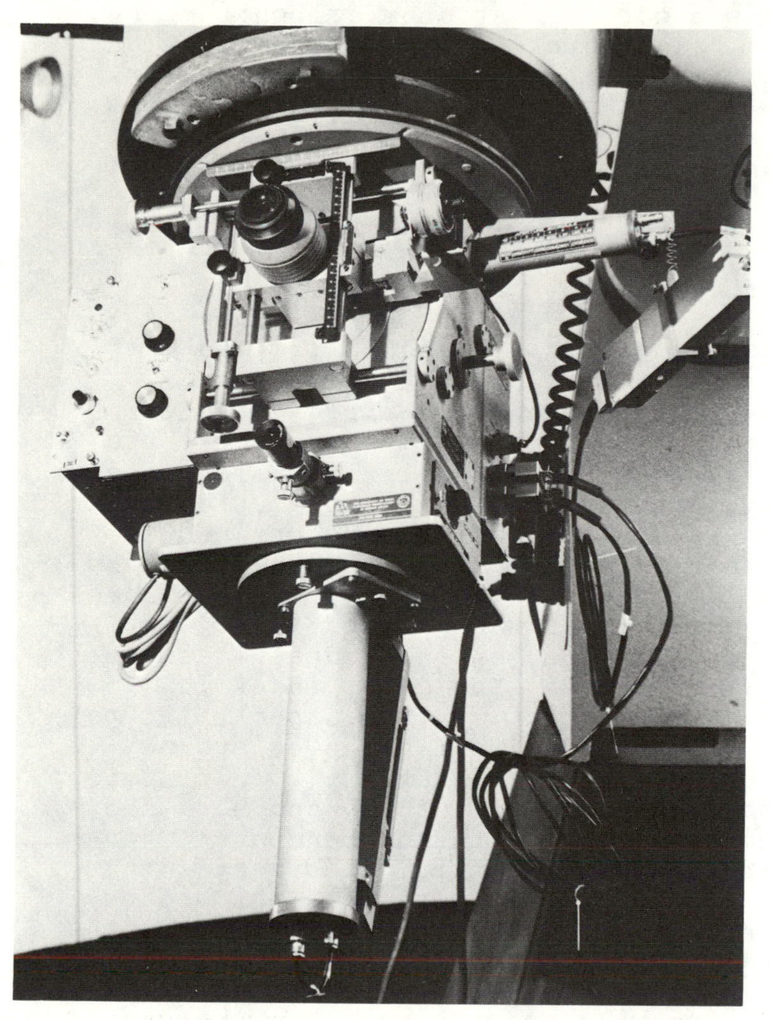

A photometer employs the accuracy and speed of a photoelectric detector to measure brightnesses and colors of stars. It can be used to gain information rapidly or to acquire data of very high precision.

OBJECTIVES

1. to understand the use of a photoelectric photometer
2. to calculate extinction and transformation coefficients and constants
3. to define the term "standard system" and explain the characteristics of the UBV system
4. to determine from your own observations the light curve resulting from a variable star, an eclipsing binary star, or an occultation

EQUIPMENT NEEDED

A telescope (12-inch aperture or larger), photoelectric photometer with auxiliary equipment, "The Arizona-Tonantzintla Catalogue."

Photometry is the science of collecting and measuring the intensity of light from a source. In astronomical photometry, the telescope is used as a "light bucket" to collect the light and direct it into the photometer which acts as the detector. Since the light is collected over a wide range of wavelengths, the intensity measurement is taken more quickly than it is with instruments employing finer spectral resolution (such as a spectrograph).

I. INTRODUCTION TO PHOTOMETRY

From the radiation laws of physics (Planck's law and Wien's law) we know that black bodies emit radiation in curves of distinctive shape with the peak intensity occurring at bluer wavelength for hotter objects (e.g., unit 11; see also figure 2 below). These variations give us the sensation of color when we look at the object. Early in this century astronomers denoted the color of a star by comparing its magnitude determined in two wavelength regions, photographic (m_{pg} in the blue region) and photovisual (m_v with maximum sensitivity near 5550 Angstroms -- like the human eye). The color index was defined as "$c = m_{pg} - m_v$". Even with careful calibration, an individual measurement was only good to about 0.1 magnitude. The color index is indicative of the surface temperature, i.e., the spectral type, of the star. The color index is positive for cooler stars, negative for hotter stars, and zero for A0 stars (by definition).

Using photoelectric photometry and carefully choosing the filter characteristics (central wavelength and bandpass), the astronomer can make a filter "system" sensitive to certain characteristics of stars. Stars are not perfect black bodies; spectral features (e.g., the Balmer jump, seen at 3646 A in the middle star of figure 2) will affect the observed magnitude. Refer to the energy distribution curves for three stars given in figure 2. In this lab unit we will use the best-known system, the broad band three-color UBV system of Johnson, Morgan, and Harris. Among the advantages of this system are:

1. The bandpasses are broad enough to allow a sufficient number of photons to be collected for fast measurements, thus one can observe very short-lived phenomena.
2. It is standardized with standard stars appearing all over the sky, in a wide range of spectral types and magnitudes.

A "standard star" is one whose magnitudes are accurately known in the desired system. It is assumed that none of the standard stars is variable.

We will use as standard stars the 1,325 bright stars published in "The Arizona-Tonantzintla Catalogue." If you were observing fainter stars (with a larger telescope in a darker site), you would need fainter standards. The Landolt list (Astronomical Journal, 1973, volume 78 page 959) contains seventh to twelfth magnitude stars located near the equator. Figure 1 shows the filter response for the UBV filters. (The blue cut-off for the U filter is atmospheric at about 3100 A due to oxygen.

The red cut-off for the V filter is determined by the long wavelength sensitivity of the photocell employed.) To duplicate the UBV system of the originators, one must remember that their system was determined not only by the type of filters but also by use of a reflector at 6,800 feet altitude (which affects the "U" response) and a 1P21 photomultiplier having a Sb-Cs (S4) cathode (which affects the "V" response). The universality of the system depends upon being able to transform your data to their system. By observing standard stars, you mathematically convert the data taken with your equipment at your location into what they would have obtained using their equipment at their location. In this manner, observations taken at many observatories around the world can be intercompared. The "natural system" using the original apparatus is given by Johnson in *Basic Astronomical Data*, chapter 11. A description of the filters is given in table 1.

Table 1. Filters

	Average Wavelength	Bandwidth
U	3600 A	700 A
B	4350	970
V	5550	850

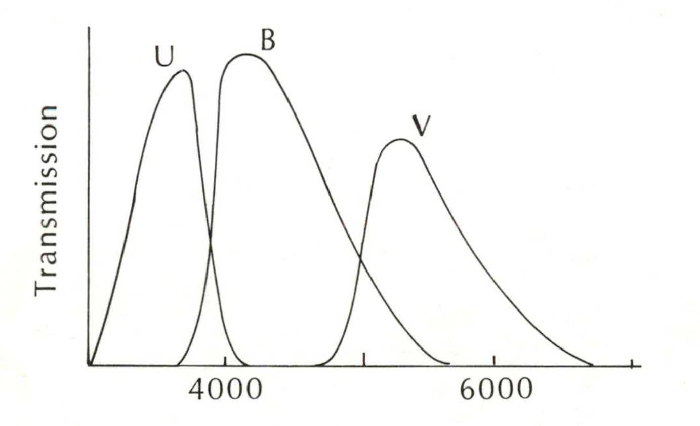

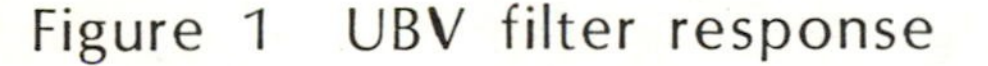
Figure 1 UBV filter response

The UBV filter response may be superimposed on the curves in figure 2. Note how different (U-B) and (B-V) colors will result.

Many types of stars can be profitably observed with UBV photometry. Many projects in the field of galactic structure involve obtaining photometry for large numbers of stars. Cluster members can be observed to obtain their age by using the cluster-fitting method. One can observe nonstellar objects, such as entire globular clusters, in order to compare their physical characteristics.

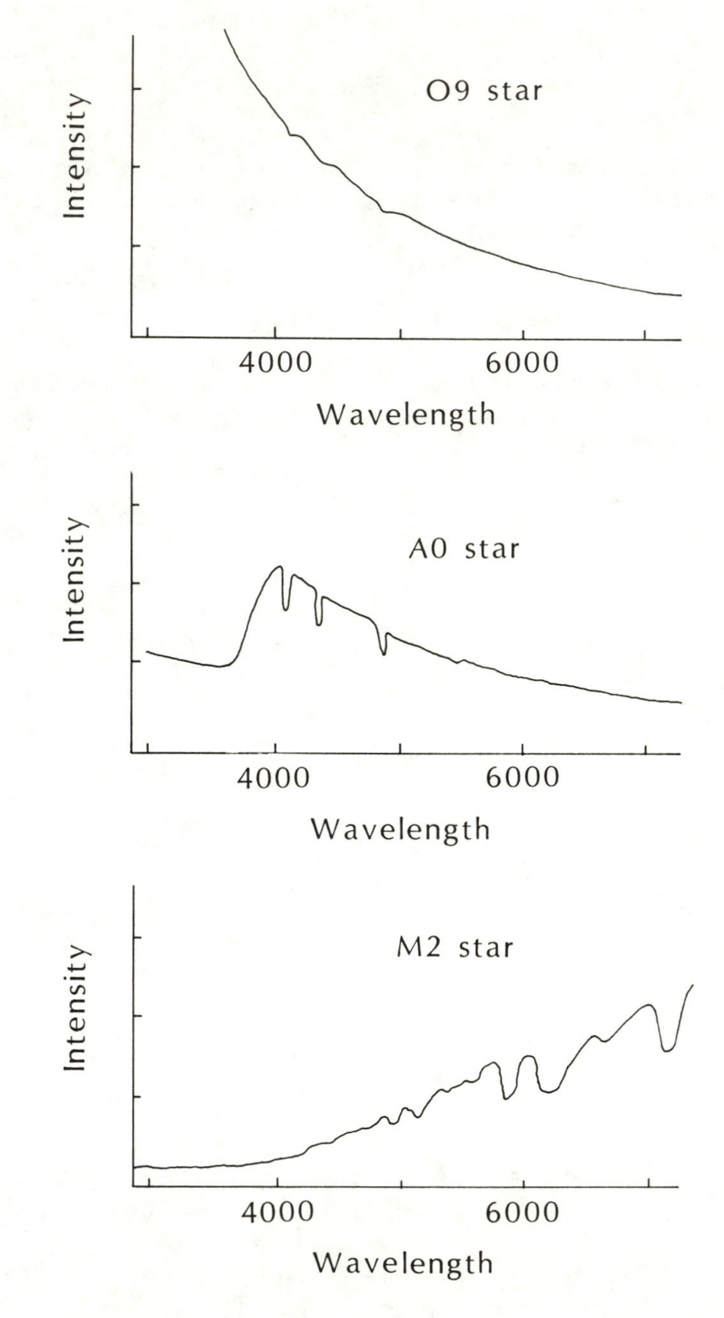

Figure 2 Stellar Intensity Distribution

One of the more interesting uses is to observe variable stars: intrinsic pulsating variables such as Delta Scuti, Mira, or RR Lyrae types; eclipsing binary stars; or exploding variables such as nova. An important specialized field is that of lunar occultations. The main focus of this lab will be to observe a variable star, removing the effects of the atmosphere and attempting a transformation to the standard system. As an alternative, you may choose to observe a lunar occultation if such an event occurs during your observing period. Occultations of bright stars are very rare. The brightness of the sky will impose a limit on the faintness of observable stars.

II. THE PHOTOMETER

Consider the simple system in the following diagram:

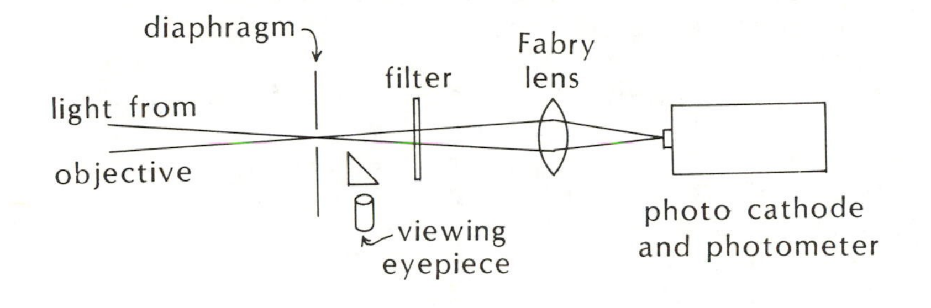

Figure 3

The diaphragm is usually one of a series of apertures of different sizes arranged on a slide with detentes for easy use. The diaphragm allows you to isolate a star with as little background sky as possible. It is desirable to use the smallest possible diaphragm. The seeing greatly affects which diaphragm aperture you chose to use on any given night. At times of poor seeing, the smallest diaphragm can not be used because some of the starlight would not pass through it.

The viewing eyepiece will check the position of the star in the diaphragm, then slide out of the way.

The filters can be arranged on a slide or a wheel. With a four-space wheel, the UBV filters can be positioned as needed, with the last space used as a "clear" position -- allowing all the light through.

The Fabry lens projects an image of the telescope objective onto the photocathode. The sensitivity of a photocathode is not uniform all over its surface, and it will show variations if the star image moves about. With the Fabry (or field) lens, the image is spread over a large area to minimize sensitivity fluctuations.

The photocathode is usually protected by a "dark slide." This slide covers the photocathode while extraneous light may be present and protects it from damage. The dark slide is physically moved out of the way when observations begin. Bright lights, such as those in the dome or a flashlight, can ruin the photocathode. Very bright astronomical objects can also damage the photocathode. Do not attempt to observe objects brighter than second magnitude. The dark slide must be re-inserted before any lights are turned on.

The spectral response of many photomultipliers changes with temperature, and these devices are often operated with a "cold box" to remedy this defect. Also, photomultiplier systems emit random surges of current, or noise, even when not observing a source of light. The amount of noise will decrease with temperature. You may use an RCA 8850 which is normally operated uncooled. It has a spectral response similar to a 1P21 but has greater signal to noise ratio.

A photomultiplier is a device combining a photo-emissive cell and an amplifier. This allows a very large multiplication of the electric current emitted by the photo-sensitive layer. The number of dynodes in the photomultiplier will affect the multiplication factor (also known as "the gain"). An incoming photon releases an electron from the photocathode; this electron is accelerated by an electric field, collides with each dynode in turn, and in each collision generates additional electrons. The result is that a photon signal at the cathode becomes amplified into a pulse of about one million electrons out of the anode. The multiplication factor is very sensitive to the voltage applied to the dynodes; thus, a very well regulated voltage supply is required. The RCA phototube is adjusted for use at 1,600 volts.

The signal from the photomultiplier is amplified before being recorded. The pulse can be measured by a variety of methods:

1. a meter which is visually read
2. chart paper recorder
3. punched paper tape
4. magnetic tape cassettes

The first two methods are less convenient since one must write in a notebook or on the chart the significant information, such as filter, diaphragm aperture, star, etc., and the reduction procedures are more involved. The latter two accomplish these "bookkeeping" tasks with switches, which allow the information to be recorded automatically with the data.

III. EXTINCTION

The earth's atmosphere absorbs starlight. Our aim is to correct our observations so that we obtain the magnitudes for U, B, and V which would have been observed if we were above the earth's atmosphere. Each measured magnitude depends on the thickness of the atmosphere through which we observe. Consider the quantity X, the air mass, measured in units of thickness at the zenith. X equals one at the zenith and increases in value as zenith distance (z) increases.

Spherical trigonometry allows us to deduce the relation between z, your latitude (ϕ), the star's declination (δ), and the hour angle (h). The hour angle is found from the time of observation and the star's right ascension (or recorded directly from the telescope readouts at the time of observation).

$$h = \text{sidereal time} - \text{right ascension of star}$$

$$\sec z = (\sin\phi \sin\delta + \cos\phi \cos\delta \cos h)^{-1}$$

To compute the air mass,

$$X = \sec z - 0.0012 \sec z (\sec^2 z - 1)$$

For our accuracy, within a zenith distance of 65 degrees, we can use X equals secant z. Our relationship between the magnitude of tne star as measured above the earth's atmosphere, m_o, and the observed magnitude, m, is:

$$m_o = m - k X \qquad (k = \text{constant})$$

Our problem is to determine the value of k for the night we are observing.

Imagine that we observe one star all night. Due to the rotation of the earth, our telescope will move to track it over its many positions in the sky. As the hour angle changes, so do the zenith distance and air mass. Question: For what star(s) is this not true?

If we plot the observed magnitude versus air mass, we obtain a plot. (See figure 4.)

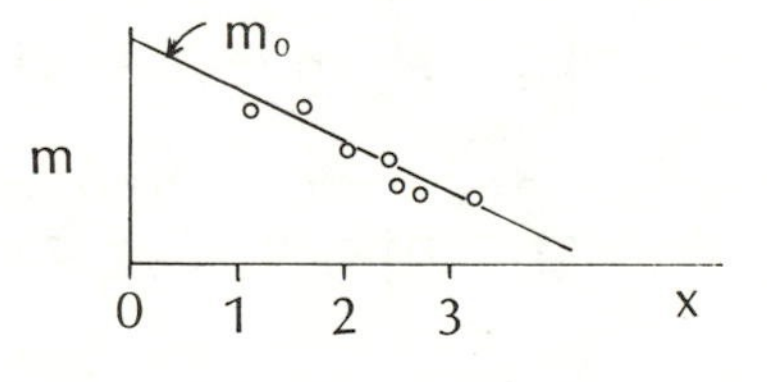

Figure 4

The slope of this line is k. The intercept of this line at X = 0 is m_o. A similar result for the color C of the star is obtained: $C_o = c - k X$. It should be noted that the atmosphere not only diminishes light but also reddens it. In other words, there is a wavelength dependence of extinction. If we made the plot in figure 4 for two stars of different colors, we would find two values of k, and our equation for magnitude would have to be corrected for this.

Traditionally this is treated as a second-order correction, so our formula would become:

$$m_o = m - k' X - k'' c X$$

where c is a color, e.g., (B - V).

Similar changes appear in the color equations:

$$C_o = c - k_c' X - k_c'' C X$$

In order to find the values of k', k_c', k'', and k_c'' one observes stars of very different color at many zenith distances.

In this lab we shall ignore these second-order effects and only consider the first-order terms. Your results in this situation will be better if you choose an extinction star which is close in color to your program star. The pertinent equations for extinction will then be:

$$v_o = v - k_v X$$

$$(b - v)_o = (b - v) - k_{bv} X$$

$$(u - b)_o = (u - b) - k_{ub} X$$

(Note: We are reserving the use of the capital letters UBV for use after reduction to the standard system.) After correction for extinction, each star will have a set of values:

$$v_o , \quad (b - v)_o , \quad (u - b)_o$$

In correcting for extinction, we have implicitly made the following assumptions:

1. extinction at the zenith is a minimum
2. for the same zenith distance, extinction is the same in all directions
3. extinction remains the same throughout the night
4. the star observed to obtain the extinction constants is nonvariable

In poor observing conditions, one or more of the above list will not be true. The condition that extinction is to remain constant throughout the night is the most easily violated condition. Choosing transformation stars at the same air mass as the program stars will help negate some of this effect. However, very poor observing conditions produce very poor data -- don't try to observe on cloudy nights! Even a slightly hazy night can be very tricky and produce margninal results.

A speedy estimate of extinction can be obtained if you observe a couple of standard stars, one near the zenith and one at a large zenith distance.

Then,

$$k = \frac{\Delta m - \Delta m_o}{\Delta X}$$

Here m_o is the known standard magnitude. Similar equations are used for the coefficients in each color.

IV. TRANSFORMATION

If your program star is a variable, and you always observe a comparison star (easily done with a two-channel photometer which observes star and sky or star and comparison star at the same time), you should immediately see the change in relative intensity. In the early days of photoelectric photometry, these relative magnitudes were often considered sufficient. They could be used to tell the period and amplitude of a pulsating variable star or to derive orbital elements for a binary star. (To first order, even extinction could be ignored.) The literature of astronomy has many such observations recorded. In attempting to combine observations from many sources, however, these relative measures are not very useful. The measurements would have been of more value if they had been transformed to a standard system. Also, if you wish to obtain any astrophysical information from the data, transformation to a standard system is necessary.

Each set of filters, telescope, and photometer system defines a "natural" system. Our aim is to relate these natural data to another "standard" system (which was originally someone else's natural system) -- the standard UBV system.

To get to the UBV system, we need to obtain transformation constants and coefficients. Using these values in the transformation equations will allow a correction for effective wavelength differences and zero-point shifts between your "natural" system and the UBV system. Observing twenty or more stars of a variety of magnitudes and colors should provide decent values of the coefficients. The wavelength coefficient should retain the same values for one site and instrument combination, with only minor adjustments. On subsequent observing nights, fewer transformation stars should be needed since only the zero-point may change. Their observation should then take only about an hour of observing time.

Now examine the transformation equations. Note that they use extinction-free values of v, (b - v), and (u - b).

$$(B - V) = \mu (b - v)_o + C_{BV}$$

$$V = v_o + \varepsilon (B - V) + C_V$$

$$(U - B) = \psi (u - b)_o + C_{UB}$$

We consider the (B −V) value first, so that we can use it in the second equation for V. The values of μ , ε , and ψ are best found by plotting. We choose to plot the variables which will be most sensitive to variations. (See figure 5.) (U − B) is plotted in a manner similar to (B − V), while V can be found more simply. (See figure 6.)

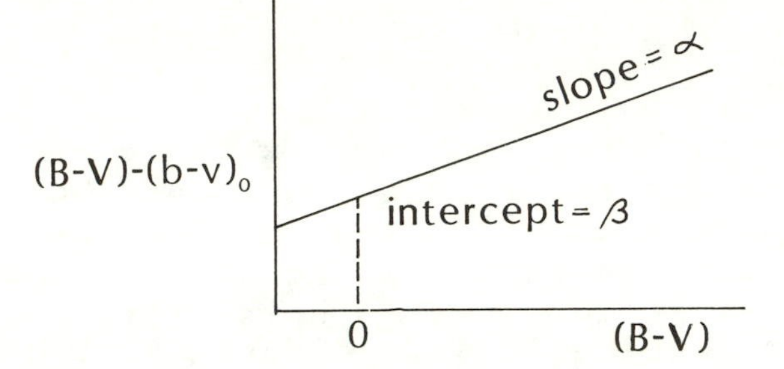

$$\mu = \frac{1}{1 - \alpha}$$

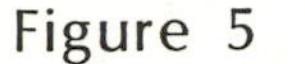
Figure 5

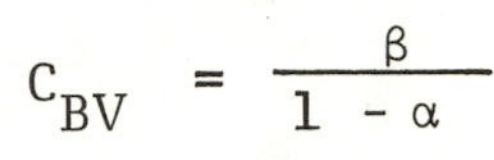

$$C_{BV} = \frac{\beta}{1 - \alpha}$$

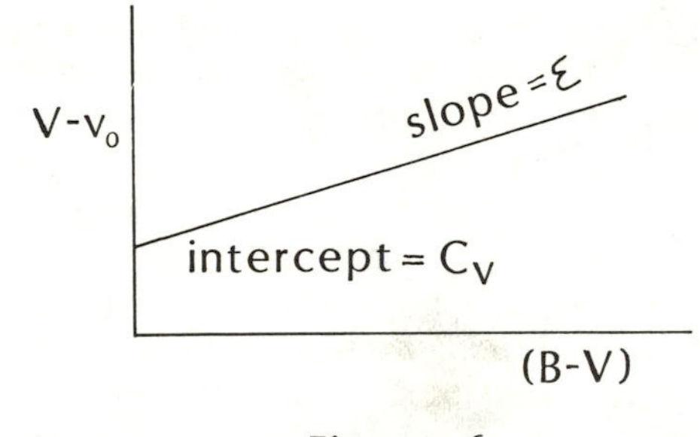

Figure 6

You will obtain better results with hand plots, since then you can SEE the data.

Now plot $(V - v_0)$ versus V, and plot the residuals in V for the standard stars versus time. The residual is obtained by applying the transformation coefficients and constants to your data and then subtracting the result from the standard value. Do not take the absolute value of the residual. What are your calculated values of the coefficients and constants? How can you tell if a given observation is bad?

If your extinction and transformation values are all right, you are ready to apply them to your program (unknown) stars. (Refer to section VI.)

V. FINDING CHARTS

With a bright star, you need only a position to locate it. Fainter stars, or bright stars in crowded fields, are a problem. The following list may prove helpful. A description of star catalogues and charts is given in "Basic Astronomical Data," appendix I, page 471. A description of the National Geographic Society-Palomar Observatory Sky Survey is also given (appendix II of the same book). Some charts are easy to copy; photographic charts are best copied with a Polaroid camera. Since the positions are given for a variety of epochs, you may need to apply a correction for the precession of the earth's axis if the epoch of the catalog is very removed from your time of observation, or if an exact position is needed to identify a faint object.

Catalog Name	Epoch	Mag Limit
Norton's Star Atlas	1950.0	6
SAO - atlas, catalog (Smithsonian Astrophysical Observatory)	1950.0	9
Becvar	1950.0	10-13
BD and extensions (Bonner Durchmusterung)	1855 +	9
Astrographic Catalog	1903-63	varies
Lick Observatory Atlas	1959 photographic	16
NGS-POSS	1950 photographic	21

VI. YOUR OBSERVING PROJECT

1. Read the guide to the photometer and reduction system.
2. Sign up for the telescope -- try to choose nights without much moonlight.
3. Choose the extinction and standard stars. Remember, you will want an extinction star that is observable most of the night over a wide range of zenith distances. You may find it convenient to record the important information about each star on a 3X5 card, and attach the cards together on a ring in order of right ascension.
4. Choose your program star(s). Make finding charts. Choose comparison stars (in the same region, about the same color, a little brighter than the program star, and nonvariable).
5. Observe ONLY extinction and transformation stars your first night. Sequence your individual observations VBUUBV, and use the average time of the set as the time.
6. Reduce your first night's data and check for errors.
7. Observe extinction, some transformation stars, and program star(s) for the rest of your observing nights. Try to observe your transformation and program stars at approximately the same air mass.

Suggested program stars include the following:

VZ Cnc	faintest mag = 8.3	period = 0.178 day
δ Scuti	5.2	0.194

Algol -- times of minimum are listed monthly in "Sky and Telescope".

The brighter members of the Pleiades. (Use Eta Tauri as your standard.)

22 Additional Projects: Where to Go from Here

Readers who have progressed to this point in the book might well appreciate references to other activities that could be used to follow up on those carried out so far. Instructors using this book might wish information on other activities that could be used to supplement and broaden their students' contact with experimental astronomy. This unit will give some guidelines on where to go from here.

There is a wide variety of astronomy activities that can be carried out. The selection in this book represents the judgment of the authors in selecting a set which illustrates important concepts with a minimum of expense and equipment. Astronomers elsewhere have also constructed such activities, although their philosophy concerning the proper design for an activity may not correspond perfectly with ours. We have assembled ours with the following goals in mind: to introduce selected important concepts in astronomy and astronomical methods through measurement activities that simulate on an introductory level the interaction of observation and theory which characterizes modern astrophysical research, and to do so with relatively simple equipment that can either be constructed from patterns or ordered at moderate cost from commercial suppliers.

Any astronomy activity will vary widely in the amount of equipment (and perhaps also prerequisite knowledge) required of the reader, and there are also wide variations in methods of presentation. Our philosophy has always been to emphasize INVESTIGATIONS in astronomy, through the process of taking measurements and interpreting the data. We have tried to avoid (with some exceptions) the presentation of data acquired by others, since it has been our experience that most readers and students coming to astronomy for the first time are more motivated to explain their own observations than they are to explain the data acquired by others.

Most of the other published astronomy activities of which we are aware rely fairly heavily on the interpretation of existing data, rather than the acquisition of new data. Once a basic interest in astronomy has been stimulated, however, there is much to be learned by examining and pondering the wide variety of astronomical observations that are available in the literature. In our discussion of activities available from other sources, we will attempt to note which of them follow our philosophy of beginning with measurements and exploration and which differ from our approach by presenting data acquired by others.

The discussion to follow will list and comment on collections of activities available commercially from other publishers which can be purchased either in book form or reprint (individually bound) format. We will divide this material into two parts:

activities approximately on the level of units 1 through 15, which can be carried out with simple equipment and a minimum of prerequisite knowledge; and activities more similar to units 16 through 21, which require either more prerequisite knowledge, more outside reading to supplement them, or more expensive or more specialized equipment.

I. INTRODUCTORY-LEVEL ACTIVITIES

A. Activities from Sky Publishing Corporation

The following activities appeared originally in "Sky and Telescope" magazine. They are available in reprint form and are generally self-contained, requiring no complicated equipment. Typically, they present observational data acquired by professional astronomers and the student is directed through a series of analyses to arrive at an understanding of the concepts being presented.

1. The Moon's Orbit. Twelve photographs of the moon illustrate its changing distance by changes in apparent size. Measuring the size leads to an understanding of the details of the orbit of the moon.

2. Spectral Classification. Thirty sample objective prism spectra are provided to classify, with supporting materials to understand the principles on which the classifications are based.

3. Rotation of Saturn and Its Rings. By measuring the doppler shift in a slit spectrogram taken across the disk and rings of Saturn, rotation periods and the planet's mass can be derived.

4. Variable Stars in M15. Eight photos of the globular cluster are used to make magnitude estimates of six RR Lyrae variable stars from which the clusters' distance and size are determined.

5. The Earth's Orbital Velocity. Doppler effects at various times of the year in a high dispersion spectrum of Arcturus are used to determine the orbital speed of the earth.

6. Proper Motion. The proper motion of 61 Cygni from data spanning 96 years is determined.

7. Pulsars. Records of radio signals from three pulsars at four frequencies are measured to determine the periods and the pulse dispersions, and, by applying a model of the pulsar phenomenon, the distance is estimated.

8. The Crab Nebula. The rate of expansion of knots and filaments in the Crab Nebula is measured from photographs 34 years apart. Combining these with radial velocity measurements from a high dispersion spectrogram yields the size and age of

the supernova remnant.

9. Hubble's Law. Redshifts for 5 elliptical galaxies are measured from their spectra and distance estimates obtained from measurements of their angular sizes on photographs, leading to a determination of the Hubble constant and an age estimate for the universe.

10. Cosmic Distance Scale. Use of the period-luminosity relation for cepheid variables leads to a distance for the Small Magellanic Cloud.

11. The Rotation of Mercury. Radar pulse spectra are measured to determine the true rotation period of the planet.

12. The Orbit of a Visual Binary. The orbit of Kruger 60 is determined from seven photographs of the system which were widely spaced in time.

B. Published Astronomy Lab Manuals

Of the collections of activities listed below, the one which comes closest to the exploratory nature of this book is by Kelsey, Hoff, and Neff.

Manual of Astronomy. R. W. Shaw and S. L. Boothroyd (Brown, Dubuque, IA, 1967), 290 pp. A large selection of exercises in workbook format.

Laboratory Exercises in Introductory Astronomy. C. M. Huffer and R. Marasso (Holt, Rinehart and Winston, New York, 1967), 255 pp. Similar to Shaw and Boothroyd.

Laboratory Exercises for General Astronomy. (Kendall/Hunt Publ,. 2460 Kerper Blvd., Dubuque, Iowa 52001, 1978.) L. Kelsey, D. Hoff, and J. Neff. Many of the activities in this book do involve students in actual data gathering. The most recent version is entitled "Activities in Astronomy."

Practical Work in Elementary Astronomy. M. G. J. Minnaert (Springer-Verlag, New York, 1969), 74 exercises. Advanced for an introductory nonmajors class, but a useful reference for instructors.

An Introduction to Experimental Astronomy. Roger B. Culver (Freeman, 1974). Well-designed activities that reach into astrophysics more than most.

Laboratory Exercises in Astronomy. Joseph Holtzinger and Michael Seeds. (Macmillan, 1976.) 38 varied activities, some with "pop-up" cutouts.

Investigations in Observational Astronomy. W. Christiansen, R. Kaitchuck, and M. Kaitchuck. (Paladin House Publ., Geneva, IL 60134.)

Astronomy - Observational Activities and Experiments. M. Gainer. (Allyn and Bacon, Boston, 1974.)

Laboratory Exercises in Astronomy. J. Safko. (University of South Carolina, 1974.)

Astronomy through Practical Investigations. (LSW Associates, P. O. Box 82, Mattituck, NY 11952), A series of 35 activities available in individual reprint form.

Projects and Demonstrations in Astronomy. H. Kruglak. (American Journal of Physics, volume 44, number 9, 1976.) This is a useful resource letter of journal articles, books, and materials dealing with astronomy activities. It is also available as Resource Letter EMAA-2 from the American Association of Physics Teachers.

II. ADVANCED ACTIVITIES

A. Investigations in Optics Using a Laser

The LASER (Light Amplification by Stimulated Emission of Radiation) produces a bright and tightly focused beam of radiation at a single wavelength. It has the additional property that all of the radiation it emits is very strictly "in phase," that is, the vibrations of the electric and magnetic fields vary in a regular way such that all parts of the beam are synchronized together. As a consequence of this property, called COHERENCE, the wave effects that light is subject to -- such as diffraction and interference -- show up more dramatically than with ordinary light. Further, the laser makes studies of polarization and viewing of holograms achievable.

Most manufacturers of small lasers also supply a kit of optics materials and a manual describing a set of activities which can be performed with the laser. The specific experiments which can be performed depend upon the materials supplied in the particular kit. We have found the following investigations to be achievable and illuminating:

1. studies of the scattering of light, as for example through a cloud of chalk dust or through water in which a precipitate is formed.
2. studies of the laws of reflection and refraction, including determinations of the index of refraction of materials, are easily carried out with simple plastic optics components. The laser beam is a natural "ray" of light.
3. diffraction from a single edge, single slit, double slit, and multi-slits shows up easily and verification of the equations of diffraction is not difficult.
4. polarization shows up well from a laser, and even the time variations and rotations of the plane of polarization are measurable.
5. any interference phenomenon in general becomes relatively easy to measure, up to and including the observation of interference fringes in a Michelson

interferometer.

6. by diffusing the beam, the viewing of commercially purchased holograms is entertaining, and some manufacturers sell kits designed to allow the users to photograph their own holograms.
7. If a photometer is available, many of the above studies can be done to considerably greater precision.

B. Programming Computers in Other Languages

An increased familiarity with computers leads naturally to the desire to increase the sophistication of the communication with them. For many years, scientific calculations have been done primarily in FORTRAN, and many versions of this language have developed over the years, from FORTRAN I through FORTRAN-77 through recent variations such as MORTRAN. PASCAL is a symbolic language of great power which has been gaining considerable popularity lately, among home computer hobbyists and professionals. Any computer language with a reasonable audience will have books and manuals designed to teach the statements of the language to prospective users. A satisfying activity is to acquire a new language and apply it to some problem in astronomy. To give just a couple of examples of astronomical applications which could be used in conjunction with practice in a new language:

1. Taking the data from the Bright Star Catalogue (D. Hoffleit, Yale University Press) for the luminous O and B stars, and using tabular data available in Allen's "Astrophysical Quantities" on the relationships between color, spectral type, and temperature, one could calculate the distances of the luminous O and B stars after correcting their observed apparent magnitudes for interstellar dust absorption by comparing their observed colors with the true colors for stars of their spectral type. Plotting these results on spherical graph paper will give some indication of the distribution of stars in the neighborhood of the sun in the galaxy, as well as just a suggestion of the spiral arm structure in the solar vicinity.
2. Using the mass-luminosity relationship, one could calculate the energy output of stars on the main sequence relative to the sun, and given the practical limitations on how faint stars can be seen at a particular location, calculate the distances to which each type of star would be visible.
3. Almost any table of astronomical data could be stored and then examined using a modern data storage and retrieval language like System 2000, an information management language. Practice exercises in such a language could be such things as cataloging a collection of slides or lab equipment. Plotting different types of stars on an H-R diagram (i.e., extending the activity introduced in unit 12) would be a natural project.
4. The mathematics of celestial navigation is a natural subject for programming on a computer.

5. Prospective teachers may find that producing a CAI (computer assisted instruction) lesson on some aspect of astronomy is a worthwhile project.

C. Personal or Self-Designed Projects

After going through some astronomy investigations, it is inevitable that your own questions will begin to come up. Investigating a question of your own formulation is one of the most intellectually challenging activities possible. However, it is not possible to be very specific here about what form such a project might take. Instructors giving courses at schools or colleges based on this book could direct students in writing new activities at the level of the first section of this book. The subject of celestial navigation mentioned above would make an excellent activity unit for students and readers on almost any level, and it could be done to a variety of levels of precision, depending on equipment available and the level of mathematical sophistication.

D. Activities Dictated by Special Equipment

Institutions that have some specialized equipment might wish to design activities using it. For example, instruments which read the density of a photographic plate or spectral plate (called microdensitometers) make precise analysis of astronomical information possible. Access to a photometer system which is set up to take observations rapidly over short time scales would allow the setup of an occultation program, through which close binary systems could be detected and an occasional stellar diameter could possibly be measured.

It should be clear from the diversity of ideas discussed above that the "sky's the limit" as far as possible activities are concerned. It is our hope that having made some progress through this book, readers will discover the joy of astronomical observations. There are astronomy clubs and societies in all parts of the world. You might enjoy continuing your activities in conjunction with a set of like-minded observers. We wish you dark skies and happy hunting in them.

Royal Astronomical Society of Canada, 124 Merton Street, Toronto, Canada M4S 2Z2

American Association of Variable Star Observers, 187 Concord Avenue, Cambridge, Mass 02138

Society of Amateur Radio Astronomers, 81 Stony Hill Road, Feeding Hills, Mass 01030

American Meteor Society, c/o Dr. David Meisel, Box 213, Geneseo, N.Y. 14454

Association of Lunar and Planetary Observers, c/o Dr. Walter Haas, Box 3A0, University Park, N.M. 88003

International Union of Amateur Astronomers c/o Mr. Ciaran, Kilbride, 16 Cedarwood Park, Ballymn, Dublin 11 Ireland

British Meteor Society, 26 Adrian Street, Dover, Kent, CT17 9AT, England.

British Astronomical Assoc, Burlington House, Piccadilly, London W1/W1VONL England.

Appendix 1. Star Maps

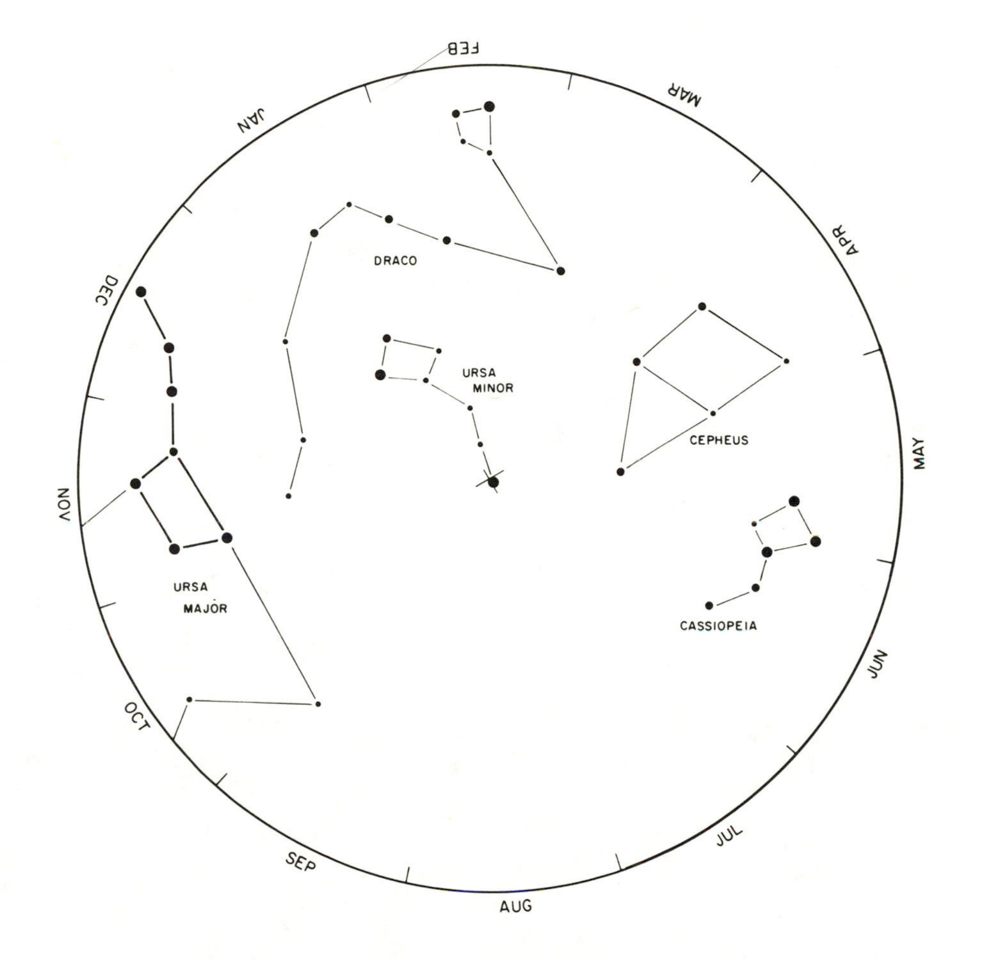

NORTH CIRCUMPOLAR CONSTELLATIONS

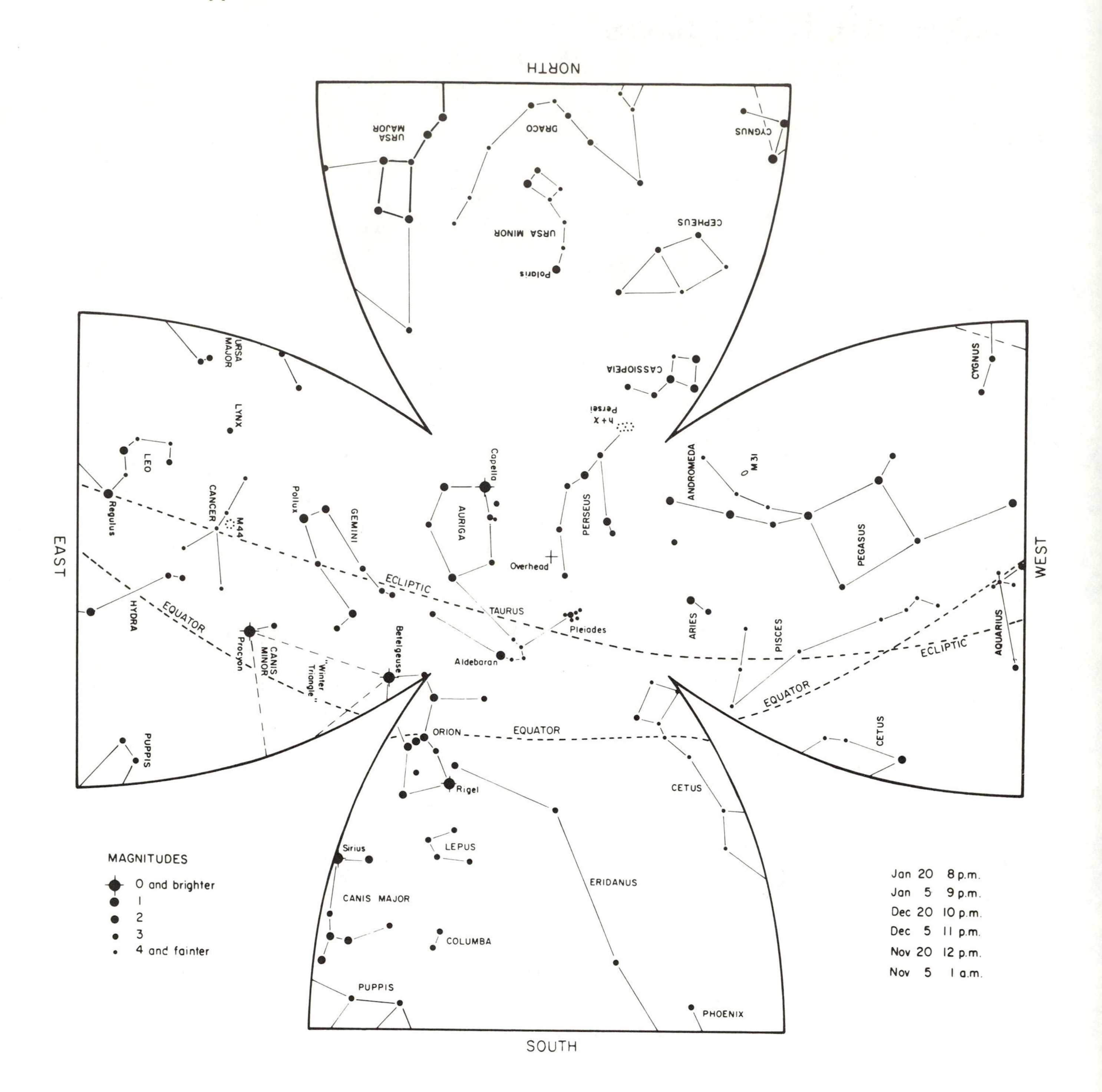
NORTH
SOUTH
EAST
WEST
URSA MAJOR
DRACO
CYGNUS
URSA MINOR
Polaris
CEPHEUS
CASSIOPEIA
h + X Persei
LYNX
LEO
Regulus
CANCER
M44
Pollux
GEMINI
Capella
AURIGA
PERSEUS
ANDROMEDA
M31
PEGASUS
Overhead
ECLIPTIC
TAURUS
Pleiades
ARIES
PISCES
AQUARIUS
HYDRA
EQUATOR
Procyon
CANIS MINOR
"Winter Triangle"
Betelgeuse
Aldebaran
ORION
Rigel
CETUS
PUPPIS
LEPUS
Sirius
CANIS MAJOR
ERIDANUS
COLUMBA
PHOENIX
MAGNITUDES
0 and brighter
1
2
3
4 and fainter
Jan 20 8 p.m.
Jan 5 9 p.m.
Dec 20 10 p.m.
Dec 5 11 p.m.
Nov 20 12 p.m.
Nov 5 1 a.m.

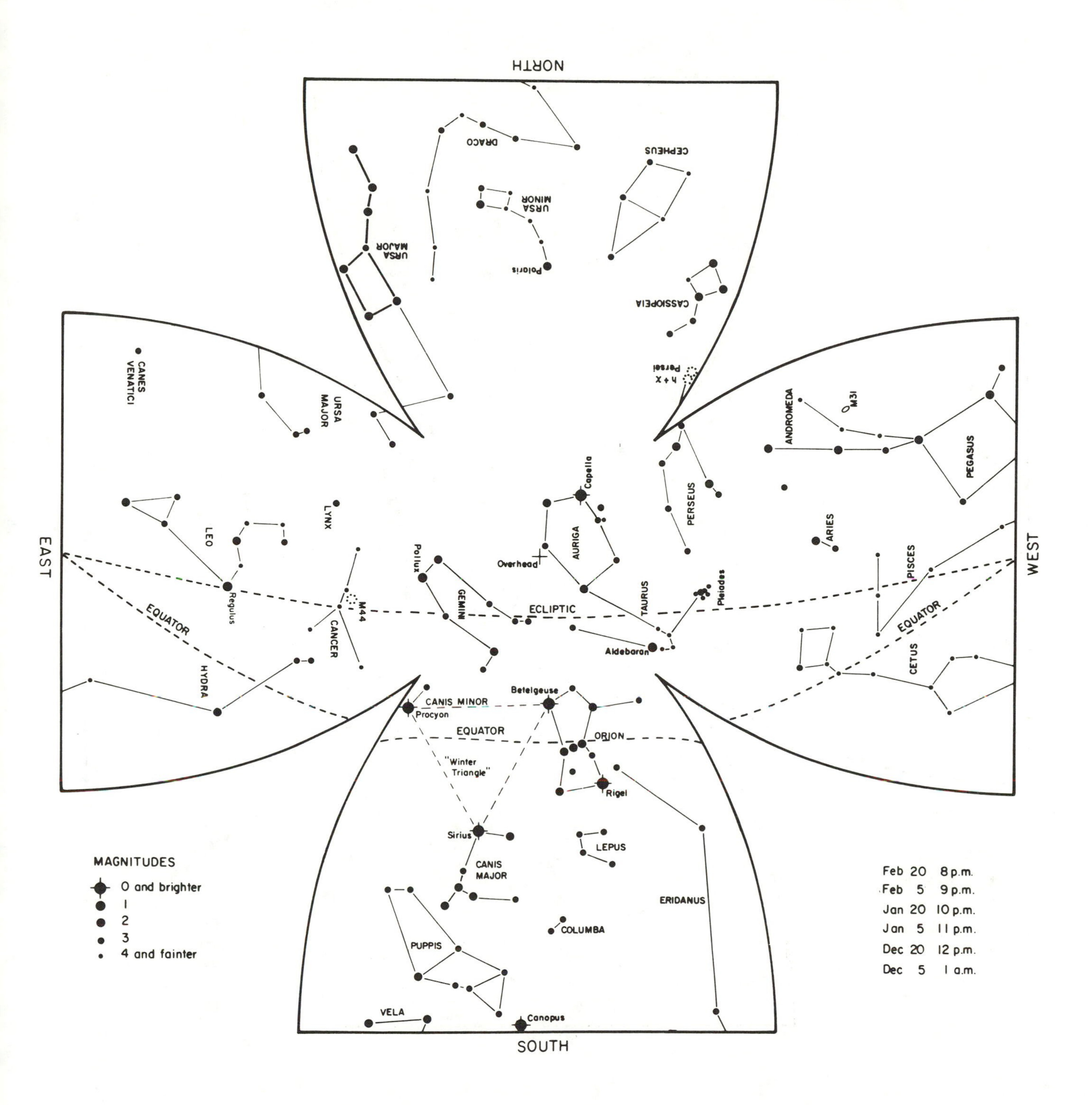
NORTH
DRACO
CEPHEUS
URSA MINOR
URSA MAJOR
Polaris
CASSIOPEIA
h + X Persei
M31
ANDROMEDA
PEGASUS
CANES VENATICI
URSA MAJOR
LYNX
LEO
Capella
AURIGA
PERSEUS
ARIES
PISCES
EAST
WEST
Pollux
GEMINI
Overhead
Regulus
M44
CANCER
ECLIPTIC
TAURUS
Pleiades
EQUATOR
Aldebaran
CETUS
HYDRA
CANIS MINOR
Procyon
Betelgeuse
EQUATOR
ORION
"Winter Triangle"
Rigel
Sirius
LEPUS
CANIS MAJOR
ERIDANUS
COLUMBA
PUPPIS
VELA
Canopus
SOUTH
MAGNITUDES
0 and brighter
1
2
3
4 and fainter
Feb 20 8 p.m.
Feb 5 9 p.m.
Jan 20 10 p.m.
Jan 5 11 p.m.
Dec 20 12 p.m.
Dec 5 1 a.m.

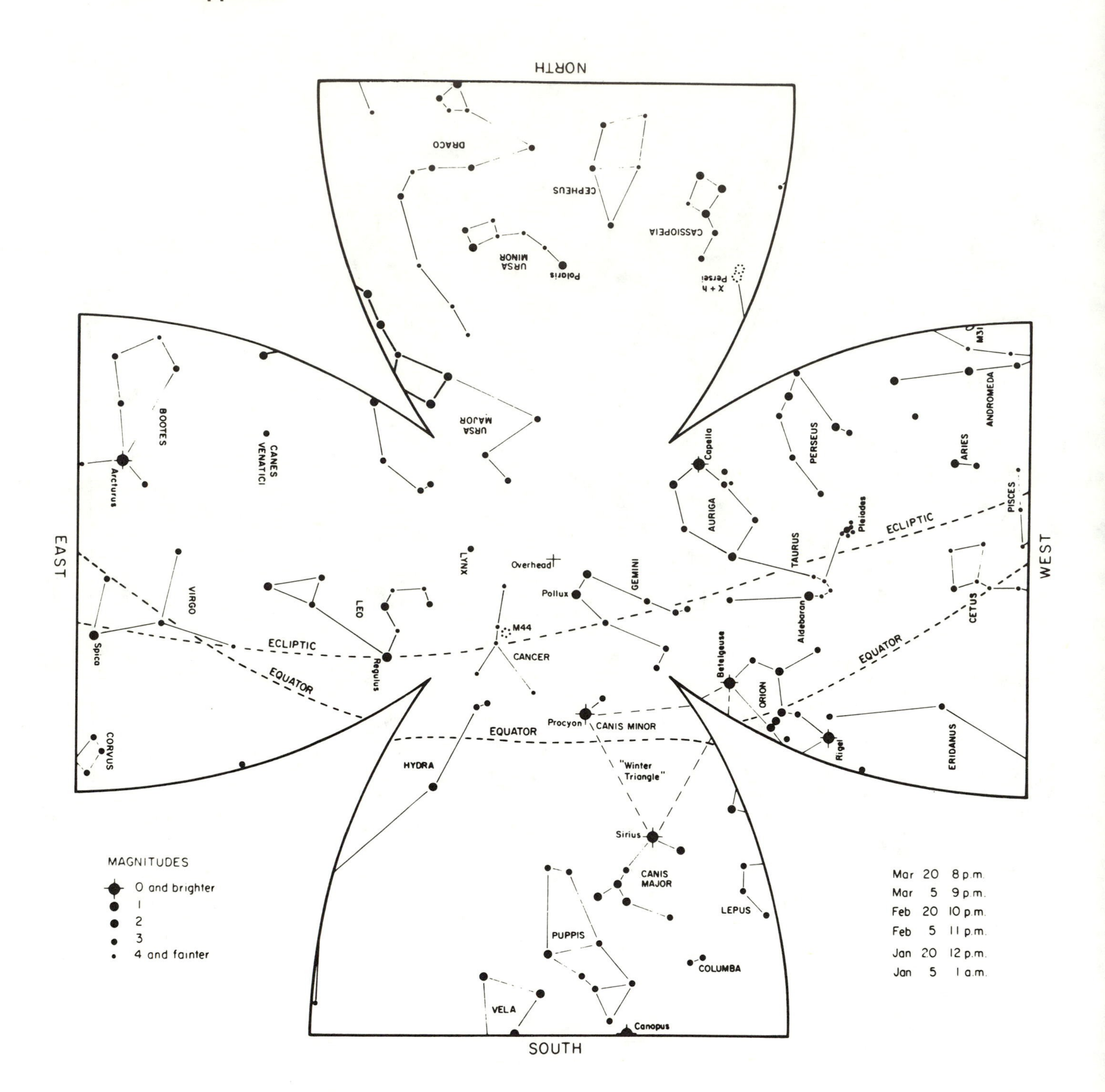

NORTH
SOUTH
EAST
WEST
DRACO
CEPHEUS
CASSIOPEIA
URSA MINOR
Polaris
h + X Persei
URSA MAJOR
BOOTES
CANES VENATICI
Arcturus
M31
ANDROMEDA
PERSEUS
Capella
ARIES
AURIGA
Pleiades
PISCES
ECLIPTIC
TAURUS
LYNX
Overhead
GEMINI
Pollux
VIRGO
LEO
CETUS
Aldebaran
M44
CANCER
Spica
Regulus
EQUATOR
Betelgeuse
ORION
Procyon
CANIS MINOR
CORVUS
HYDRA
"Winter Triangle"
Rigel
ERIDANUS
Sirius
CANIS MAJOR
LEPUS
PUPPIS
COLUMBA
VELA
Canopus
MAGNITUDES
0 and brighter
1
2
3
4 and fainter
Mar 20 8 p.m.
Mar 5 9 p.m.
Feb 20 10 p.m.
Feb 5 11 p.m.
Jan 20 12 p.m.
Jan 5 1 a.m.

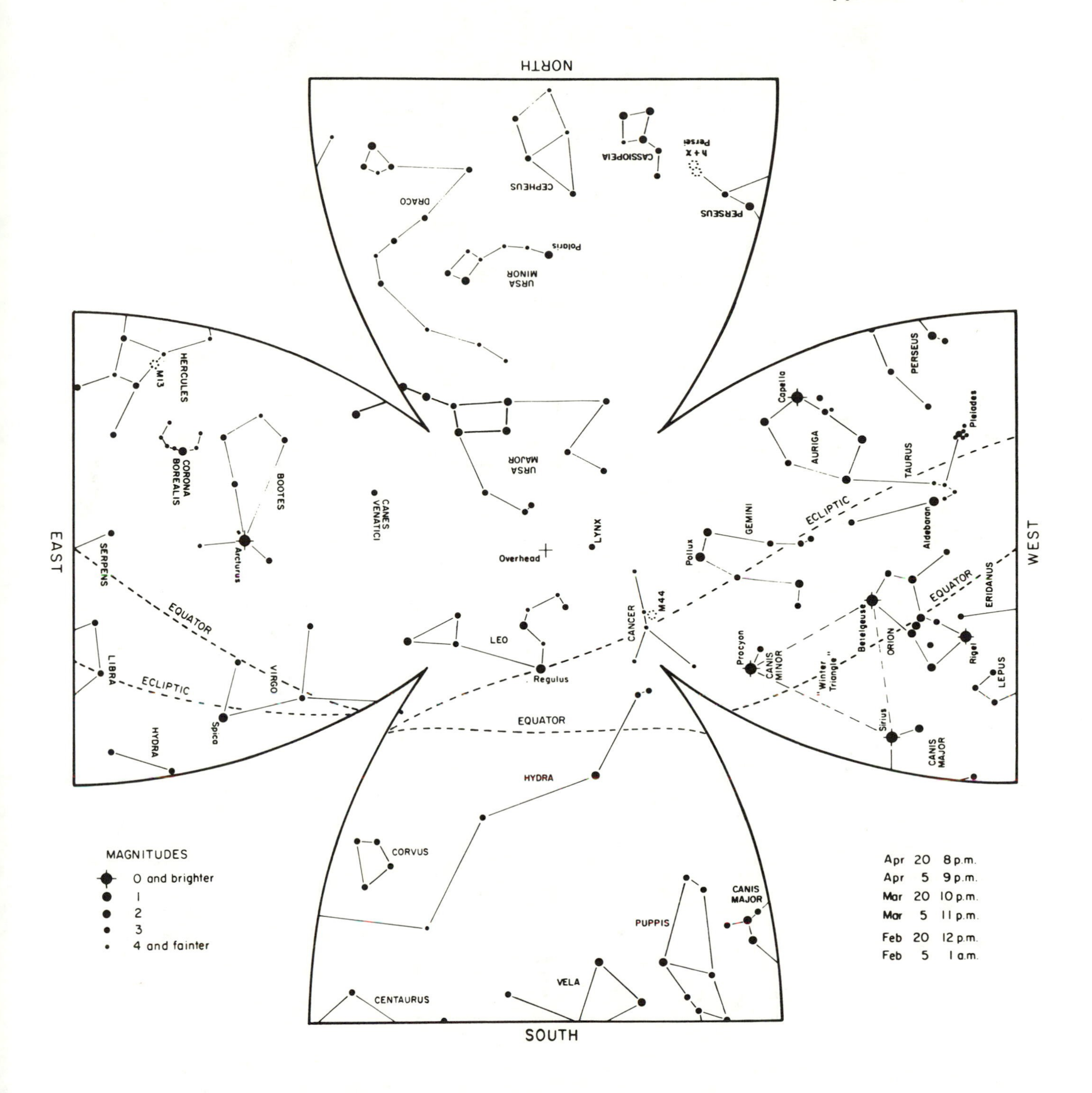

NORTH
SOUTH
EAST
WEST
CASSIOPEIA
h + χ Persei
CEPHEUS
DRACO
PERSEUS
Polaris
URSA MINOR
HERCULES
M13
CORONA BOREALIS
BOOTES
URSA MAJOR
CANES VENATICI
LYNX
Overhead
Arcturus
SERPENS
EQUATOR
ECLIPTIC
LIBRA
VIRGO
Spica
HYDRA
LEO
Regulus
CANCER
M44
Capella
AURIGA
TAURUS
Pleiades
Aldebaran
GEMINI
Pollux
Procyon
CANIS MINOR
"Winter Triangle"
Betelgeuse
ORION
Rigel
ERIDANUS
LEPUS
Sirius
CANIS MAJOR
CORVUS
PUPPIS
VELA
CENTAURUS
MAGNITUDES
0 and brighter
1
2
3
4 and fainter
Apr 20 8 p.m.
Apr 5 9 p.m.
Mar 20 10 p.m.
Mar 5 11 p.m.
Feb 20 12 p.m.
Feb 5 1 a.m.

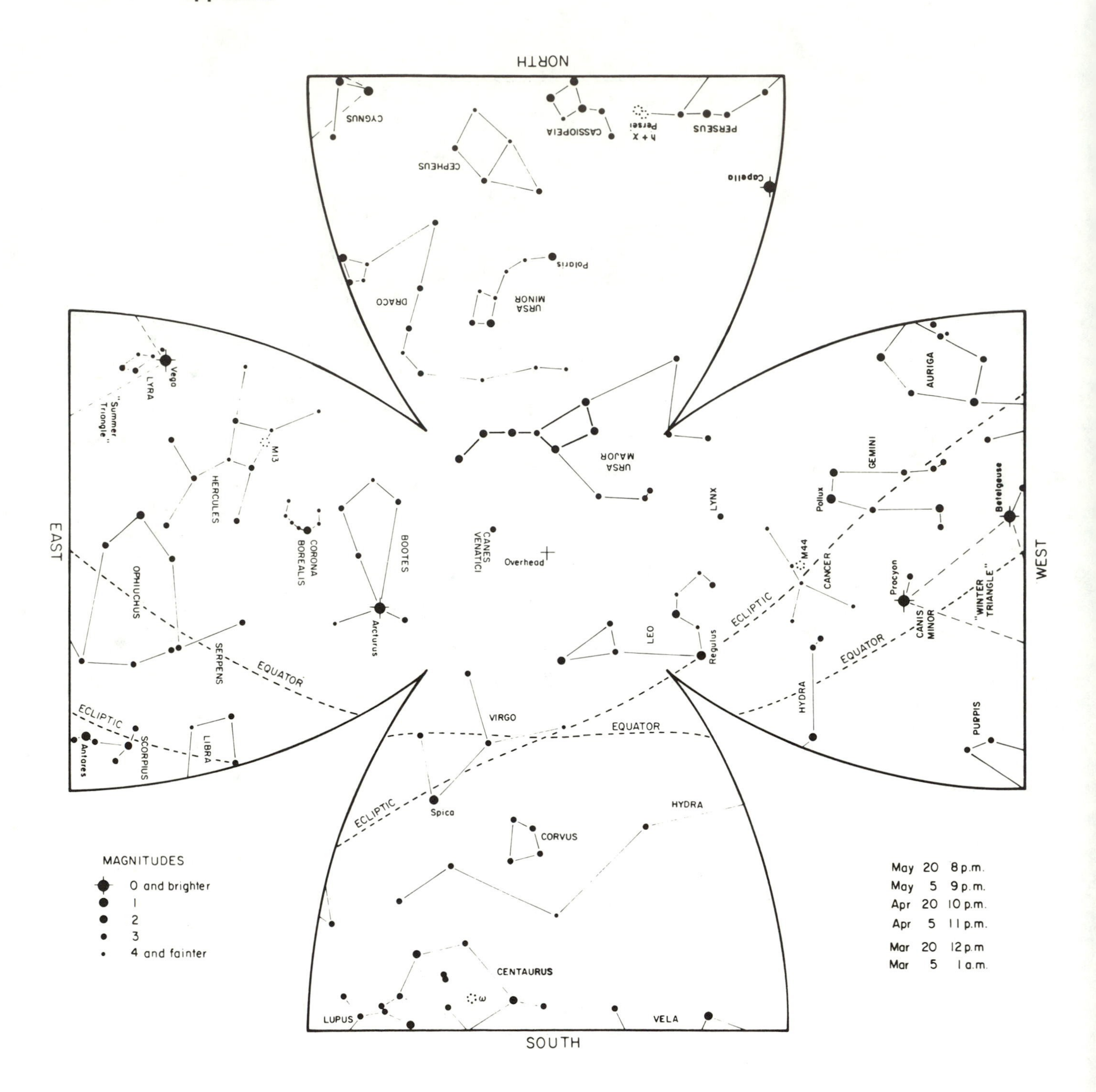
NORTH
SOUTH
EAST
WEST
CYGNUS
CASSIOPEIA
h + X Persei
PERSEUS
CEPHEUS
Capella
Polaris
DRACO
URSA MINOR
URSA MAJOR
LYRA
Vega
"Summer Triangle"
M13
HERCULES
CORONA BOREALIS
BOOTES
CANES VENATICI
Overhead
LYNX
AURIGA
GEMINI
Pollux
Betelgeuse
M44
CANCER
Procyon
CANIS MINOR
"WINTER TRIANGLE"
OPHIUCHUS
SERPENS
EQUATOR
Arcturus
LEO
Regulus
ECLIPTIC
HYDRA
PUPPIS
VIRGO
ECLIPTIC
SCORPIUS
Antares
LIBRA
Spica
CORVUS
CENTAURUS
LUPUS
VELA
MAGNITUDES
0 and brighter
1
2
3
4 and fainter
May 20 8 p.m.
May 5 9 p.m.
Apr 20 10 p.m.
Apr 5 11 p.m.
Mar 20 12 p.m
Mar 5 1 a.m.

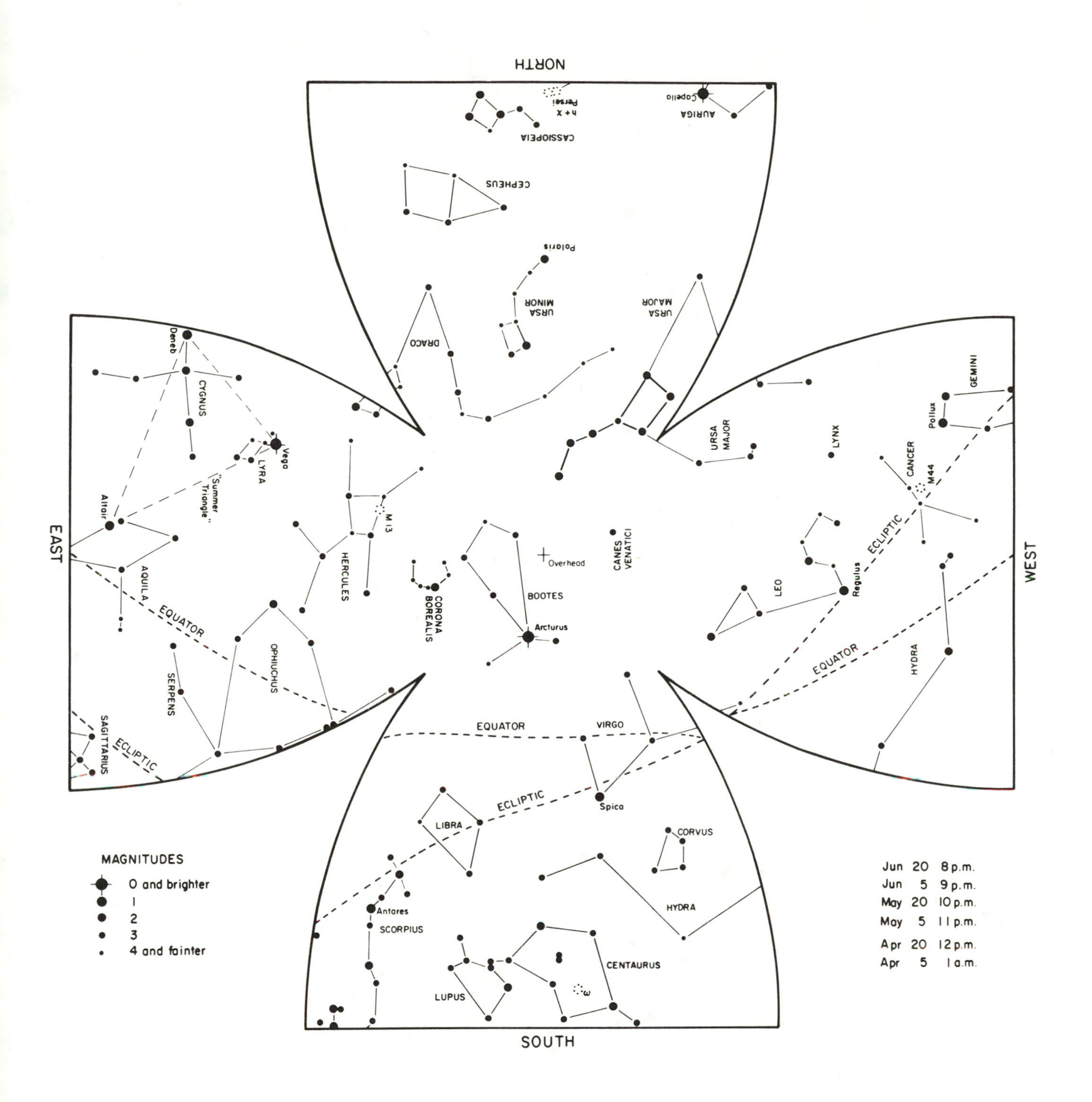
NORTH
SOUTH
EAST
WEST
CASSIOPEIA
h + χ Persei
AURIGA
Capella
CEPHEUS
Polaris
URSA MINOR
URSA MAJOR
DRACO
Deneb
CYGNUS
Vega
LYRA
"Summer Triangle"
Altair
AQUILA
M 13
HERCULES
CORONA BOREALIS
BOOTES
Overhead
CANES VENATICI
Arcturus
OPHIUCHUS
SERPENS
SAGITTARIUS
EQUATOR
ECLIPTIC
GEMINI
Pollux
LYNX
CANCER
M44
LEO
Regulus
HYDRA
VIRGO
Spica
LIBRA
CORVUS
Antares
SCORPIUS
CENTAURUS
LUPUS
ω
MAGNITUDES
0 and brighter
1
2
3
4 and fainter
Jun 20 8 p.m.
Jun 5 9 p.m.
May 20 10 p.m.
May 5 11 p.m.
Apr 20 12 p.m.
Apr 5 1 a.m.

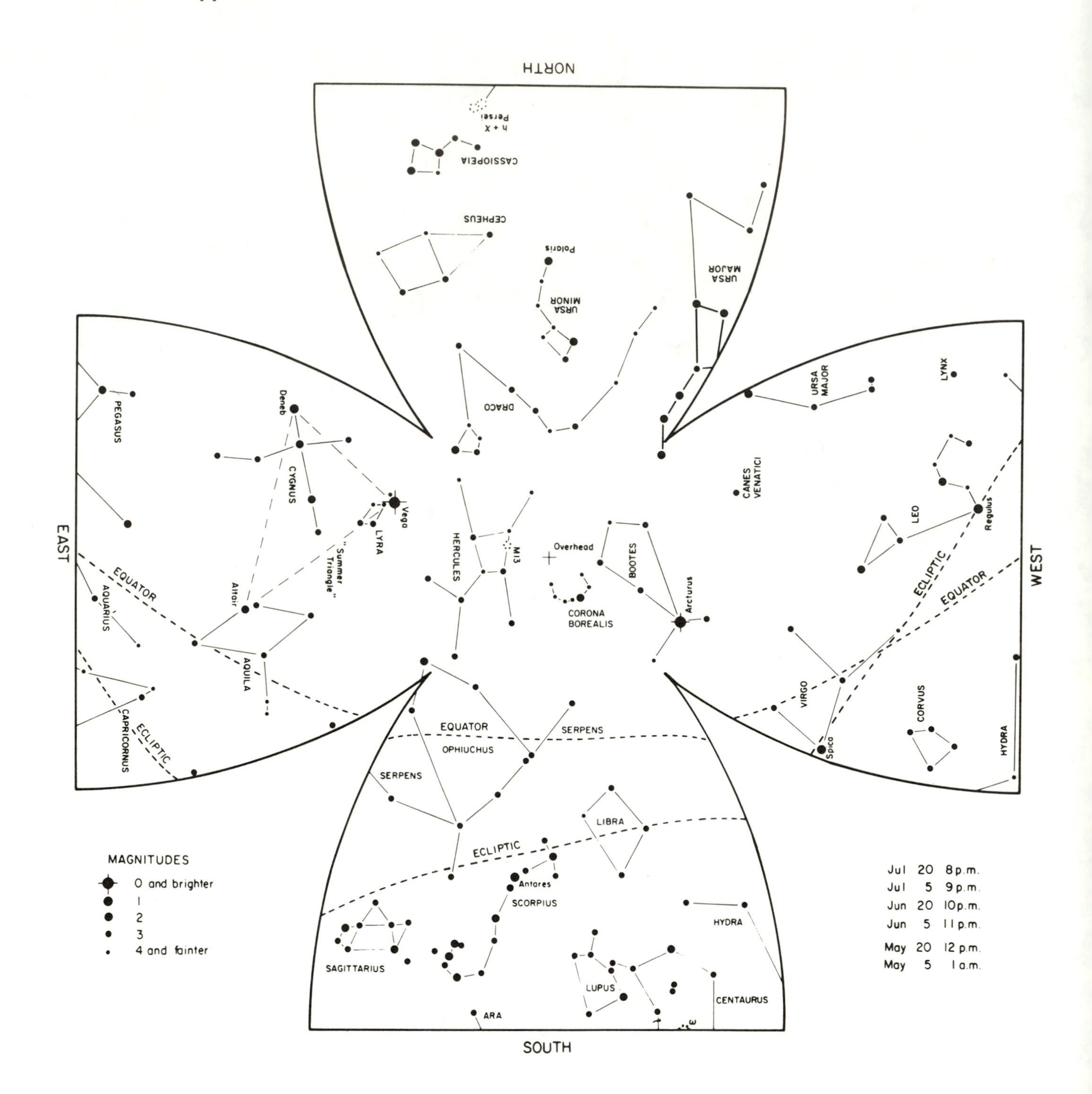
NORTH
SOUTH
EAST
WEST
h + X Persei
CASSIOPEIA
CEPHEUS
Polaris
URSA MINOR
URSA MAJOR
DRACO
PEGASUS
Deneb
CYGNUS
Vega
LYRA
"Summer Triangle"
Altair
AQUILA
AQUARIUS
CAPRICORNUS
EQUATOR
ECLIPTIC
HERCULES
M13
Overhead
CORONA BOREALIS
BOOTES
Arcturus
CANES VENATICI
LYNX
LEO
Regulus
VIRGO
Spica
CORVUS
HYDRA
OPHIUCHUS
SERPENS
LIBRA
Antares
SCORPIUS
SAGITTARIUS
LUPUS
CENTAURUS
ARA
MAGNITUDES
0 and brighter
1
2
3
4 and fainter
Jul 20 8 p.m.
Jul 5 9 p.m.
Jun 20 10 p.m.
Jun 5 11 p.m.
May 20 12 p.m.
May 5 1 a.m.

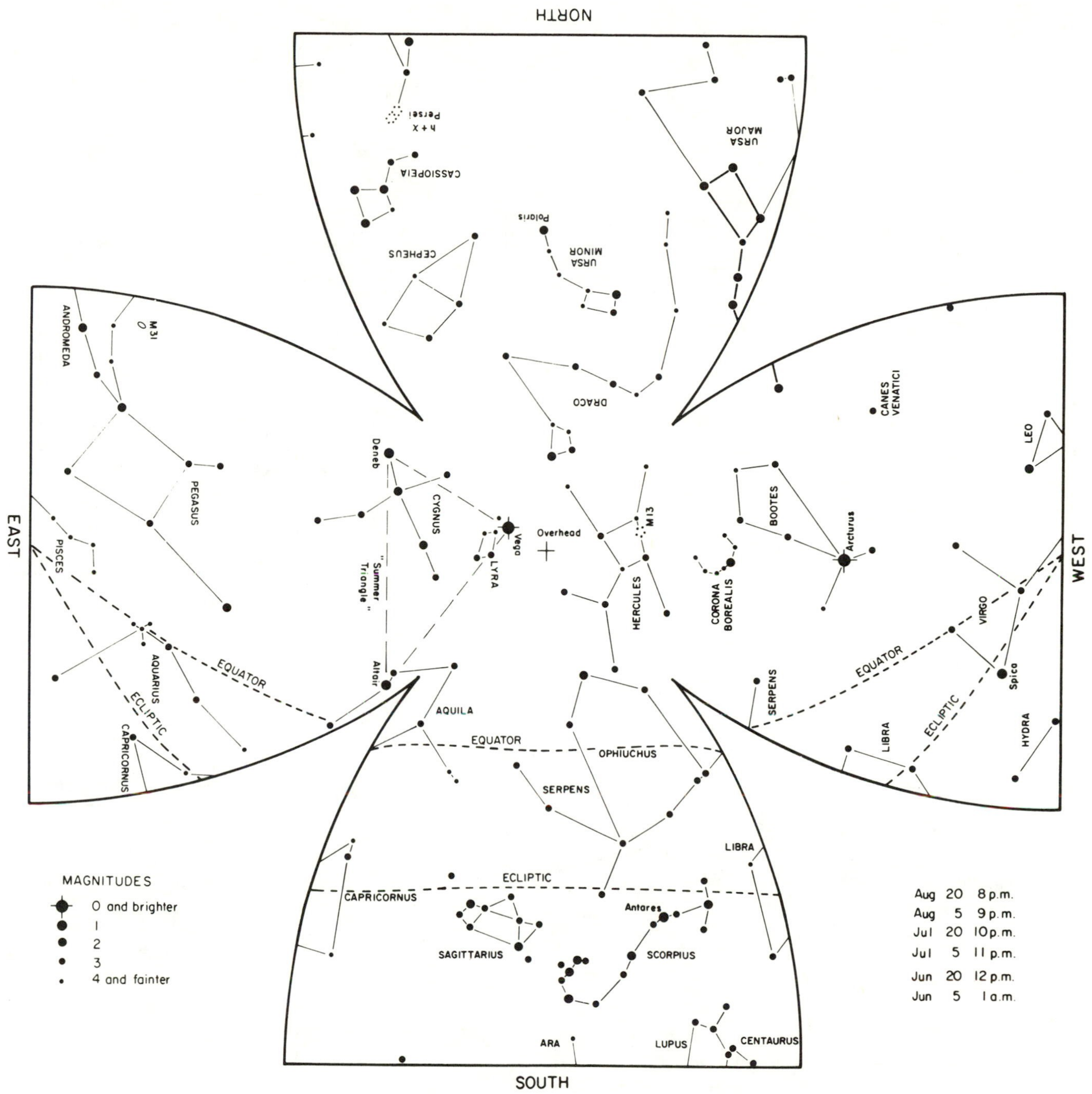
NORTH
SOUTH
EAST
WEST
Overhead
URSA MAJOR
CASSIOPEIA
CEPHEUS
URSA MINOR
Polaris
DRACO
ANDROMEDA
M31
PEGASUS
PISCES
AQUARIUS
CAPRICORNUS
EQUATOR
ECLIPTIC
Deneb
CYGNUS
Vega
LYRA
"Summer Triangle"
Altair
AQUILA
HERCULES
M13
CORONA BOREALIS
BOOTES
Arcturus
CANES VENATICI
LEO
VIRGO
Spica
LIBRA
HYDRA
SERPENS
OPHIUCHUS
SAGITTARIUS
Antares
SCORPIUS
ARA
LUPUS
CENTAURUS
MAGNITUDES
0 and brighter
1
2
3
4 and fainter
Aug 20 8 p.m.
Aug 5 9 p.m.
Jul 20 10 p.m.
Jul 5 11 p.m.
Jun 20 12 p.m.
Jun 5 1 a.m.

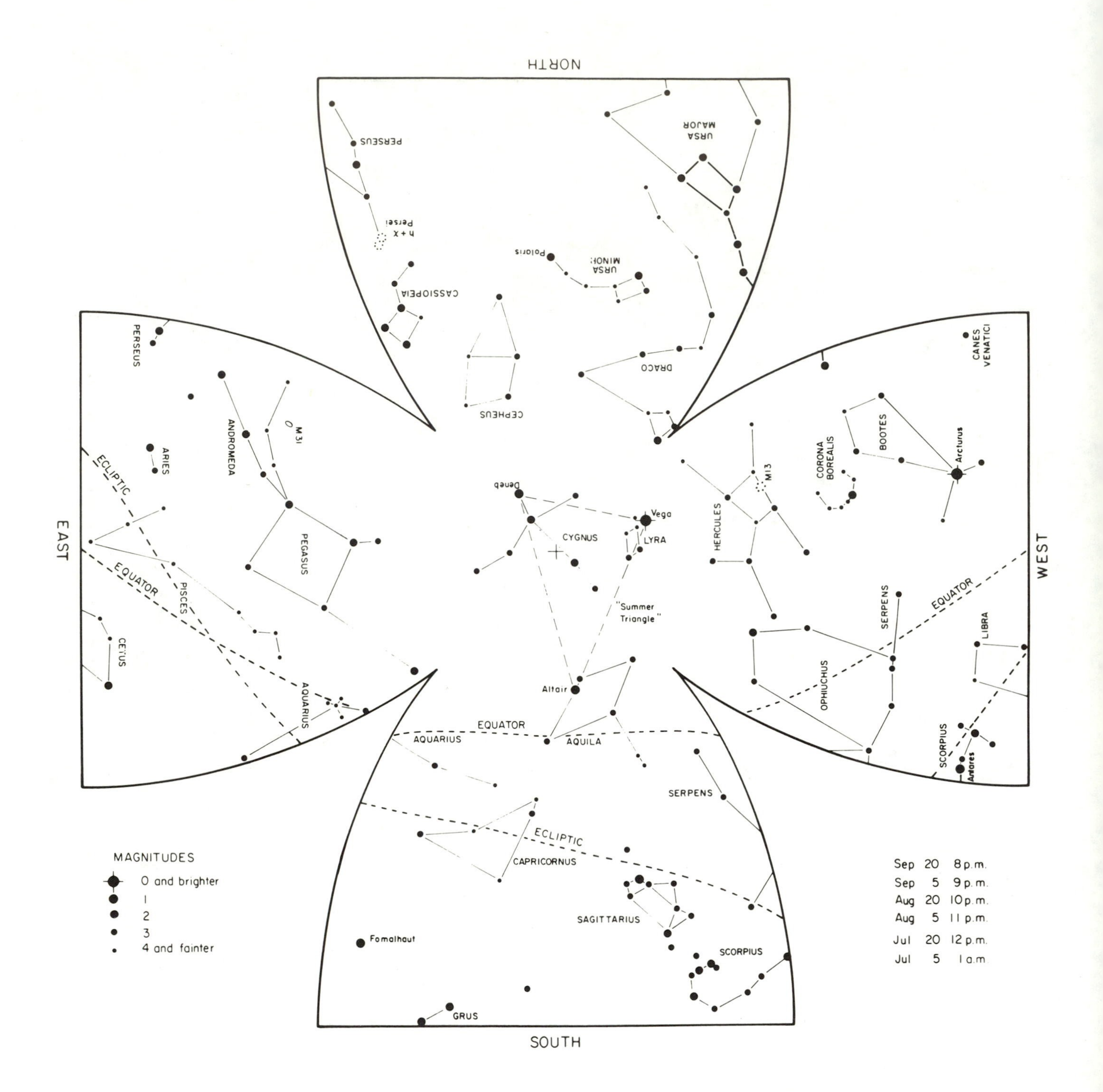

NORTH
SOUTH
EAST
WEST
PERSEUS
URSA MAJOR
h + X Persei
URSA MINOR
Polaris
CASSIOPEIA
DRACO
CEPHEUS
CANES VENATICI
ANDROMEDA
M31
ARIES
ECLIPTIC
EQUATOR
PISCES
PEGASUS
CETUS
AQUARIUS
Deneb
CYGNUS
Vega
LYRA
"Summer Triangle"
Altair
AQUILA
HERCULES
M13
CORONA BOREALIS
BOOTES
Arcturus
SERPENS
OPHIUCHUS
LIBRA
SCORPIUS
Antares
CAPRICORNUS
SAGITTARIUS
Fomalhaut
GRUS
MAGNITUDES
0 and brighter
1
2
3
4 and fainter
Sep 20 8 p.m.
Sep 5 9 p.m.
Aug 20 10 p.m.
Aug 5 11 p.m.
Jul 20 12 p.m.
Jul 5 1 a.m.

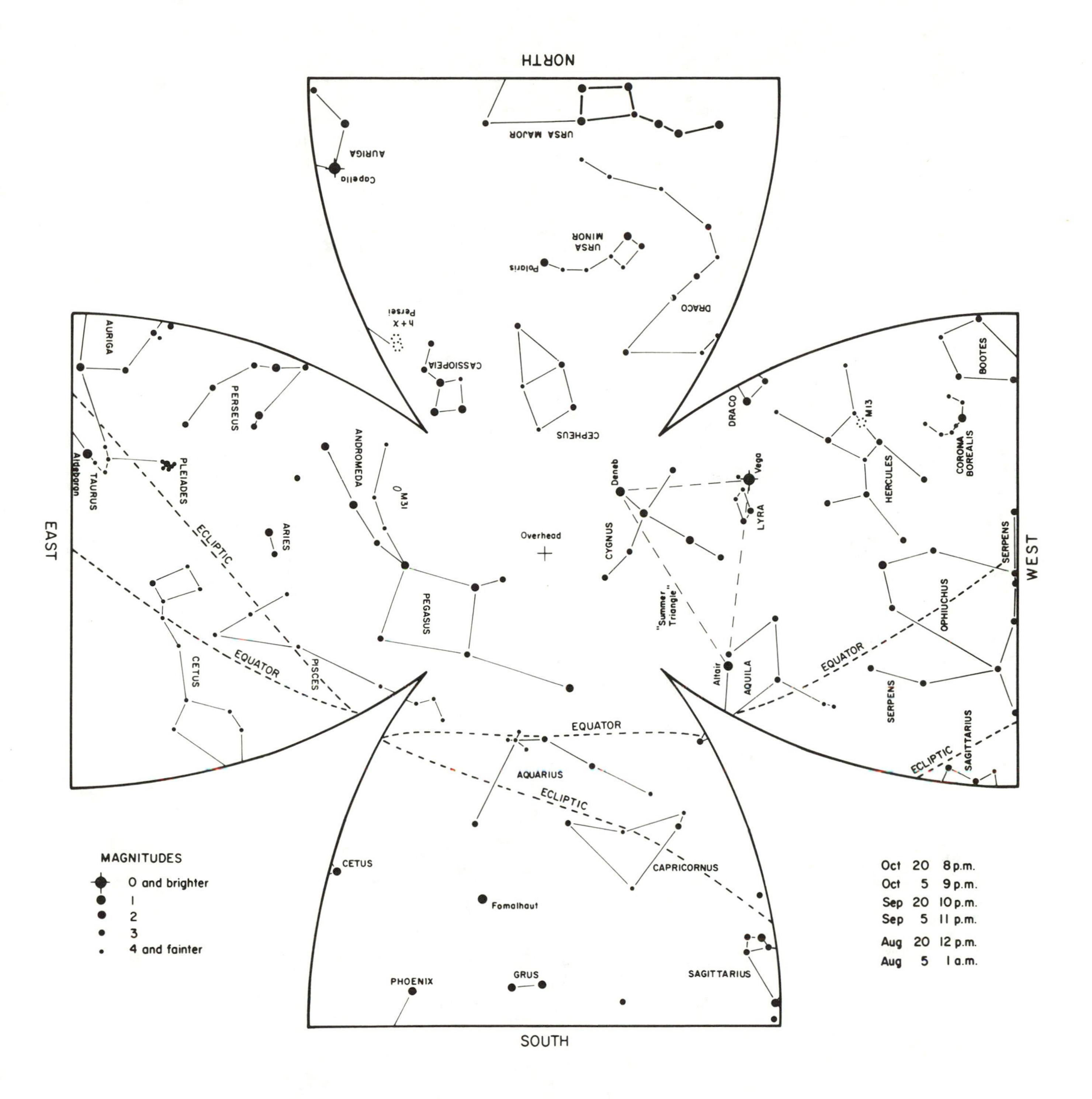
NORTH
SOUTH
EAST
WEST
URSA MAJOR
AURIGA
Capella
URSA MINOR
Polaris
DRACO
h + X Persei
CASSIOPEIA
CEPHEUS
PERSEUS
PLEIADES
Aldebaran
TAURUS
ANDROMEDA
M31
ARIES
ECLIPTIC
EQUATOR
CETUS
PISCES
PEGASUS
Overhead
CYGNUS
Deneb
Vega
LYRA
"Summer Triangle"
Altair
AQUILA
DRACO
M13
HERCULES
BOOTES
CORONA BOREALIS
SERPENS
OPHIUCHUS
SAGITTARIUS
AQUARIUS
CAPRICORNUS
Fomalhaut
PHOENIX
GRUS
MAGNITUDES
0 and brighter
1
2
3
4 and fainter
Oct 20 8 p.m.
Oct 5 9 p.m.
Sep 20 10 p.m.
Sep 5 11 p.m.
Aug 20 12 p.m.
Aug 5 1 a.m.

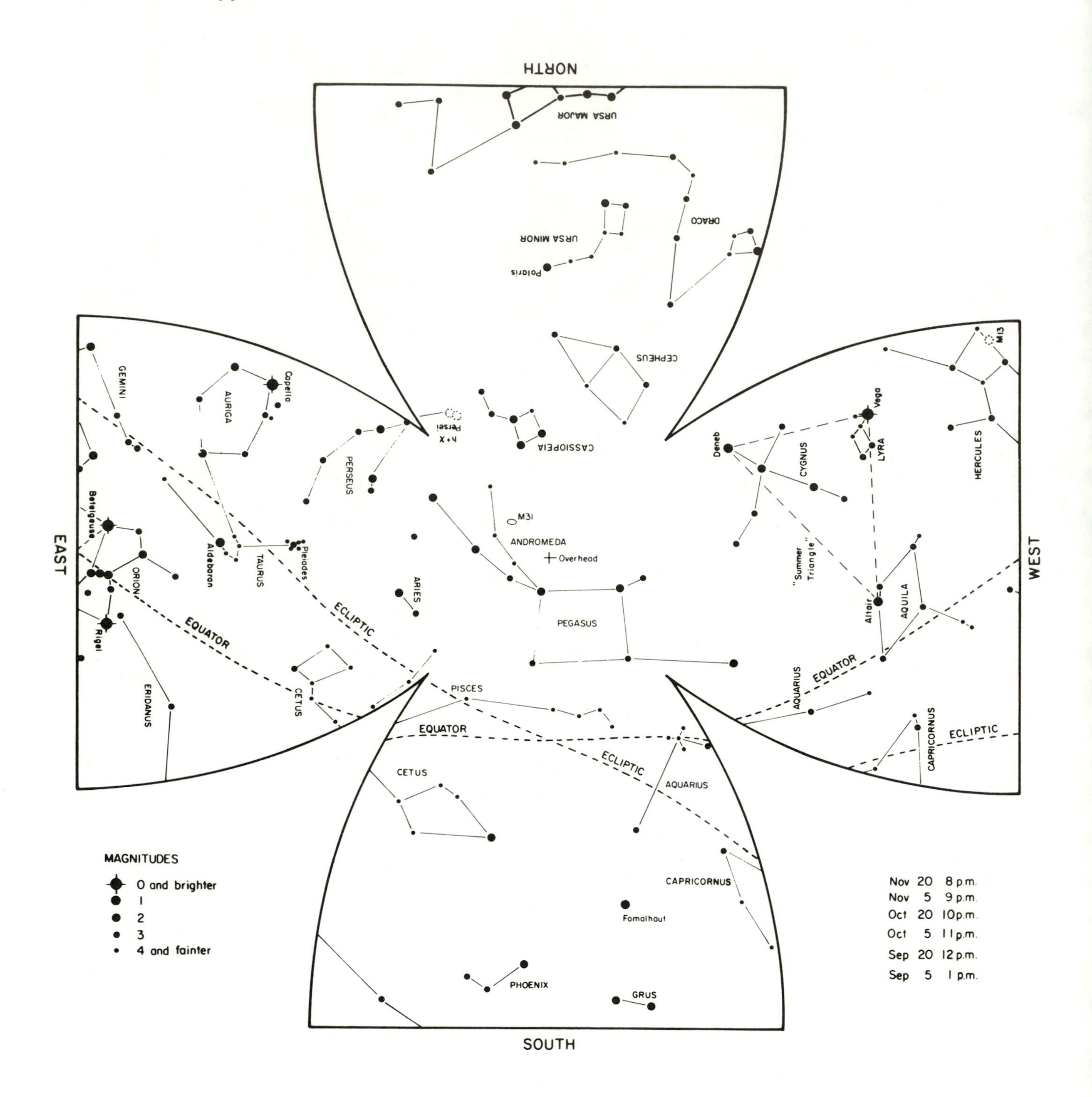
NORTH
URSA MAJOR
DRACO
URSA MINOR
Polaris
CEPHEUS
CASSIOPEIA
h + X Persei
M13
HERCULES
Vega
LYRA
Deneb
CYGNUS
"Summer Triangle"
Altair
AQUILA
GEMINI
Capella
AURIGA
PERSEUS
Betelgeuse
ORION
Aldebaran
TAURUS
Pleiades
Rigel
M31
ANDROMEDA
Overhead
ARIES
EAST
WEST
PEGASUS
EQUATOR
ECLIPTIC
ERIDANUS
CETUS
PISCES
AQUARIUS
CAPRICORNUS
AQUARIUS
CETUS
CAPRICORNUS
Fomalhaut
PHOENIX
GRUS
SOUTH
MAGNITUDES
0 and brighter
1
2
3
4 and fainter
Nov 20 8 p.m.
Nov 5 9 p.m.
Oct 20 10 p.m.
Oct 5 11 p.m.
Sep 20 12 p.m.
Sep 5 1 p.m.

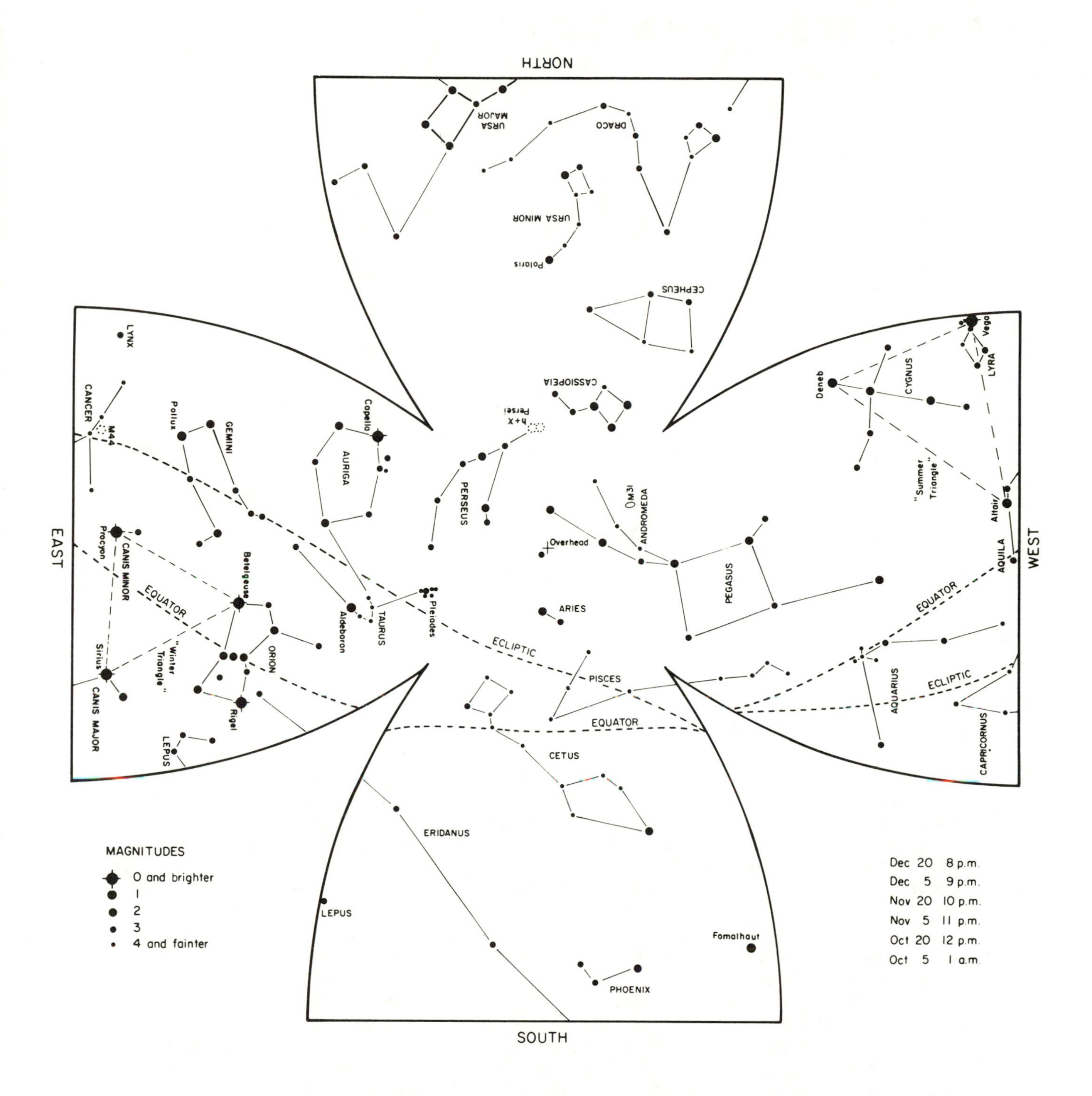

Dec 20	8 p.m.
Dec 5	9 p.m.
Nov 20	10 p.m.
Nov 5	11 p.m.
Oct 20	12 p.m.
Oct 5	1 a.m.

Appendix 2. Using a Sextant

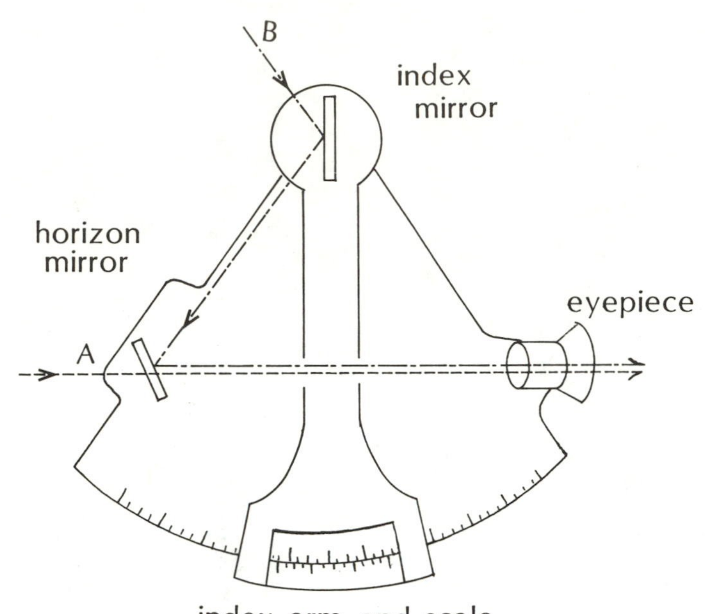

Figure 1

The sextant was and is an important navigational instrument for sailors. The model you will use can measure angles to an accuracy of 2' (1/30 degree). Figure 1 shows a diagram of the sextant. Notice that light can arrive at the eyepiece from two directions. The direct beam of light from an object in the direction A comes straight into the eyepiece from the object, just bypassing the horizon mirror but passing near it. The light from the object in direction B hits the index mirror and then the horizon mirror and then enters the eyepiece. The observer adjusts the movable arm of the sextant until the object at A and the object at B are seen side by side in the eyepiece, and the angle between them is read off the scale at the bottom.

To make your most accurate measurements with the sextant, you need to be able to read the vernier scale. In figure 2, the scale of the sextant is illustrated as set on a certain angle. To read the indicated angle, go first to the index marker, the zero on the bottom scale. Notice that it rests between 43 and 44 on the top scale. This means that angle is between 43 and 44 degrees. To find out WHERE between these two angles it is set, you compare the two sets of angle markings (top and bottom scales) and find two marks that are exactly lined up. Then reading the bottom scale at this lined-up point gives the number of minutes of arc. The matched lines are indicated in figure 2; since the marks are each two minutes of arc apart on the bottom scale, the total value of the angle indicated is 43 degrees 26 minutes of arc.

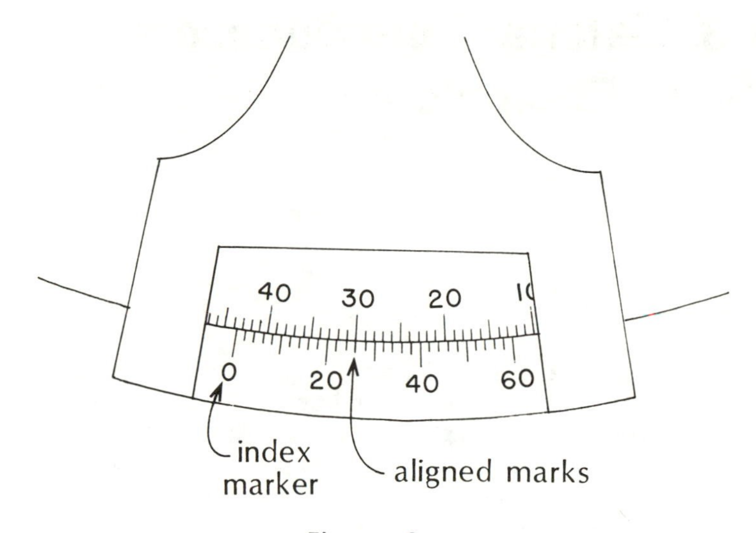

Figure 2

Take the sextant outside, set 0 on the bottom scale against 1 degree on the other scale, hold the sextant HORIZONTALLY so that the long dimension of the horizon mirror is parallel to the ground, and sight a large building. Notice that you can see the building both in the horizon mirror and above it but that the two images are slightly offset from each other. Adjust the sextant by moving the index arm with your other hand, making small corrections while continuing to look through the eyepiece, until the two images are exactly lined up with each other. At this point, the two zeros on the two scales should be very nearly coincident. If they are not, the sextant has a ZERO-POINT ERROR that must be corrected for. You are in effect measuring the angle between an object and itself, which must by definition by zero. The small difference from zero that the sextant shows must be an inherent error in the sextant. Read the value of the error off of the vernier scale (or use both scales if it is greater than one degree). This is called a zero-point error, and it will be included in every measurement which you take. For maximum accuracy, it must be taken into consideration in your calculations. What is the zero-point error of your sextant? How should you correct your sextant readings for the error?

Now aim the sextant at another building or object and measure its angular size. Adjust the index arm until you see the right-hand side of the building in the mirror lined up with the left-hand side of the building as seen by sighting just above the mirror. You can now see how the sextant can be used to measure the angle between two objects, such as the sun and the horizon. You simply sight one directly and move the index arm until you can simultaneously sight the other object through the mirror system. With care, you could also measure the angles between stars at night. One defect of the cross-staff was that, when the angle between two objects got too large, you could no longer sight it. The sextant can conveniently measure angles up to 100 degrees.

Note that, if you measure the sun, you should use the shades provided on the sextant to protect your eyes.

Appendix 3. Materials and Suppliers for the Course Equipment

This appendix contains a listing of certain sources for the equipment needed in the various units. The rather arbitrary listing is based upon our experience at the University of Texas; obviously, the information is not meant to be all-inclusive. For example, there are many other suppliers of telescopes besides the few listed here. You could even make your own telescope. Addresses of these suppliers are given at the end of the appendix.

Unit 1

cross-staff, quadrant patterns	insert in this book
meterstick, wood, scissors, string, thumbtacks	local

Unit 2

cross-staff, quadrant	unit 1
nomograph	page 1-3
SCl chart	Sky Publishing
protractor, compass, ruler terrestrial map	local
star map	appendix 1
transparent celestial globe with horizon	Fisher Scientific, Hubbard, Southern Biological, MMI Corporation

Unit 3

binoculars	local
small telescope	Edmund Scientific, local
lunar globe	MMI Corporation, Fisher Scientific

Unit 4

cross-staff	unit 1
SCl chart	op. cit. (unit 2)
sextant (optional)	Fisher Scientific

Unit 5

cross-staff	unit 1
sextant	op. cit. (unit 4)
SC1 chart	op. cit. (unit 2)
string, thumbtacks, SLR camera (optional)	local

Unit 6

gnomon, paper, protractor, box	local
celestial globe	op. cit. (unit 2)

Unit 7

gnomon	unit 6
bottle, thermometer, tape, ruler	local
celestial globe	op. cit. (unit 2)
sextant	op. cit. (unit 4)

Unit 8

optical bench	Central Scientific, Science Kit, Fisher Scientific
lenses (a variety of focal lengths)	Edmund Scientific
concave mirror	Edmund Scientific
ray box and accessories	Sargent-Welch
fluorescent bulb, ruler	local

Unit 9

SLR camera, camera lens, film	local
tripod, cable release	local
darkroom and equipment	local

Unit 10

small telescope	op. cit. (unit 3)
binoculars	local
telescope with equatorial drive	Edmund Scientific, Celestron, Criterion Scientific
Norton's Star Atlas	Sky Publishing

Unit 11

spectrometer pattern	insert in this book
double-edged razor blade	local
cardboard, glue, meterstick	local
transmission diffraction grating	Edmund Scientific
spectrum emission tubes	local (e.g., neon from advertising sign maker), Central Scientific

Unit 12

cross-staff	unit 1
sextant	op. cit. (unit 4)

Unit 13

NGS-POSS prints	Palomar Survey Cal Tech
magnifier	local, Edmund Scientific
wedge scale	page 13-11
reticle	Edmund Scientific
galaxy photo cards (optional)	Fisher Scientific

Unit 14

NGS-POSS prints	op. cit. (unit 13)
magnifier with reticle	Edmund Scientific
small ruler	local
transparent overlay	pattern on page 14-13

Unit 15

computer	e.g., Apple, TRS-80, Pet, Atari,...

Unit 16

solar telescope, H-alpha filter	Carson Astronomical Instruments
grating, accessory lenses	Edmund Scientific
meterstick	local

Unit 17

none

Unit 18

telescope with equatorial drive	op. cit. (unit 10)
SLR camera	local
film and equipped darkroom	local
astronomical film (103a-O, 103a-F)	Astro-Cards

Unit 19

large telescope camera for that telescope	sources as for unit 10
sheet film	local (possibly special order at camera-store)
Schmidt camera	Celestron International
equipped darkroom	local

Unit 20

telescope (30-cm or larger)	sources as for unit 10
spectrograph	Optomechanics
equipped darkroom	local

Unit 21

telescope (12-inch or larger)	sources as for unit 10
photoelectric photometer	Pacific Photometric, Optec, Inc.
Arizona-Tonantzintla catalogue	Sky Publishing

Unit 22

laser	Metrologic

SOURCES FOR EQUIPMENT

Astro-Cards
P.O. Box 35
Natrona Heights, Pa. 15065

Carson Astronomical Instruments
120 Erbbe
Albuquerque, N. M. 87112

Central Scientific Company = Cenco
6901 E 12 Street
Tulsa, Okla. 74112

Celestron International
2835 Columbia Street
Torrance, Calif. 90503

Criterion Scientific Instruments
620 Oakwood Avenue
West Hartford, Conn. 06110

Edmund Scientific Company
101 E Gloucester Pike
Barrington, N. J. 08007

Fisher Scientific Company
4301 Alpha Road
Dallas, Texas 75234

Hubbard
1946 Raymond Drive
Northbrook, Ill. 60062

Metrologic Instrument
143 Harding Avenue
Bellmawr, N. J. 08031

MMI Corporation
2950 Wyman Parkway
Baltimore, Md. 21211

Optec, Inc.
199 Smith
Lowell, Mich. 49331

Pacific Photometric
5675 Landregan Street
Emeryville, Calif. 94608

Palomar Survey Prints for Laboratory Exercises (six-pack)
Bookstore
California Institute of Technology
Pasadena, Calif. 91109

Sargent-Welch Scientific Company
7300 North Linder Avenue
Skokie, Ill. 60076

Science Kit, Inc.
777 E Park Drive
Tonawanda, N. Y. 14150

Sky Publishing Company
49 Bay State Road
Cambridge, Mass. 02238

Southern Biological Supply
P.O. Box 68
McKenzie, Tenn. 38201

Appendix 4. How These Materials Are Organized into a Course at the University of Texas at Austin

I. THE KELLER-METHOD SELF-PACED APPROACH AND HOW IT IS GRADED

This course is not an encyclopedic survey of astronomy; it concentrates on the methods and reasoning processes involved in an observational science, using actual observations and discovery activities to give a researchlike exposure to the subject. It is a "do-it-yourself" course, and you proceed through the various activities at whatever rate you choose. Your grade in the class will be determined by how many of the units you successfully complete. There are no lectures in the class; the class periods exist as times for you to do the observational activities, obtain guidance from the instructor when you have questions, and take the tests which demonstrate your mastery of the concepts.

Mastery is the key word in a course constructed in this fashion. In a conventional lecture class, a considerable amount of material is covered, and a certain percentage of comprehension (perhaps as low as 50 percent, depending upon the curve adopted by the instructor) is considered passing. In a Keller-method class, a lesser amount of material is covered, but a high degree of understanding is required. The course is structured to facilitate acquiring a complete comprehension of the materials covered.

To this end, when you take a quiz over a particular unit, there are only two grades you can earn on it -- PASS (which indicates a perfect or near perfect score) and REPEAT (indicating you need to prepare some more and try again). There is no stigma attached to "failing" a test and no penalties are assigned to it. You simply continue working on the materials until the required degree of comprehension is achieved, and your final grade depends upon how many unit tests you pass. Thus in a real sense, you are rewarded only for your successes and not penalized for your failures. Courses designed in this fashion have historically been found to be satisfying both to teach and to take and to result in a better understanding of the material than conventional classes. As a student, you will have a thorough and complete understanding of its concepts, not a 50 percent understanding. Until you achieve such understanding of its concepts, you do not continue on to other material, simply because the course is moving along at a certain rate. If you work diligently and do not procrastinate, there is no reason why you should not be able to earn an A in the class; some students will achieve their A before the semester is over. To earn an A in the class, you must complete 8 of the regular units (units 1 through 15) and 1 of the advanced units (units 16 through 21). To earn a B, complete 8 regular units. For a C, complete 6 regular units; and for a D, 4 units. Unit 22 contains a variety of activities that can be carried out, some introductory and some advanced.

II. CHOOSING THE ACTIVITIES THAT YOU WILL DO

The activities you select in the class will depend on your interests and time available. Units 1 and 2 are required of everyone and are prerequisite to all the other activities, but past this point, there are various tracks you can follow through the course materials. For example, if you were most interested in aspects of astronomy that you could continue to follow as a naked-eye hobby the rest of your life, you might choose to do units 3 and 4 (the moon), unit 5 (the planets), and units 6 and 7 (the sun). If you have a strong interest in stars and their life cycles, you would want to do units 11, 12, 13, and 14.

You should read through the descriptions of the activities at the beginning of the book, and begin to plan your path through the course materials. As you do, pay careful attention to the prerequisites listed for several of the units; these requirements will direct your path through the concepts necessary to achieve your goals at the end of the semester. When you have selected a main path for your studies, you will then choose other activities to fill out the number required for the grade you wish to achieve. Your choices may also depend upon weather patterns and available equipment.

This class does not assume any background in astronomy, physics, or mathematics and has no formal prerequisites except upper-division standing. Most of the early units are written for students on an introductory level. However, some students may enter the course with a more complete background in physics and math and would prefer to do activities more commensurate with their level of skills. Students who have completed the introductory calculus-based physics sequence for science or engineering majors may not obtain credit for units 1 and 8. Such students are encouraged to do only 6 of the regular activities plus 2 of the advanced units (for an A) or any other combination agreeable to the instructor.

III. HOW TO PROCEED THROUGH THE CLASS

Obtain and keep a bound notebook for recording your observations and writing your conclusions from them. The best type has lined paper on one side and graph paper on the other. As you do your observations and take measurements, you will enter them into this notebook, and your instructor will check it from time to time to verify that your progress is satisfactory. Record everything you do in this book, and attempt to organize your comments and data entry in such a way that the instructor can read it and understand what you have done. For example, before taking a series of measurements, construct a table in which the data can be entered. Taking the time to do this in advance will actually save you time, because it will insure that you have an organized understanding of what you are doing and save you from wasting time doing unnecessary observations. Be sure to record the time and date of all observations and any other information necessary to allow you to reconstruct later what you did and what you learned.

A. Choose a unit (in consultation with your instructor if necessary) and carefully read through its introduction, prerequisites, and equipment list. If you decide to work on this unit, notify your instructor for the class record, and sign out any necessary equipment.

B. Proceed through the directions in the unit, performing any measurements and answering any questions it may ask, recording everything in your notebook. If you have any problems or questions, take them to your instructor when they occur. The instructor is familiar with the types of difficulties that occur in going through the materials and is prepared to lead you through them if necessary. Frequent checks and consultations is a good idea in general, because such conferences can be used to uncover misconceptions about ideas and inefficiencies in procedures before they waste a lot of time.

C. In some of the units, it may be convenient, necessary, or desirable to work with a partner. Such an arrangement gives moral support and stretches the equipment supplies. However, it will not necessarily speed up progress, and care should be taken that both partners take their own data and do their own unique interpretation of that data. Having one partner DO the observations and other WRITE them down is not acceptable.

D. When you are finished with the duties required by the unit, you should take all of your measurements and conclusions to the instructor for the final checking. You may need to repeat some measurements at this time.

E. Before taking the quiz on the unit, you should give a mini-lecture to another student in the class (not your partner, if you had one) concerning what you have done. These talks are a valuable part of the class experience; they give you an opportunity to organize your ideas on the concepts of the unit and finalize your understanding of it. They are a valuable preparation for the quiz or they should be properly prepared before delivery; that is, take a few minutes to outline what you will say before starting to talk. The feedback you obtain from your listener may help you to uncover areas of the activity you have not adequately understood.

These mini-lectures are also valuable to the listener. They will increase the listener's understanding of the material if the listener has completed the unit previously, and enrich the listener's class experience if the lecture is on a unit that he or she has not done. The listener should ask questions and if necessary comment on parts of the presentation that do not seem clear. If you thoroughly understand the material, you should be able to present a lucid and understandable discussion of it to your listener. When mini-lectures are given and/or heard, this should be reported to the instructor to be recorded in the class records-book. You must listen to as many mini-lectures as you give to complete the course.

F. Finally, take the quiz on the unit. Some are oral quizzes and some are written, some open-book and some closed, depending upon the type of activity completed. If you do not pass the quiz, return to the material and study it further, and take another quiz at another time. You may take a quiz as often as necessary to pass the unit, but you may try the quizzes on a particular unit only twice in one class period. After passing the quiz, proceed to the next unit of your choice.

IV. FURTHER INFORMATION ABOUT THE CLASS

A. In general, you will be allowed to work on two units at a time, one outdoor unit for good weather and one indoor unit for poor weather. To work on more than two at once has been found by past experience to be distracting and harmful to your rate of progress. The one exception to this rule is if you choose to do the two long term activities in the class, units 4 and 7. These require observations over a lengthy time interval and, if you are working on these, you many continue other activities while doing them.

B. No incompletes will be given as grades in this class, except in well-documented emergency situations. Consult with your instructor as soon as such situations develop, not when it is too late to take some corrective action.

The only way to make a poor grade in this class is to procrastinate. If you feel you cannot discipline yourself adequately to maintain a proper rate of progress in a course of this type, drop it immediately and enroll in another astronomy class. You may consult with the instructor at any time to check whether your rate of progress is adequate to achieve the grade you desire. Many students finish the class well before the end of the semester, thus freeing themselves for finals in their other classes.

During the summer nine-week semester, you must complete slightly more than one unit per week to earn an A, slightly under for a B. In a regular 15-week semester, a uniform rate of about 3/4 unit per week will earn an A. For your own sanity and that of your instructor, do not attempt to pass a large number of units during the last week of class. It takes time to check and grade a unit, and if a logjam of students develops, you will have some long waits and may have to repeat some work. Since you work only on two units at a time, there is no benefit whatsoever to holding finished work back from being graded.

C. If you wish to do the long-term projects in class, you must make sure to get them started in time. The last meeting of the class is the last opportunity for presenting data and taking quizzes. There is no final exam. Allow plenty of time to finish your work; for projects dependent on the weather, allow extra time. In any particular semester, consult your instructor for the latest dates at which time-critical projects can be started.

Index